T0327632

SOLUTIONS IN LIDAR PROFILING OF THE ATMOSPHERE

SOLUTIONS IN LIDAR PROFILING OF THE ATMOSPHERE

VLADIMIR A. KOVALEV

WILEY

Published by John Wiley & Sons, Inc., Hoboken, New Jersey
Published simultaneously in Canada

Library of Congress Cataloging-in-Publication Data:

Kovalev, Vladimir A.
Solutions in lidar profiling of the atmosphere / Vladimir A Kovalev.
1 online resource.
Includes index.
Description based on print version record and CIP data provided by publisher; resource not viewed.
ISBN 978-1-118-96327-2 (ePub) – ISBN 978-1-118-96328-9 (Adobe PDF) –
ISBN 978-1-118-44219-7 (cloth) 1. Atmosphere–Laser observations. 2. Atmosphere–Remote sensing. 3. Meteorological optics. I. Title.
QC976.L36
551.5028′7–dc23

2014024758

Printed in the United States of America

10 9 8 7 6 5 4 3 2 1

CONTENTS

PREFACE

Modern atmospheric science meets extremely complex and challenging problems, and thousands of researchers are enthusiastically looking for ways to find their solutions. The experimental methods and data are considered the basis for answering the scientific questions of human interest. However, the interpretation of the data remains an issue.

The most common way to interpret the experimental data in atmospheric science is by solving a number of related equations or a single equation with a number of unknowns. Unfortunately, any method of interpretation of the experimental data obtained in the real atmosphere requires assumptions. Many such assumptions have been transformed into implicit premises and now often go unmentioned. Reliable interpretation of the experimental data can have place only when all assumptions and implicit premises are met; this is where the devil hides.

The total uncertainty of the measured quantity of interest depends on how large the uncertainties in the involved parameters are. Each measurement task can only be solved within the limits established by the uncertainty of the involved elements. No way exists for determining the exact value of any atmospheric quantity of interest, only some likely value can be found. An accurate estimate of the uncertainty of an unknown quantity could only be found if the casual fluctuation of this quantity during its measurement obeyed some relatively simple laws so that existing error propagation theories can be applied. Generally, the total uncertainty includes two independent components, namely random uncertainty and systematic uncertainty. Until now, there has been no proper method for estimating systematic uncertainty, including that resulting from the often mandatory implementation of the *a priori* assumption or assumptions. Therefore, researchers do their best to, in some way, minimize systematic distortions before or during the experiment; this approach allows them to focus on the random uncertainty. Analysis of the random phenomena still remains the main method for estimating the uncertainty in atmospheric studies. It is generally assumed that the laws governing the chance phenomena of interest are fixed in nature, and these laws are ultimately determined. In other words, it is assumed that the uncertainties obey simple "rules of the game." The distribution of a random variable as a symmetrical or a nonsymmetrical bell-shaped graph is a typical example. The so-called normal distribution, named after the German mathematician Carl Friedrich Gauss (1777–1885), and the Poisson distribution, named after the French mathematician and physicist Simeon Denis Poisson (1781–1840), have remained the fundamental theoretical basis for the investigation of atmospheric processes for more than 200 years. Nothing new, at least with the same level of importance, has been proposed for error analysis since the nineteenth century.

Unfortunately, nature does not obey our relatively simple formulas. It obeys its own and much more complicated laws. All of our formulas are surrogates; they only approximate the atmospheric processes, proposing simplified solution schemes for these processes. It is quite rare when an accurate approximation is achieved by a simple formula, like the one for gravitational interaction or the famous Einstein formulas. A simple formula is generally valid only when the process is governed by a small number of influential parameters. In such cases, the influence of all other parameters is minor and even their significant variations do not change the essential characteristics of the process. Unfortunately, most processes in the atmosphere depend on a great number of parameters, whose variations during a measurement may have nothing in common with the relatively simple laws used in applied statistics. The actual fluctuations, and hence, the uncertainty distributions of the involved parameters are often unpredictable. No strict mathematics exist that would permit an exact evaluation of the reliability of the solutions obtained in the presence of nonstatistical uncertainties. Therefore, the assumption that relatively simple statistical laws govern chance phenomena is the compelling issue in atmospheric sciences. Meanwhile, statistical estimates may only be true under certain limited conditions, which are quite often not properly met in real atmospheres. In atmospheric physics, it is quite difficult to establish whether the phenomenon under investigation meets these conditions, and accordingly, to estimate the reliability of the applied statistics. The fluctuations of the involved unknowns may vary in an unpredictable way, often far from the assumed simple laws. The inappropriate use of statistics yields wrong conclusions, which unfortunately, often look extremely plausible and mislead both their authors and readers.

The simplest example of the doubtful use of statistics is when using temporal averaging of lidar signals during the vertical profiling of the atmosphere. The real atmosphere cannot be considered as horizontally homogeneous even in the statistical sense because the variations of the optical parameters do not obey any predictable statistical distribution. To overcome this issue, the more rigid assumption of a "frozen" atmosphere is commonly used. The time during which the atmosphere should remain "frozen" may change from some seconds to half an hour and more. This assumption is so common in lidar profiling of the atmosphere that it is rarely even mentioned in the publications. In other words, such an assumption is now one among other implicit premises.

The principles of estimating uncertainties based on purely statistical models and conventional error propagation theory are inappropriate for investigating atmospheric processes with lidar. The conventional theoretical basis for random error estimates is very restrictive and requires rigid conditions, which are rarely satisfactorily met in real atmospheres and real lidar signals. First, the uncertainties of the involved parameters are often large, preventing the conventional transformation from differentials to finite differences used in standard error propagation. Second, the random errors of such parameters cannot always be accurately described by some simple distribution, such as Gaussian or Poisson. Third, the quantities used in lidar data processing can be correlated; the level of correlation often changes with the measurement range, and no reliable methods exist to determine the actual behavior of the uncertainty. Fourth, the measured atmospheric parameters may not be constant during the measurement period because of atmospheric turbulence, particularly for large averaging times used by deep atmospheric sounders. Apart from this, the total uncertainty can include any number of systematic errors of an unknown sign, constant, or variable that may cause large and often hidden distortions in the retrieved atmospheric profile.

The harsh reality is that instead of having truly concrete methods for uncertainty analyses, researchers are often playing a variant of the DADT game, "Don't ask about the systematic errors, – don't tell about these." A common justification for such play is the excuse that the statistical methods we have are the best that we have ...

Actually, the lidar researcher does not measure the optical parameters of the atmosphere; he measures only the sum of the backscatter signal and the background component. Therefore, today many lidar specialists avoid using the word "measurement" in their papers, or at least, in the title of their papers. Readers of scientific literature related to lidar searching of the atmosphere should notice that instead of using the term "measurement," many authors prefer using terms such as "monitoring," "profiling," "retrieving profiles," or "performing observations" of profiles of interest when describing their experimental results. These terms cloud reality, and I believe this is the proper time to dot our "i's" and cross our "t's." Truthfully, one should admit that today's lidar data processing technique implements more and more elements of a simulation rather than a measurement. In other words, lidar solutions should be considered as models, that is, simplified reflections of reality, which represent physical processes in some general way. As is known, modeling is typically used when it is impossible to create conditions in which one can accurately measure the parameter of interest. Models use assumptions and accumulated statistics while true measurements do not. Accordingly, numerical estimates in lidar observations made by using model dependencies will always be much less accurate than direct measurements, and this fact should be freely admitted. Only after such an admission can the appropriate elements of modeling technique in lidar searching be openly discussed. Triggering such a discussion is the basic goal of this book.

The term "profiling" can be defined as a reconstruction of a particular optical parameter of the atmosphere using the characteristics of the backscattered signal and some *a priori* assumptions based on statistics or sometimes just educated guesses.

The difference between measurement and profiling is that, unlike measurement, profiling gives only some general idea of the shape of the parameter of interest rather than precise details.

Comprehensive analysis shows that in any type of lidar profiling, the most significant errors occur during signal inversion, when the optical parameters of the atmosphere are extracted from the lidar signals using a number of implicit premises and *a priori* assumptions. Inverting the lidar signal, the researcher actually builds some simulation based on past lidar observations, some assumptions, implicit premises, some statistics, and finally, on the researcher's intuition and common sense. Under such conditions, the researcher can obtain only an estimate of the atmospheric profile of interest with an uncertainty that cannot be accurately quantified. In this book, instead of using the long phrase "simulation based on past lidar observations," I will also use the shorter phrase "a posteriori simulation."

Some methods used for profiling the aerosol atmosphere with lidar, as discussed in this book, have no rigid mathematical foundation; they are generally based, as the author believes, on common sense. Unfortunately, in the practice of atmospheric investigations, this is often the only way to interpret physical processes in the atmosphere in a meaningful way. Common statistics perform extremely poorly, for example, in smoky atmospheres, and this deficiency forced me to look for alternative ways for processing lidar data. Using alternative methods to invert lidar signals allows comparing results and estimating the credibility of different methods. The accuracy of the retrieved results cannot be estimated as with data that obey statistical laws. However, the use of alternative solutions gives one an estimate of how reliable the retrieved data are. This is the central premise of this book.

Fortunately, apart from the standard error-propagation procedure for statistical random errors, two alternative methods exist that allow investigating the effects of systematic and random errors without relying on common statistical laws. The first is a sensitivity study in which expected uncertainties in the involved quantities or likely signal distortions are used in numerical simulations in order to evaluate the distortion level in the output parameter of interest. In these simulations, a virtual lidar operates in a synthetic atmosphere, and its synthetic corrupted signals together with the selected *a priori* assumptions are used to retrieve the optical parameters of the atmosphere. Such a method may be used, for example, to analyze how an overestimated or underestimated backscatter-to-extinction ratio influences the accuracy of the extinction-coefficient profile extracted from elastic lidar data. To use this method, an analytical dependence may be obtained by combining and solving two inversion equations. The first equation is derived for the actual backscatter-to extinction ratio, used in the simulation, and the second is the solution obtained for the assumed incorrect ratio. Such an investigation is especially useful when making an error analysis for the case where large random or systematic errors are involved. This method provides a reasonable estimate of the total measurement uncertainty; it allows avoiding common underestimation of uncertainty when systematic distortions are ignored. The other method may be used when investigating the real lidar data, for example, the influence of the particular parameter taken *a priori*. This method may also be used to understand how an overestimated or underestimated backscatter-to-extinction ratio influences the extinction-coefficient profile extracted from the real noisy signal in the real atmosphere under investigation.

Some recommendations in this book, which follow from such nonstandard methods of error estimation, cannot be unanimously justified. One cannot claim, for example, 68% confidence in retrieved data that includes uncertainty not treatable statistically. Considering the problem of combining random and systematic errors, Taylor (1997) wrote: "No simple theory tells us what to do about systematic errors. In fact, the only theory of systematic error is that they must be identified and reduced… However, this goal is often not attainable… There are various ways to proceed [the total uncertainty calculation]. None can really be rigorously justified… Because the errors… are surely independent…, using the quadratic sum [of random and systematic uncertainties] is probably reasonable. The expression cannot really be rigorously justified… Nonetheless, it does at least provide a reasonable estimate of our total uncertainty, given that our apparatus has systematic uncertainties we could not eliminate."

Defending the approaches and methods proposed in this book, I can only paraphrase Taylor by saying that there are no rigorous justifications for these methods except common sense. This principle of performing error analysis and estimation based on common sense is unavoidable and will remain the center of the author's attention in this book.

The book consists of three chapters. In Chapter 1, the basic issues of elastic-lidar-data inversion are discussed considering this task as a typical ill-posed problem. Chapter 2 discusses the specifics and the issues in separating the backscatter and transmission terms in the lidar equation. Chapter 3 considers the specifics of profiling the atmosphere with scanning lidar that operates in a multiangle mode. This book is intended for the users of atmospheric lidar, particularly newcomers who are starting their lidar investigations. The author believes that this book will allow them to see the real situation in remote sensing and current impassable restrictions in this area of atmospheric investigation.

An attentive reader will notice that the book contains a lot of repetition. The author has included such repetition deliberately. From his long experience, he knows that most readers of scientific books have neither the time nor the desire to read the book from cover to cover. Generally, they focus only on sections, in which the subject of their interest is discussed. Taking this into account, the author has tried to make the chapters and sections of the book as self-contained as possible. Therefore, the most specific and the most important points discussed in the book may be repeated in different sections, so that the reader has no need to jump from section to section to understand the points discussed in the section relevant to his or her interest.

ACKNOWLEDGMENTS

The author wishes to acknowledge the assistance of the USDA Forest Service Missoula Fire Sciences Laboratory for making this book possible. The author is deeply indebted to his colleagues, especially to the members of the lidar team, namely Cyle Wold and Alexander Petkov, who enthusiastically gathered and processed the lidar experimental data, which became the basis for many concepts in this book. Finally, I would like to thank the staff of the publisher, John Wiley & Sons Ltd., for their collaboration during the production of the book.

DEFINITIONS

α	Angstrom exponent
β	Total (molecular and particulate) scattering coefficient, $\beta = \beta_m + \beta_p (\text{m}^{-1}, \text{km}^{-1})$
β_m	Molecular scattering coefficient (m^{-1}, km^{-1})
β_p	Particulate scattering coefficient (m^{-1}, km^{-1})
β_π	Total (molecular and particulate) backscatter coefficient, $\beta_\pi = \beta_{\pi,m} + \beta_{\pi,p}$ (m^{-1} steradian^{-1})
$\beta_{\pi,m}$	Molecular backscatter coefficient (m^{-1} steradian^{-1})
$\beta_{\pi,p}$	Particulate backscatter coefficient (m^{-1} steradian^{-1})
β_{on}	Total (molecular and particulate) scattering coefficient at the DIAL wavelength λ_{on}
β_{off}	Total (molecular and particulate) scattering coefficient at the DIAL wavelength λ_{off}
$\beta_{\pi,\text{on}}$	Total (molecular and particulate) backscatter coefficient at the DIAL wavelength λ_{on}
$\beta_{\pi,\text{off}}$	Total (molecular and particulate) backscatter coefficient at the DIAL wavelength λ_{off}
$\beta_{\pi,R}$	Inelastic backscatter coefficient at the Raman shifted wavelength λ_R
δ_P	Distortion component of the lidar-signal multiplicative factor, $(1 + \delta_P)$, caused by non-ideal transformation of the backscattered light into the output electrical signal

δ_{mult}	Ratio of the multiple scattering signal to the single scattering signal
ΔB	Remaining offset in the backscatter signal after removing the estimated constant, $\langle B \rangle$ from the total signal
$\Delta\sigma = \sigma_{on} - \sigma_{off}$	Differential absorption cross section of ozone for the *on* and *off* wavelengths
$\epsilon(h_{max})$	Criterion for the selection of the optimal maximum height, h_{max}, for the signal inversion
$\epsilon(\Delta h_i)$	Criterion for equalizing the alternative piecewise optical depths within a restricted interval, Δh_i
θ	Azimuthal angle of the scanning lidar
κ	Total (molecular and particulate) extinction coefficient, $\kappa = \kappa_m + \kappa_p$ (m^{-1}, km^{-1})
$\kappa_{p,0}$	Particulate extinction coefficient at the wavelength λ_0 of the emitted light of the Raman lidar
$\kappa_{p,R}$	Particulate extinction coefficient at the Raman shifted wavelength λ_R
$\kappa_{m,0}$	Molecular extinction coefficient at the wavelength λ_0 of the emitted light of the Raman lidar
$\kappa_{m,R}$	Molecular extinction coefficient at the Raman shifted wavelength λ_R
κ_{gr}	Total extinction coefficient at ground level
$\kappa^{(dif)}$	Total extinction coefficient determined through numerical differentiation
$\kappa_p^{(dif)}$	Particulate extinction coefficient determined through numerical differentiation
$\kappa_p^{(i)}(h)$	Particulate piecewise extinction coefficient within a segmented interval versus height
$\kappa_{p,j}^{(i)}$	Piecewise particulate extinction coefficient in slope direction
κ_w	Transformed extinction coefficient in the elastic lidar solution
λ	Wavelength of the emitted and backscattered light of the elastic lidar
λ_0	Wavelength of the emitted light of the Raman lidar
λ_R	Wavelength of the Raman shifted signal
λ_{on}	Wavelength within the enlarged absorption spectrum of ozone used in the DIAL profiling of the ozone concentration
λ_{off}	Wavelength outside the enlarged absorption spectrum of ozone used in the DIAL profiling of the ozone concentration
Λ	Criterion for minimizing the difference between the alternative transmission profiles within a segmented interval
Π_p	Particulate backscatter-to-extinction ratio (steradian^{-1})

Π_m	Molecular backscatter-to-extinction ratio (steradian^{-1})
$\Sigma\tau_p^{(0,R)}$	Total of the particulate optical depths at the wavelength, λ_0, emitted by the laser, and at the Raman shifted wavelength, λ_R
Σw_j	Sum of the low- and high-frequency noise components in the lidar signal
$\Sigma w_{j,\text{low}}$	Sum of the low frequency noise components that remains in the signal after its smoothing
$\Sigma\kappa_p^{(0,R)}$	Sum of the particulate extinction coefficient at the wavelength, λ_0, emitted by the laser, and at the Raman shifted wavelength, λ_R
$\frac{d\sigma_{\pi,R}}{d\Omega}$	Range-independent differential Raman backscatter cross section
ς	Correction factor in Raman lidar equation corresponding to the variable Angstrom coefficient
ς_0	Correction factor in Raman lidar equation corresponding to the assumed constant Angstrom coefficient
τ	Total (molecular and particulate) optical depth, $\tau = \tau_m + \tau_p$
τ_m	Molecular optical depth
τ_p	Particulate optical depth
τ_0	Total (molecular and particulate) optical depth at the emitted laser wavelength, λ_0
τ_R	Total (molecular and particulate) optical depth at the Raman shifted wavelength, λ_R
τ_{90}	Total (molecular and particulate) optical depth determined directly in zenith; $\tau_{90} = \tau_{p,90} + \tau_{m,90}$
τ_{mod}	Model optical depth used for the extrapolation of the optical depth derived with lidar down to the ground level
τ_{vert}	Total (molecular and particulate) optical depth in the vertical direction determined from the multiangle data of scanning lidar
τ_{sh}	Shaped total optical depth, which increments versus height or range are either positive or equal to zero
$\tau_p^{(i)}$	Particulate piecewise optical depth
$\tau_{p,\text{up}}$	Estimated upper limit of the shaped optical depth, $\tau_{p,\text{sh}}$
$\tau_{p,\text{low}}$	Estimated lower limit of the shaped optical depth, $\tau_{p,\text{sh}}$
υ	Distortion component of the optical depth originated in additive and/or multiplicative components in the lidar signal
υ_{rand}	Random noise component in the distorted optical depth profile
υ_{sys}	Systematic distortion component in the distorted optical depth profile
φ	Elevation angle of vertically scanning lidar
χ	Fixed levels of the ratio function, $R_Y(h)$ (in zenith profiling) or $R_{\theta,\max}(h)$ (in multiangle profiling) used for determining maximal height of the atmospheric layer with increased backscattering

a_{π}	Ratio of the particulate to the molecular lidar ratio, S_p/S_m
$A(h)$	Interception point of the linear fit, $Y(h)$ with y-axis at the height, h
B	Range-independent offset in the recorded lidar signal, which estimated value is $\langle B\rangle$
C	Lidar solution constant, its estimated value is $\langle C\rangle$
$EF[\delta f(x), \delta x]$	Error factor which is defined as the absolute value the ratio of the fractional error of the output function, $\delta f(x)$, to the fractional error of the input element, δx
$f_m(h)$	Temperature and pressure dependent attenuation factor for the Cabannes spectrum in the HSRL equation
f_p	Rejection ratio of the scattered light from particulates in the HSRL equation
h	Height from ground level to the scattering volume
h_{min}	Minimum height used for the lidar signal inversion
h_{max}	Maximum height used for the lidar signal inversion
h_b	Reference height for which an assumed boundary condition for the lidar equation solution is taken
h_s	Middle point of the interval from h to $h + s$, where s is the range resolution used for numerical differentiation
$I(r_a, r)$	Integral of the square range-corrected signal
n_{ozone}	Ozone concentration (ppb)
$N_R(T, p)$	Atmospheric number density of the Raman scattering molecules as a function of temperature (T) and pressure (p)
P_{Σ}	Signal recorded by lidar, which is a sum of a backscatter signal P and a constant offset, B
$P(r)$	Backscatter signal not distorted by a multiplicative and/or an additive component
$\langle P(r)\rangle$	Distorted backscatter signal used for the inversion
P_{on}	Backscatter signal of the DIAL measured at the wavelength λ_{on}
P_{off}	Backscatter signals of the DIAL measured at the wavelength λ_{off}
P_{90}	Backscatter signal in the zenith direction
$P_{\Sigma,90}$	Total signal in the zenith direction
P_m	Backscatter signal at the output of the molecular channel of High Spectral Resolution Lidar
P_j	Backscatter signal measured under the slope angle φ
P_R	Raman signal at the shifted wavelength, λ_R
q	Overlap function of the emitted laser light beam and the cone of the receiver telescope field of view
q_{eff}	Effective overlap function determined in the multiangle mode

r	Range at which the lidar signal is considered
r_0	Distance from the lidar to the nearest point of the complete overlap area
r_b	Boundary (reference) point for the lidar signal inversion
r_{min}	Minimum range used for the lidar signal inversion
r_{max}	Maximum range used for the lidar signal inversion
s	Range resolution in numerical differentiation
SOR	Signal-to-offset ratio
S_m	Molecular lidar ratio (steradian)
S_p	Aerosol lidar ratio (steradian)
$\overline{S_p(h_i, h)}$	Column-integrated particulate lidar ratio over the altitude range h_i–h
$S_p^{(i)}$	Piecewise range-independent particulate lidar ratio within a restricted interval
$T_\Sigma^2(0, h)$	Two-way total (molecular and particulate) transmittance from ground level to the height, h; $T_\Sigma^2(0, h) = T_p^2(0, h)T_m^2(0, h)$
$T_{90}^2(0, h)$	Two-way total (molecular and particulate) transmittance in the zenith direction
$T_m^2(0, h)$	Two-way molecular transmittance from ground level to the height h
$T_p^2(0, h)$	Two-way particulate transmittance from ground level to the height h
$T_0(0, r)$	One-way total (molecular and particulate) transmittance versus range in the Raman measurement at the laser wavelength, λ_0
$T_R(0, r)$	One-way total (molecular and particulate) transmittance versus range at the Raman shifted wavelength, λ_R.
$T_{\Sigma,j,\mathrm{vert}}^2(0, h)$	Two-way total vertical transmittance determined from the signals measured along the set of slope angles, φ
$\overline{T_{\Sigma,\mathrm{vert}}^2(0, h)}$	Average two-way total transmittance in the vertical direction determined from the multiangle data
$\overline{T_{p,\mathrm{vert}}^2(0, h)}$	Average two-way particulate transmittance in the vertical direction determined from the multiangle data
$w(r_m, r_n)$	Weight function for the calculation of the extinction coefficient within the overlapping interval r_m–r_n using alternative two-way transmittance profiles
$w(h_i, h_{i+1})$	Weight function for the calculation of the extinction coefficient within the interval h_i–h_{i+1} using two alternative optical depth profiles
x	Independent variable in the Kano–Hamilton solution, uniquely related with the slope direction, φ [$x = (\sin\varphi)^{-1}$]

y_i	Natural logarithm of the square-range-corrected signal measured under slope direction, φ
y_{90}	Natural logarithm of the square-range-corrected signal measured in the zenith direction
$Y(x, h)$	linear fit for the data points y_i taken versus independent variable, x
z	Independent variable in the transformation of the recorded signal into the function $y^*(r)$
Z_w	Transformed square-range-corrected elastic backscatter signal

1

INVERSION OF ELASTIC-LIDAR DATA AS AN ILL-POSED PROBLEM

1.1 RECORDING AND INITIAL PROCESSING OF THE LIDAR SIGNAL: ESSENTIALS AND SPECIFICS

Before starting a detailed consideration of the basic principles of lidar-data analyses, the principal issue of atmospheric remote sensing should be clearly stated. Any atmospheric formula used in practice - the lidar equation being no exception - is some surrogate of reality. Atmospheric laws are extremely complicated, and, in practical terms, their realization can be analyzed only in some simplified form. It follows from this fact that the commonly used lidar equation considered below is nothing but the mathematical description of a simplified model of the real backscattered signal. Therefore, it is quite sensible to compare the formulas for ideal and real lidar backscatter signals.

1.1.1 Lidar Equation and Real Lidar Signal: How Well Do They Match?

In the classical form, the single backscatter signal $P_j(r)$ at the range r, recorded by elastic lidar in the two-component atmosphere is written as

$$P_j(r) = C_j \beta_\pi(r) q(r) r^{-2} T_\Sigma^2(0, r). \tag{1.1}$$

where $C_j = \eta P_{0,j}$ is the lidar constant, which contains all instrumental range-independent parameters, the emitted laser power $P_{0,j}$ and the efficiency factor η. The latter includes the total efficiency of optical components, such as the telescope and filters, and the light-voltage conversion factor; $\beta_\pi(r)$ is the total backscatter coefficient which includes the molecular and particulate components, that is, $\beta_\pi(r) = \beta_{\pi,p}(r) + \beta_{\pi,m}(r)$; $T^2_\Sigma(0,r)$ is the total, particulate and molecular, two-way transmittance over the distance from the lidar location to the range r, that is

$$T^2_\Sigma(0,r) = T^2_p(0,r)T^2_m(0,r) = \exp\left[-2\int_0^r \kappa_p\left(r'\right)dr'\right]\exp\left[-2\int_0^r \kappa_m\left(r'\right)dr'\right],$$

where $\kappa_p(r)$ and $\kappa_m(r)$ are particulate and molecular extinction coefficients, respectively. In the general case, both $\kappa_p(r)$ and $\kappa_m(r)$ can include scattering and absorption components.

The term $q(r)$ in Eq. (1.1) is an overlap function of the emitted laser light beam and the cone of the receiver telescope field of view. At distances close to lidar, the overlap is incomplete. In this near zone, only a part of the backscattered light created by the laser beam is "seen" by the telescope and reaches the photoreceiver. At these distances, the function $q(r)$ monotonically increases with the range until reaching its maximum level; the nearest point r_0 where this takes place and the telescope "sees" the whole laser beam, is considered the minimum distance of complete overlap. Over the ranges $r > r_0$, the laser beam fully remains within the telescope field of view and allows one to simplify the signal inversion by using the condition $q(r) = \text{const}$. Generally, $q(r)$ is normalized to unity, so that for ranges larger than r_0, the overlap function reduces to $q(r) = 1$. Note also that Eq. (1.1) shows the intensity of single-scattered lidar return; multiple scattering is assumed absent. Accordingly, this equation is not valid for highly polluted atmospheres, clouds, dense smokes, etc., if special corrections are not taken.

In practical measurements, no single backscatter signal is processed. Instead, the sum (or the average) of a number of backscattered signals N_{sign} is determined, that is,

$$P(r) = \sum_{j=1}^{N_{\text{sign}}} P_j(r). \tag{1.2}$$

The sum of the signals accumulated during the selected measurement time is then used for inversion of the signal $P(r)$ into the profile of the required optical parameters of the atmosphere.

There is an extremely important implicit premise behind the summation in Eq. (1.2). It is assumed, that during measurement, the atmosphere is "frozen". This term implies that when the set of signals $P_j(r)$ is recorded, the profiles of $\beta_\pi(r)$ and $T^2_\Sigma(0,r)$ do not vary, that is, they both are "frozen." This is a very stringent requirement, especially when the time for collecting the required number of shots is large. The details of this issue are considered in Section 1.3.2.

In most cases, the real signal recorded by the lidar instrument also includes the additive range-independent component B, whose value depends on the background luminance and the electric offset due to signal amplification and digitization. Accordingly, the recorded lidar signal is

$$P_{\Sigma}(r) = Cq(r)\beta_{\pi}(r)r^{-2}T_{\Sigma}^{2}(0, r) + B \tag{1.3}$$

The determination of the lidar backscatter signal includes two independent operations: recording the profile of the total lidar signal, which is the sum of the backscattered light and background component (Eq. 1.3), and separating the profile of the backscattered signal. Then the backscatter signal is square-range-corrected, and this corrected signal is used for the next inversions. The consequent operations used for obtaining the square-range-corrected lidar signal are shown in Fig. 1.1. If the lidar has a multichannel receiver, such operations are performed with the recorded signal from each channel.

The first and the second procedures, shown in the gray filled blocks 1 and 2, include establishing the levels of the initial temporal smoothing of the recorded signals, that is, determining the recording time and the number of shots per unit time to obtain the lidar signal $P_{\Sigma}(r)$. The next two procedures deal with the calculation of the square-range-corrected signal. The procedure in the block 3 removes the offset B from the total signal $P_{\Sigma}(r)$. This procedure results in obtaining the electrical signal versus time, proportional to the backscatter light power. The signal is square-range-corrected, and the profile of the range-corrected backscatter signal $P(r)r^2$ is then analyzed and inverted into an atmospheric profile of interest.

The schematic of the ideal lidar signal transformation before its inversion is given in Fig. 1.2. The quantity I_{bsc} in the input block is the intensity of the backscatter light and I_{bgr} is the background luminance.

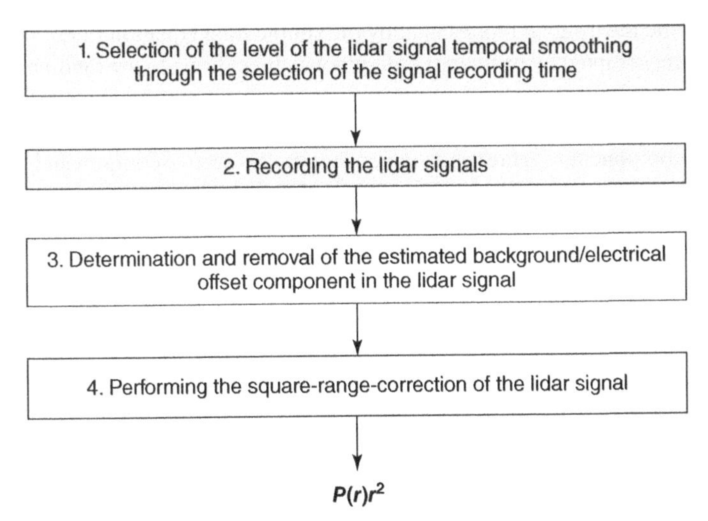

Fig. 1.1 Schematic of determining the profile of the square-range-corrected backscatter signal.

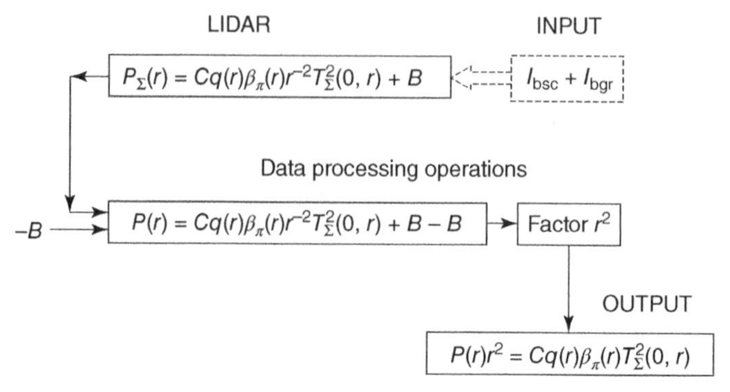

Fig. 1.2 Schematic of the ideal transformation of the light energy on the photodetector into the square-range-corrected backscatter signal.

1.1.2 Multiplicative and Additive Distortions in the Lidar Signal: Essentials and Specifics

The ideal backscatter signal not corrupted by noise and systematic distortion obeys the formula

$$P(r) = Cq(r)\beta_\pi(r)r^{-2}T_\Sigma^2(0, r). \tag{1.4}$$

Unfortunately, there are no grounds to expect that one can transform the backscattered light energy into a digitized electrical signal and then separate the backscatter signal of interest from the sky background without adding additional distortion. In other words, the actual signal recorded by the lidar receiver is much more complicated than the ideal signal represented by the simple lidar equation in Eq. (1.4). First, the signal is corrupted by ordinary random noise. In addition to the random noise, the actual measured signal always has some nonzero distorting components, which generally cannot be properly estimated. The signal recorded by lidar generally includes two distortion components: (i) the range-dependent multiplicative component $[1 + \delta_P(r)]$, which produces some nonlinear distortion in the signal during its passing from the input to the output of the receiver and (ii) the sum of the random noise components $\sum w_j(r)$ in a wide frequency spectra where some components do not obey common statistics and cannot reliably be filtered out. Thus, the actual total signal recorded by the lidar receiver can be written as

$$P_\Sigma(r) = Cq(r)\beta_\pi(r)r^{-2}T_\Sigma^2(0, r)[1 + \delta_P(r)] + \sum w_j(r) + B. \tag{1.5}$$

In the short form, it can be written as a combination of the ideal backscatter signal $P(r)$ and the signal distortion components, that is,

$$P_\Sigma(r) = P(r)[1 + \delta_P(r)] + \sum w_j(r) + B. \tag{1.6}$$

For general analysis of signal distortions and their influence on the inverted optical parameters, the multiplicative component is written in the form $[1 + \delta_P(r)]$, representing the backscatter signal distorted by the multiplicative component as the product $P(r)[1 + \delta_P(r)]$. The interaction of the noise with the backscatter signal is extremely complicated. For simplicity, we consider the noise as a sum of random noise components in the wide spectra $\sum w_j(r)$, which can be treated as another additive component in the lidar signal, that is, as the electromagnetic noise superimposed on the signal. The wide spectra noise $\sum w_j(r)$ can be generated by either the signal circuits or external sources; the latter, appearing to be random, is often synchronized to an external event. Typically, such electromagnetic noise is a slowly decaying ringing voltage induced after triggering the light pulse (Ahmad and Bulliet, 1994).

One should point out that, in spite of the complicated form of the signal in Eq. (1.5), this formula is also a simplified model of the recorded lidar signal, but closer to reality than Eq. (1.3). The simplified models of the multiplicative and additive distortions in Eqs. (1.5) and (1.6) are convenient enough for general analyses. They allow one to avoid going deeply into the specific details of the numerous sources of lidar signal distortion.

Before the lidar signal inversion can be made, the component B in Eq. (1.6) should be estimated and removed, as shown in block 3 in Fig. 1.1. Obviously, it is unrealistic to expect that the constant offset B can be estimated with zero uncertainty. In practical terms, the estimate of this offset, which we define as $\langle B \rangle$, will always somewhat differ from the actual B, so that the difference $\Delta B = B - \langle B \rangle$ will not necessarily be equal zero. As shown in Fig. 1.3 (a) and (b), the remaining offset ΔB can be either positive or negative. In both plots of this figure, the thin solid curves show the total signal, which is the sum of the backscatter signal $P(r)$ and the constant component B; the true value of the latter is shown as a solid horizontal line. The estimated constant component $\langle B \rangle$ is shown as a thin dashed horizontal line, so that the magnitude of the distorted backscatter signal used for inversion is $\langle P(r) \rangle = P_{\Sigma}(r) - \langle B \rangle$. For clarity, other distortion components in Eq. (1.6), $\sum w_j(r)$ and $\delta_P(r)$, are not shown in these plots.

Before the backscatter signal $\langle P(r) \rangle$ is inverted into atmospheric optical profiles, it is generally smoothed. Signal smoothing is an extremely delicate operation. Both the backscatter signal $P(r)$ and the random noise $\sum w_j(r)$ consist of wide-spectra frequency components, and generally, no reliable separation of the signal and noise components is possible. Here, the use of assumptions and guesses are the only methods available to the researcher. The common implicit premise is that all the high-frequency noise components are normally (lognormally, etc.) distributed, so these can be easily separated and removed by simple filtering. All the bumps and concavities remaining in the backscatter signal after such filtering are assumed to be the true details of the backscatter signal. Such an implicit premise, commonly used during data inversion, is too simple to provide truly reliable inversion results. Indeed, any type of filtering used in the lidar system reduces the impact of noise on the backscatter signal. However, the filtering procedure adds its own characteristic distortion to signal inversion results, which depend on the unavoidable issue of

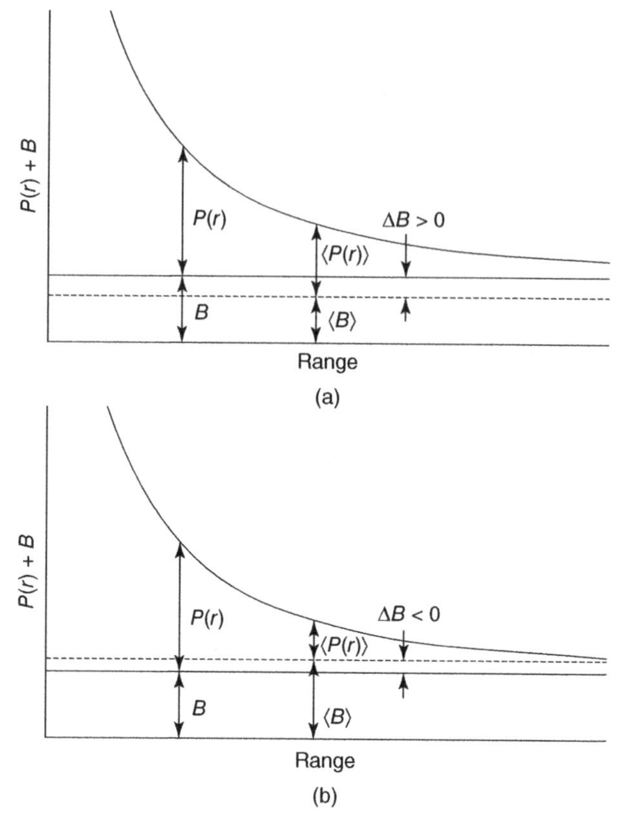

Fig. 1.3 (a) The thin solid curve shows the sum of the backscatter signal $P(r)$ and the constant offset B; the latter is shown as a solid horizontal line. The estimated constant $\langle B\rangle$ is shown as a thin dashed horizontal line. The remaining offset ΔB is positive. (b) The same as in (a) but the remaining offset ΔB is negative.

determining at what frequency to cut off the spectra of the recorded signal. One should also keep in mind that low-frequency noise components cannot be properly removed from the recorded signal by simple filtering; such an operation may be accompanied by unacceptable distortion in the backscatter signal.

Even moderate signal smoothing transforms the term $\sum w_j(r)$, removing or reducing high-frequency components in it. Therefore, after the smoothing operation, the term $\sum w_j(r)$ should be reduced to the term, we symbolize as $\sum w_{j,\mathrm{low}}(r)$. This term denotes the low-frequency components, which were not filtered out during the smoothing operation. Thus, after subtracting the estimated offset $\langle B\rangle$ from the recorded signal $P_{\Sigma}(r)$ and spatial smoothing, the real backscatter signal $\langle P(r)\rangle$ can be written as

$$\langle P(r)\rangle = P(r)[1 + \delta_P(r)] + \sum w_{j,low}(r) + \Delta B. \tag{1.7}$$

The multiplicative distortion factor $[1 + \delta_p(r)]$ may be produced by many sources, such as nonlinear transformation of the backscattered light into the electrical signal, the receiver's restricted frequency bandwidth, the presence of multiple scattering (see below), etc. Numerous analyses performed by different researchers have shown that the signals in the near zone, where their magnitude sharply changes over short time intervals, can be strongly impacted by this factor. In addition, the spatial nonuniformity of the photomultiplier photocathode can also contribute to the factor. This distortion occurs due to the displacement of the scattering volume image on the photoreceiver (Simeonov et al., 1999; Agishev and Comeron, 2002). The influence of the distortion component $\delta_P(r) \neq 0$ in the near end of the lidar measurement range is especially significant when the lidar has an extended incomplete overlap zone, which extends to some kilometers (Berkoff et al., 2003, 2005; Eloranta et al., 2004). The stability requirements for the shape of the overlap function $q(r)$ are extremely stringent; if these requirements are not met, the influence of the distortion factor becomes significant (see Section 1.8).

Additional signal distortion can come from the so-called signal-induced noise (or signal-induced background) at the far end of the measurement range. This effect can take place in lidars that use photomultiplier tubes for transforming backscattered light into electrical signals (Shimizu et al., 1985; Zhao, 1999; Sunesson et al., 1994; Lee et al., 1990; Fiorani et al., 1997). Presumably, it is caused by exposure of the photomultiplier tube to intense lidar returns from the lower atmosphere. It can be considered an afterpulsing effect, which occurs in photomultiplier tubes after the abrupt termination of a short light pulse (Hunt and Poultney, 1975; Bristow, 2002). Afterpulsing depends on many factors, such as ambient and internal temperature, the laser pulse energy, the background level. As follows from the study by Campbell et al. (2002), no empirical relationship can be directly derived to address this issue, and therefore, afterpulsing calibration must be done frequently, approximately every month.

In addition to the aforementioned distortion sources, multiple scatter will also implement a nonzero distortion term $\delta_P(r)$. The lidar equation in its classic form [Eq. (1.4)] assumes that lidar profiling is made in the atmosphere where only single scattering takes place. The presence of additional multiscattering components in the recorded signal can be treated as the presence of the multiplicative factor $\delta_P(r) > 0$. Indeed, assuming a nonzero ratio of the multiple-scattering component $P_{\text{mult}}(r)$ to the single scattering component, that is,

$$\delta_{mult}(r) = \frac{P_{mult}(r)}{P(r)} \neq 0$$

one can write the corresponding backscatter signal as (Veretennikov and Abramochkin, 2009)

$$\langle P(r) \rangle = P(r)[1 + \delta_{\text{mult}}(r)]. \tag{1.8}$$

In other words, when the single-scattering lidar equation is used, the presence of the multi-scattering component in the lidar signal can be considered a signal distortion component, that is, $\delta_P(r) = \delta_{mult}(r)$. While in general, the range-dependent function $\delta_P(r)$ in Eqs. (1.6) and (1.7) may be either positive or negative over the lidar operative range, the multiple-scattering component $\delta_{mult}(r)$ is always positive.

The reliable estimation of the multiple-scattering factor $\delta_{mult}(r)$ remains a significant issue because of the extremely simplified model solutions used to invert the signal in dense atmospheres. Initially, a simple correction factor was proposed (Platt, 1973, 1979). As multiple scattering is additive, it results in more backscattered light on the photodetector than there would be in a single-scattering atmosphere. This effect, in turn, reduces the inverted optical depth when the single-scattering equation is used, especially over distant ranges. Platt proposed to correct this decrease by implementing a so-called "effective" optical depth, the product of the true optical depth multiplied by η_{mult}, where the constant factor $\eta_{mult} < 1$. The question that immediately follows is, "Under what conditions would the assumption of the range-independent factor be practical and how should such a factor be specified?" This question has no satisfactory answer. A number of more complicated strategies for the inversion of lidar signals corrupted by multiple scattering have been proposed (e. g., Bissonnette, 1995; Bissonnette et al., 2002; Eloranta, 1998; Zege et al., 1995; Zuev et al., 1978; Kovalev, 2003). However, the reliability of all the existing signal inversion methods remains questionable. Unfortunately, there is no comprehensive method which would allow one to estimate the influence of multiple scattering, using either the relatively simple models and solutions, such as those proposed by Weinman (1976), Eloranta (1998), Bissonnette (1995, 1996), etc., or the significantly more sophisticated models. Any such methodology resorts to modeling rather than presenting a true measurement method.

As there is no current methodology for profiling dense atmospheres, some hardware solutions to solve this issue have been proposed, which would presumably decrease the factor $\delta_{mult}(r)$. The most common attempt to reduce the relative level of multiple scattering in recorded lidar signals has been to decrease the field of view of the receiving telescope. Unfortunately, this creates other problems. The cost of such a solution is an extremely extended incomplete overlap zone. For example, the angular field-of-view of high spectral resolution lidar (HSRL) developed at the University of Wisconsin for unattended operation in Arctic is only 45 μrad; the lidar signals measured in zenith depart from the standard $1/r^2$ dependence for distances up to 3 km (Razenkov et al., 2002). Such an extended incomplete overlap zone requires additional calibrations and other measures to avoid defocusing the optical system, which may occur, for example, due to temperature changes. A similar issue is also inherent to the commercial micropulse lidar (MPL) (Berkoff et al., 2003, 2005).

There are also many other factors which may worsen the lidar signal inversion results. Such error sources may be related to specifics of photon-counting statistics, to the photomultiplier afterpulsing effect, and to the uncertainty in the determination of the calibration parameters, including the profile of the overlap function, which is significant when the signals in the incomplete overlap zone are inverted into the optical parameters of the sounded atmosphere.

The most vexing factor in lidar profiling of the atmosphere is that the distorted backscatter signal rather than the ideal one has to be square-range-corrected before the required inversion can take place. Therefore, the schematic of the transformations of the lidar signal shown in Fig. 1.2 is very far from reality. The schematic of the transformation of the real lidar signal before its inversion is given in Fig. 1.4, where a larger number of quantities are involved as compared to the idealized schematic in Fig. 1.2. Taking into account Eqs. (1.5) and (1.7), one can present the real square-range-corrected signal $r^2\langle P(r)\rangle$ as

$$r^2\langle P(r)\rangle = Cq(r)\beta_\pi(r)T^2_\Sigma(0,r)[1+\delta_P(r)] + r^2\sum w_{j,\mathrm{low}}(r) + r^2\Delta B. \tag{1.9}$$

Note that the first term in the right side of the formula, which contains the information of interest, generally decreases with range, whereas the two distortion components, ΔB and $\sum w_j(r)$, increase. This effect significantly restricts profiling of the atmosphere at distant ranges.

The theoretical estimates of the lidar-signal-inversion accuracy are generally focused on the three obvious sources of measurement uncertainty: (i) the random signal noise, which dramatically increases at distant ranges after the measured signal is range corrected (Sasano and Nakane, 1984; Sasano et al., 1985; Bissonnette, 1986; Kunz and Leeuw, 1993; Kunz 1996; Durieux and Fiorani, 1996; Rocadenbosch, 1998; Whiteman, 1999); (ii) the uncertainty in the assumed behavior of the backscatter-to-extinction ratio profile, generally taken *a priori* when processing the elastic-lidar data; and (iii) the distortion of the atmospheric profile caused by an inaccurate estimation of the lidar-equation constant C due to, for example, an inaccurate selection of a boundary value in the lidar-equation solution. As shown in this section, there are a number of additional sources of the distortions, which should be taken into account to avoid overestimating the accuracy of the derived profiles. However, the evaluation of the errors created by systematic distortions in the lidar signal is an issue which is difficult to address. When profiling the atmosphere, the possibility of properly evaluating the errors in the derived atmospheric profiles is extremely small. Most systematic distortions are not detectable, especially when measurements are made in one-directional mode.

There is no practical and reliable way for estimation of the distortion components $\delta_P(r)$, $\sum w_{j,\mathrm{low}}(r)$, and ΔB in the real lidar signal, and their influence on

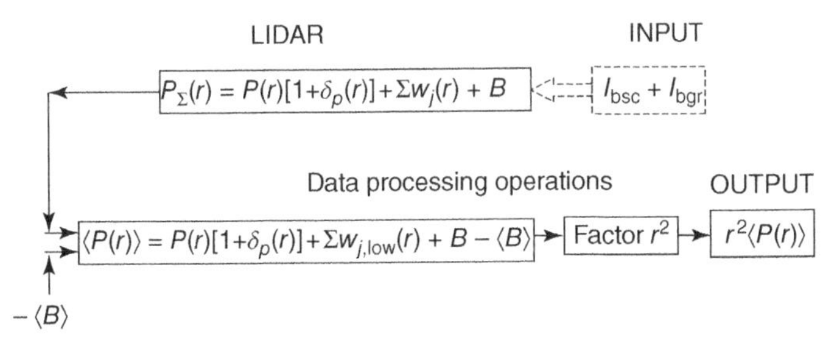

Fig. 1.4 Schematic of the transformation of the nonidealized photodetector signal into the square-range-corrected backscatter signal.

the optical profile derived from the signal. The most effective way to investigate the influence of these distortion components on the inversion results is by performing numerical simulations. In such a simulation, the synthetic atmospheric profile $f(x)$ is compared with the distorted profile $\langle f(x)\rangle$, retrieved from the signal of a virtual lidar, obtained in accordance with the schematic in Fig. 1.4. The error in the profile retrieved from the virtual lidar signal is determined relative to the initial (test) profile, rather than relative to some statistical average. Such a simulation does not require transition to the infinitesimal quantities as the standard error propagation methodology. Therefore, it can be used in the situations when large errors are involved, as it happens during profiling of the atmosphere.

Applying this principle to the "true" backscattering signal $P(r)$ and its distorted profile $\langle P(r)\rangle$ defined in the schematic in Fig. 1.4, one can obtain from Eq. (1.7) a simple formula for the relative error in the distorted lidar signal:

$$\delta P(r) = \frac{\langle P(r)\rangle}{P(r)} - 1 = \delta_P(r) + \frac{\sum w_{j,\mathrm{low}}(r)}{P(r)} + \frac{\Delta B}{P(r)}. \tag{1.10}$$

As follows from Eq. (1.10), over distant ranges, where $P(r)$ is small, the low-frequency random noise and the nonzero component ΔB become the most significant issues for determining the accurate backscatter signal. The analysis made in the next section shows that even a minor remaining offset ΔB is extremely destructive over ranges where it becomes comparable with $P(r)$.

Let us summarize. The shape of the optical depth and extinction-coefficient profiles obtained through the inversion of the square-range-corrected signal will significantly depend on the assumptions and the implicit premises used when performing the initial procedures in the blocks 1, 2, and 3 in Fig. 1.1. Even the initial selection of the data accumulation time, which assumes a stationary (frozen) atmosphere during the lidar sounding time, is erroneous when the signal data are accumulated in a nonstationary atmosphere. It will be shown in Section 1.3 that the violation of this requirement makes the resulting profile of the atmosphere dependent not only on the atmospheric characteristics but also on the selected averaging time. Unfortunately, the selection of the data averaging time is not the only procedure related to the implicit premises and the assumptions used for the inversion of the backscatter signal into an atmospheric profile of interest. Additional implicit premises and assumptions are related to the estimation and removal of the constant offset from the recorded signal, with spatial averaging of the backscatter signal, and with the selection of the minimum and maximum operative ranges. Disregarding other factors, such as multiple scattering in the recorded lidar signal, yields additional uncertainty in the inversion results, which are virtually impossible to address.

In Table 1.1, the assumptions and implicit premises which may cause significant distortions in the signal inversion results are briefly tabulated.

TABLE 1.1 General Assumptions and Implicit Premises When Processing the Elastic-Lidar Signal and Separating the Backscatter Component

	Data Processing Procedure	Assumptions/Implicit Premises
1.	Temporal averaging of the recorded lidar signal	The atmosphere during the lidar signal recording time is stationary ("frozen"), or the backscatter signals fluctuate slightly and randomly about some mean value, and their fluctuations obey an assumed statistical distribution
2.	Estimating and removing the constant offset in the recorded signal	(a) The low-frequency noise components and the multiplicative distortion component do not influence the estimation of the constant offset
		(b) The remaining offset in the backscatter signal is minor and can be ignored as compared with other errors
3.	Spatial averaging of the backscatter signal	(a) The level of signal spatial averaging is adequate for the actual behavior of the backscattering profile; the shape of the averaged backscatter profile is neither overly smoothed, nor over fitted
		(b) The low-frequency noise components in the lidar signal are minor, and the variations in the backscatter signal originate in the real spatial variations of the backscattered light

1.2 ALGORITHMS FOR EXTRACTION OF THE EXTINCTION-COEFFICIENT PROFILE FROM THE ELASTIC-LIDAR SIGNAL

1.2.1 Basics

As follows from Eq. (1.4), the elastic-lidar equation for the single-scattering atmosphere includes three key parameters, a constant C and two range-dependent functions, the backscatter coefficient $\beta_\pi(r)$ in the scattering volume and the two-way total attenuation $T_\Sigma^2(0, r)$ along the path from the lidar to the scattering volume. Note that the classic lidar equation does not describe the recorded signal, it deals only with the backscatter component, that is, with the signal obtained after subtraction of the constant offset B from the recorded signal. The implicit premise hidden in the classic lidar equation is that the signal constant component is precisely estimated and removed, that is, $\Delta B = 0$.

Even in such a simplified form, the backscatter lidar signal contains insufficient information to separate the range-dependent components, $\beta_\pi(r)$ and $T_\Sigma^2(0, r)$, and extract from the signal the particulate profiles of interest, the extinction coefficient or the particulate optical depth. This task is intractable without using some assumptions or auxiliary data from independent remote sensing instrumentation.

In most cases, such auxiliary data are not available, so accurate separation of the two range-dependent components in the lidar equation can only be made if the relationship between the particulate extinction and backscatter components, $\kappa_p(r)$ and $\beta_{\pi,p}(r)$, is known. Unfortunately, this relationship is generally unknown. The simplest, but not the most reliable way of separating the extinction and backscatter components in the lidar equation is based on the assumption of either a range-independent particulate backscatter-to-extinction ratio (that is, $\beta_{\pi,p}(r)/\kappa_p(r) = \text{const.}$), or on some analytical dependence between the particulate extinction and the particulate backscattering (Fernald, 1982, 1984; Klett, 1981).

Let us consider the basic inversion algorithm for single-wavelength elastic lidar. In accordance with Eq. (1.4), the square-range-corrected backscatter signal of the zenith-directed lidar can be written in the form,

$$P(h)h^2 = CT_{\Sigma}^2(0, h_0)[\Pi_p(h)\kappa_p(h) + \Pi_m\kappa_m(T^{(h)}p^{(h)})]$$
$$\exp\left\{-2\int_{h_0}^{h}\left[\kappa_m\left(T^{(h')}p^{(h')}\right) + \kappa_p(h')\right]dh'\right\}, \qquad (1.11)$$

where $T_{\Sigma}^2(0, h_0)$ is the two-way total transmission term for the altitude range from ground level to the minimal operative height h_0, selected somewhere in the complete overlap zone, that is, where the overlap function has reached its maximum value, $q(r) = 1$; $\kappa_p(h)$ is the particulate extinction coefficient at the height h; the term $\kappa_m(T^{(h)}, p^{(h)})$ is the molecular extinction coefficient, which is a function of temperature ($T^{(h)}$) and pressure ($p^{(h)}$) at the height h. In lidar profiling of the atmosphere, the molecular temperature- and pressure-dependent extinction coefficient $\kappa_m(T^{(h)}, p^{(h)})$ is converted into the height-dependent function $\kappa_m(h)$. Thus, the extraction of the particulate extinction coefficient $\kappa_p(h)$ from the lidar equation requires knowledge of the molecular extinction profile within the lidar-searched zone. Sometimes, such data are obtained from balloon measurements; if these are not available, the most appropriate statistical models are used. In other words, the profile $\kappa_m(h)$ is never determined directly along the laser beam path during lidar sounding. Therefore, the actual uncertainty in the estimated profile $\kappa_m(h)$ cannot be precisely estimated. Fortunately, this uncertainty is generally minor. The particulate and molecular backscatter-to-extinction ratios, $\Pi_p(h)$ and Π_m, are defined as

$$\Pi_p(h) = \frac{\beta_{\pi,p}(h)}{\kappa_p(h)}, \qquad (1.12)$$

and

$$\Pi_m = \frac{\beta_{\pi,m}(h)}{\kappa_m(h)} = \frac{3}{8\pi} = \text{const.} \qquad (1.13)$$

To extract the profile of the particulate extinction coefficient, the square-range-corrected lidar signal in Eq. (1.11) should be transformed into a function that has only one variable $y(x)$; that is, the transformed signal should have the structure

$$Z(x) = C' y(x) \exp\left\{-2\int y\left(x'\right) dx'\right\}.$$

To achieve this form, the square-range-corrected signal $P(h)h^2$, should be transformed into the special function (Kovalev and Eichinger, 2004)

$$Z_w(h) = Y_w(h)[P(h)h^2], \tag{1.14}$$

where

$$Y_w(h) = \frac{1}{\Pi_p(h)} \exp\left\{-2\int_{h_0}^{h} \kappa_m\left(h'\right)[a_\pi(h') - 1]dh'\right\}, \tag{1.15}$$

The variable $a_\pi(h)$ is defined as

$$a_\pi(h) = \frac{\Pi_m}{\Pi_p(h)} = \frac{S_p(h)}{S_m}, \tag{1.16}$$

where the reversed backscatter-to-extinction ratios, $S_p(h)$ and S_m, are the so-called particulate and molecular lidar ratios, respectively. Unlike $\Pi_p(h)$ and $\Pi_m(h)$, which determine the fraction of backscattered light in the total backscattering, the notion of the lidar ratio has no proper physical context; however, in some cases, the reversed backscatter-to-extinction ratios are more convenient for practical use.

Calculating the transformed function $Z_w(h)$ requires knowledge of the profile $\Pi_p(h)$, which is unknown, and therefore, should be selected *a priori*. The only practical solution to this issue is the *a priori* selection of Π_p = const. The selected Π_p should always be taken as a nonzero positive value, even in aerosol-free areas, where $\kappa_p(h) = 0$. In this case, $\beta_{\pi,p}(h) = \Pi_p \kappa_p(h) = 0$, so any arbitrarily selected positive Π_p will be consistent for any aerosol-free area.

The use of the analytical dependence between $\kappa_p(r)$ and $\beta_{\pi,p}(r)$ in the form $\beta_{\pi,p}(r) = C_p[\kappa_p(r)]^a$, proposed by Klett (1981), does not solve the problem, as it requires an *a priori* selection of the exponent, a. The selection of $a = 1$, commonly used in practice, reduces this power-law dependence to the selection of constant Π_p.

In practical terms, it is not possible to invert the elastic-lidar signal into the atmospheric profile without using some *a priori* assumption (or assumptions) and adding uncertainty to the inversion result. The need of the *a priori* selection of the backscatter-to-extinction ratio is the weakest point in the elastic-lidar solution, as it transforms the analytical solution into a solution model. There is no simple and reliable way to check the validity of the assumed relationship between $\kappa_p(r)$ and $\beta_{\pi,p}(r)$ in the atmosphere searched by lidar. In many cases, such an *a priori* assumption can yield a highly inaccurate inversion result and in heterogeneous atmospheres, such errors may be extremely large (Kovalev and Eichinger, 2004).This issue is not a specific of lidar profiling of the atmosphere, it is a typical issue when attempting the transformation of an equation with two unknowns into an equation with one unknown.

The transformed signal $Z_w(h)$ can be written in the form

$$Z_w(h) = C'\kappa_w(h)\exp\left\{-2\int_{h_0}^{h} \kappa_w\left(h'\right)dh'\right\}, \tag{1.17}$$

where C' is a constant, and the variable $\kappa_w(h)$ is related to $\kappa_p(h)$ and $\kappa_m(h)$ in the form

$$\kappa_w(h) = \kappa_p(h) + \alpha_\pi(h)\kappa_m(h). \tag{1.18}$$

The variable function $\kappa_w(h)$ can be formally considered as a sum of two weighted extinction coefficients, where the weight of the molecular component is equal to $a_\pi(h)$ and that of the particulate component is unity. The purely theoretical solution for Eq. (1.17) can be written in the simple form (Kaul, 1977; Zuev et al., 1978)

$$\kappa_w(h) = \frac{Z_w(h)}{2\int_h^{\infty} Z_w(h')dh'}. \tag{1.19}$$

For practical application, the solution for the particulate extinction coefficient $\kappa_p(h)$ can be rewritten in the form (Barret and Ben-Dov, 1967; Viezee et al., 1969; Collis, 1969; Davis, 1969; Zege et al., 1971; Browell et al., 1985)

$$\kappa_p(h) = \frac{Z_w(h)}{CT_\Sigma^2(0,h_1) - 2\int_{h_1}^{h} Z_w(h')dh'} - a_\pi(h)\kappa_m(h), \tag{1.20}$$

where $h_1 \geq h_0$ is a starting near-end point where the extinction coefficient is determined, and

$$T_\Sigma^2(0,h_1) = \exp\left\{-2\int_0^{h_1} \left[\kappa_p\left(h'\right) + \kappa_m(h')\right]dh'\right\}. \tag{1.21}$$

The solution in Eq. (1.20) is difficult to apply in practice because of the necessity of knowing both the constant C and the two-way transmittance over the altitude range from ground level to the height h_1. Therefore, to derive the extinction coefficient $\kappa_p(h)$ from the elastic-lidar signals, more practical solutions, known as the *boundary-point* and *optical-depth* solutions, are commonly used.

1.2.2 Fernald's Boundary-Point Solution

Before analyzing the Fernald boundary-point solution, commonly termed *Klett-Fernald's* solution, some clarification is required. This solution was not proposed in the well-known work by Klett (1981). In that study, the power-law relationship between particulate backscattering and extinction was used. Currently, such a solution is considered obsolete. The solution in its modern form was initially proposed by Fernald (1982), and then analyzed in the studies by Sasano and Nakane (1984), Fernald (1984), and Klett (1985).

For zenith-directed lidar, the boundary-point solution can be rewritten from Eq. (1.20) in the form

$$\kappa_p(h) = \frac{Z_w(h)}{\frac{Z_w(h_b)}{\kappa_w(h_b)} - 2\int_{h_b}^{h} Z_w(h')dh'} - a_\pi(h)\kappa_m(h), \tag{1.22}$$

where h_b is a height at which the quantity $\kappa_w(h_b)$ should be someway established. Its value is determined similarly to Eq. (1.18), that is,

$$\kappa_w(h_b) = \kappa_p(h_b) + \alpha_\pi(h_b)\kappa_m(h_b). \tag{1.23}$$

If h_b is selected close to the near end of the lidar measurement range, the less stable near-end solution is used (Kovalev and Eichinger, 2004). The selection of h_b closer to the far end provides more stability, and in a polluted atmosphere, a more accurate solution (Klett, 1981). In this case, $h_b > h$, and accordingly, the sum of the two terms rather than their difference is determined in the denominator of Eq. (1.22).

To find the profile of the extinction coefficient $\kappa_p(h)$, the local particulate and molecular extinction coefficients, $\kappa_p(h_b)$ and $\kappa_m(h_b)$, and the ratio $a_\pi(h)$ over the entire operative altitude range should be assumed; in practice, there is no reliable method for grounded establishing these. The selection of the appropriate boundary condition in the troposphere is possible, but is always an issue. The assumption of an aerosol-free atmosphere at high altitudes is the most common assumption when the far-end solution is used. However, this assumption should be applied with caution. Recent HSRL observations discovered that even at heights of 9 km, a nonzero level of background particulates can be observed, and the ratio of the particulate backscatter coefficient $\beta_{\pi,p}$ to the molecular backscatter coefficient $\beta_{\pi,m}$ can reach levels of 10% or more (Hair et al., 2008). A poor selection of $\kappa_p(h_b)$ compared to the real optical characteristics at the boundary point can significantly distort the derived extinction-coefficient profile (Russell et al., 1979; Bissonnette, 1986; Rocadenbosch and Comerón, 1999). However, one should keep in mind that this is only a secondary issue. The main issue - even when inverting an ideal, not corrupted, elastic-lidar signal - is the selection of the appropriate backscatter-to-extinction ratio profile. This selection, even when based on statistical estimates, is the most critical issue in the elastic lidar profiling of the atmosphere (Sasano and Nakane, 1984; Ansmann et al., 1992; Ferrare et al., 1998).

Thus, the solution of the relatively simple lidar equation actually includes a number of implicit premises and explicit assumptions. The Fernald's solution would allow *measuring* the extinction coefficient with lidar if the backscatter-to-extinction ratio over the profiling range and the extinction coefficient at the reference point r_b could someway be measured, and accordingly, their uncertainty known. However, in practice, this is unrealistic, so assumptions about the numerical values of the above parameters are the only way to apply Eq. (1.22). Such assumptions are nothing but the selection of the solution model for determining the profile of the extinction coefficient in the examined atmosphere. There is no real way to estimate the actual uncertainty of the extracted extinction coefficient $\kappa_p(h)$. Therefore the determination of its profile is eventually a *simulation* based on the specifics of the measured lidar-signal profile and on the selected solution model.

1.2.3 Optical Depth Solution

The variant of the elastic-lidar solution, known as the optical depth solution, was initially considered in the studies by Fernald et al. (1972), Uthe and Livingston (1986), Balin et al. (1987), and Weinman (1988). The theoretical solution, which follows from Eq. (1.19), can be written in the form

$$\kappa_p(h) = \frac{0.5Z_w(h)}{I_w(h_1,\infty) - \int_{h_1}^{h} Z_w(h')dh'} - a_\pi(h)\kappa_m(h), \tag{1.24}$$

where h_1 is a starting near-end point of the operative range where the extinction coefficient can be determined and

$$I_w(h_1,\infty) = \int_{h_1}^{\infty} Z_w(h')dh'. \tag{1.25}$$

The lidar operative range is always restricted to heights from $h_{\min}$ to $h_{\max}$, and the optical depth solution in Eq. (1.24) can be rewritten in the more practical form

$$\kappa_p(h) = \frac{0.5Z_w(h)}{\frac{\int_{h_{\min}}^{h_{\max}} Z_w(h')dr'}{1-V_w^2(h_{\min},h_{\max})} - \int_{h_{\min}}^{h} Z_w(h')dh'} - a_\pi(h)\kappa_m(h), \tag{1.26}$$

where

$$V_w^2(h_{\min},h_{\max}) = T_\Sigma^2(h_{\min},h_{\max})\exp\left\{-2\int_{h_{\min}}^{h_{\max}} \kappa_m\left(h'\right)[a_\pi(h') - 1]dh'\right\}, \tag{1.27}$$

and

$$T_\Sigma^2(h_{\min},h_{\max}) = \exp\left\{-2\int_{h_{\min}}^{h_{\max}} \left[\kappa_m\left(h'\right) + \kappa_m(h')\right] dh'\right\}. \tag{1.28}$$

This solution is valid within the complete overlap zone, so that for vertically directed lidar, $h_{min} \geq r_0$. For practical use of Eq. (1.26), the term $T_{\Sigma}^2(h_{\min}, h_{\max})$ should be estimated by some means and some constant backscatter-to-extinction ratio within the interval $h_{\min} - h_{\max}$ assumed. The determination of the two-way transmittance term $T_{\Sigma}^2(h_{\min}, h_{\max})$, however, is not a trivial task. If the operative range of the zenith-directed lidar is enough extended, it may be estimated using sun-photometer data (Fernald et al., 1972; Balis et al., 2000 and 2003; Marenco et al., 1997; Cuesta et al., 2008). However, the sun photometer measures the atmospheric transmittance $T_{\Sigma,\text{sun}}(0, \infty)$ through the whole atmosphere, whereas the operative altitude range of lidar is always restricted from $h_{\min}$ to $h_{\max}$. Obviously, when auxiliary sun-photometer data are available, the acceptable accuracy of the profiling of the atmosphere with Eq. (1.26) can be achieved only if the discrepancy between the quantities $T_{\Sigma,\text{sun}}(0, \infty)$ and $\sqrt{T_{\Sigma}^2(h_{\min}, h_{\max})}$ is someway minimized. This condition means that before processing the lidar data, the optical depth outside the lidar operative range, that is, over the distant zone outside $h_{\max}$ and within the lidar near-end zone, from ground level to $h_{\min}$, should be someway estimated. If such an estimation is done, the value of $T_{\Sigma}(h_{\min}, h_{\max})$ can be calculated from the obvious formula

$$T_{\Sigma}(h_{\min}, h_{\max}) = \frac{T_{\Sigma,\text{sun}}(0, \infty)}{T_{\Sigma}(0, h_{\min})T_{\Sigma}(h_{\max}, \infty)}. \tag{1.29}$$

When the maximal lidar measurement range of the ground-based and zenith-directed lidar extends up to 8–10 km heights, the optical depth of the upper atmosphere in clear-sky conditions can often be ignored as compared to the optical depth $\tau(h_{\min}, h_{\max})$ within the lidar operative range; in other words, under such a condition, the quantity $T_{\Sigma}(h_{\max}, \infty) \approx 1$. The other component $T_{\Sigma}(0, h_{\min})$ cannot be ignored because the aerosol pollution near ground is generally maximal. In some cases, especially if the lidar has an extended incomplete overlap zone, the optical depth $\tau(0, h_{\min})$ may even be comparable to $\tau(h_{\min}, h_{\max})$. In such atmospheres, $\tau(0, h_{\min})$ may be roughly estimated using the data of ground-based instrumentation and the assumption of a homogeneous atmosphere within the layer from ground level to $h_{\min}$ (Ferguson and Stephens, 1983; Marenco et al., 1997). More sophisticated methods for estimating the optical depth $\tau(0, h_{\min})$ are considered in Section 2.5.3.

Let us summarize. The use of the classic lidar-equation solutions (more correctly – the solution models) for inverting the signals of zenith-directed elastic lidar always requires some *a priori* assumption about the vertical profile of the backscatter-to-extinction ratio over the lidar total operative range. In addition to this principal requirement, each aforementioned solution has its own issues and obligatory assumptions. The solution in Eq. (1.20) requires the knowledge of the constant C and the total transmittance within the near zone from $h = 0$ to $h = h_1$. The Fernald's solution [Eq. (1.22)] does not need knowledge of these parameters; however, if the assumption of an aerosol-free atmosphere at reference point h_b is not valid, the particulate extinction coefficient and the lidar ratio at this point should be taken *a priori*.

Finally, the optical depth solution [Eq. (1.26)] requires knowledge of the two-way transmission over the height interval from h_{min} to h_{max}. However, as previously stated, the most important issue for determining the atmospheric profiles with elastic-lidar data is the need of some assumption on the profile of $\Pi_p(h)$ over the searched atmosphere. As a result, each researcher selects the numerical value for Π_p that he or she believes is most likely for the atmosphere of interest. Such a state of affairs may result in significant differences in the optical profiles obtained by different researchers from the same lidar data, especially when profiling of the atmosphere is made during complex meteorological conditions. In fact, no researcher can definitely prove that his or her results are true for the real atmospheric situation. The only way to validate how well the results obtained track reality is by making use of measurements from additional instrumentation. However, in practice, such observations are rarely possible, and what is worse, in most cases, the additional instrumentation introduces additional sources of uncertainty.

1.2.4 Implicit Premises and Mandatory Assumptions Required for Inversion of the Elastic Lidar Signal into the Atmospheric Profile

As shown in Section 1.1, the assumptions and implicit premises used during the elastic-lidar signal recording and its initial processing can significantly worsen the accuracy of the extracted optical profiles of the atmosphere. The next signal inversion procedures, illustrated in Fig. 1.5 can dramatically worsen the situation. In the schematic, only the most important specifics of the operations are briefly given; note that the consequence of the operations may be different when different solution algorithms are used. The selection of the retrieval algorithm is the first step (block 1), after which additional smoothing of the signal, if necessary, is made (block 2). This operation involves an important implicit premise. As is known, inadequate smoothing can result in either losing important details in the inverted profile or in so-called overfitting. Extraordinary smoothing can result in the loss of real local variations in the output function of interest. Overfitting occurs when the smoothed function shape is formed by noise instead of the underlying noise-corrupted function of interest (Hawkins, 2004; Vorontsov, 2006). When selecting the level of spatial signal smoothing, the implicit premise is that the low-frequency noise components $\sum w_{j,\mathrm{low}}(r)$ remaining after initial signal filtering, are negligible and can be ignored.

The next step (block 3) includes the selection and the adaptation (when possible) of the available auxiliary data obtained by independent instrumentation. This operation requires establishing the way in which such data will be used. The auxiliary instrumentation may include *in situ* instrumentation at ground level (i.e., an integrating and a backscatter nephelometer) located at the lidar sounding site and the sun photometer, wavelengths of which either coincide or are close to the lidar operational wavelength. The information obtained from balloon measurement of the vertical molecular density and temperature profiles can be used instead of applying statistical profiles. Such independent data from auxiliary instrumentation allows putting helpful restrictions on the inversion result. On the other hand, the use of auxiliary data in lidar profiling

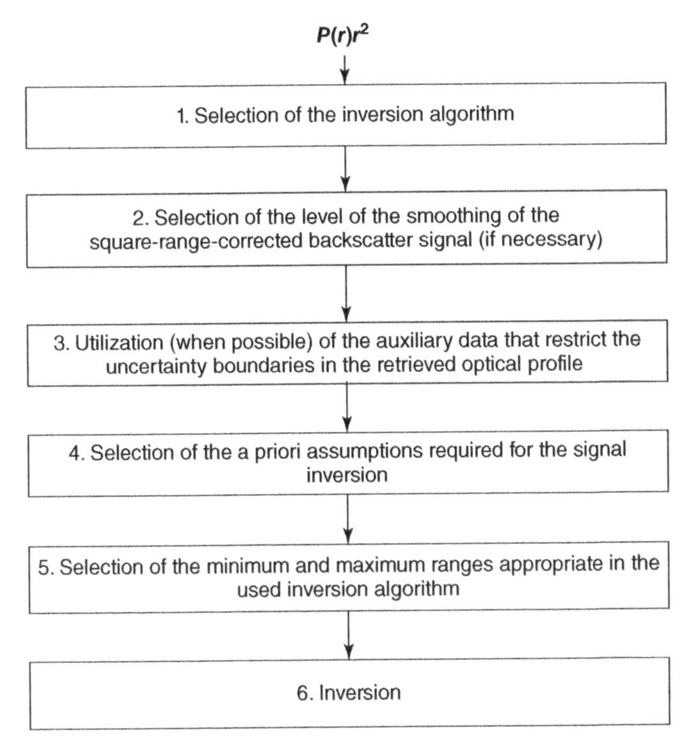

Fig. 1.5 General schematic of the square-range-corrected lidar signal.

of the atmosphere may be accompanied by the use of additional implicit premises or even additional assumptions.

The next step (block 4), in which the *a priori* assumptions, mandatory for the utilized inversion algorithm, are selected, is an extremely critical step in the inversion procedure. As mentioned above, the most important issue when inverting the elastic-lidar signal is the selection of the backscatter-to-extinction ratio. Another concern, which is inherent to the boundary-point solution, is the selection of the height where the assumption of the aerosol-free atmosphere, which allows the application of the far-end solution, is valid.

Taking into account the level of the remaining signal noise and the length of the incomplete overlap zone, the minimum and maximum ranges, r_{min} and r_{max}, within which the inversion algorithm will be applied, should be established (block 5). In some cases, for example, for the optical depth and far-end solutions, the selection of r_{max} should be done extremely cautiously. Proper selection of the lidar operative range reduces the influence of additive and multiplicative signal distortions caused by the nonzero offset and the low-frequency noise remaining in the signal. Finally, *a posteriori* analysis of the inversion result can be quite useful: it can allow the researcher to decide whether the inversion result is acceptable or has to be redone after selecting, for example, a reduced r_{max}. The assumptions and the implicit premises used for the inversion of the elastic-lidar signals are briefly listed in Table 1.2.

Thus, lidar profiling of the atmospheric parameter of interest, such as the optical depth or the extinction coefficient, requires the evaluation of a number of parameters, atmospheric and instrumental, not related directly to the investigated optical parameter. The accuracy of the inversion result may strongly depend on the assumptions that the researcher considers sensible, whereas the adequacy of these for the real atmospheric situation during lidar searching remains unknown.

Summarizing the above inversion objectives, one should state that lidar profiling of the atmosphere yields some probable shape of the optical profile of interest but not necessarily the most probable shape in the sense of the maximal likelihood. In other words, instead of a true mean profile with its standard deviation, one obtains the profile that meets the assumptions used for signal inversion. The uncertainty of the inverted optical profile cannot be strictly quantified, and no information can be extracted on how precise the details of this profile are.

As shown in Chapters 2 and 3, an assumption or a number of assumptions are needed practically for any lidar-data processing algorithm. Such assumptions may be related to the wavelength dependence of the particulate extinction coefficient (in differential absorption lidar (DIAL) and Raman measurements), the selection of different weighting schemes (Cheng et al., 2004), the specifics of the used technique

TABLE 1.2 Assumptions and the Implicit Premises Used for the Inversion of the Elastic-Lidar Signal

	Inversion Procedure of the Elastic-Lidar Signal	Assumptions/Implicit Premises
1.	Determination of the particulate extinction coefficient using Fernald's solution [Eq. (1.22)]	(a) *A priori* selected constant value of the particulate backscatter-to-extinction ratio is adequate for the actual optical situation (b) Height fluctuations of actual backscatter-to-extinction ratio are minor and do not cause distortion in the extracted profile of the extinction coefficient; (c) The assumed boundary value at the reference height is close to the actual value
2.	Determination of the particulate extinction coefficient using the optical depth solution [Eq. (1.26)]	(a) *A priori* selected constant value of the particulate backscatter-to-extinction ratio is appropriate for the actual optical situation (b) A difference between the estimated two-way transmittance over the total operative range and the actual two-way transmittance over this range does not result in significant distortion in the retrieved profile of the extinction coefficient

for the numerical differentiation, etc. Any such assumptions, inaccurately addressed, may cause significant distortions in the derived atmospheric profile.

1.3 PROFILING OF THE OPTICAL PARAMETERS OF THE ATMOSPHERE AS A SIMULATION BASED ON PAST OBSERVATIONS

1.3.1 Definitions of the Terms

According to the basics of the error analysis, two characteristics should generally be taken into consideration when estimating a measurement result: its accuracy and its precision. The term "accuracy" is defined as a measure of how close the result of the experiment is to the true value; the term "precision" defines the measure of how well the result has been determined (Bevington and Robinson, 2003). The measurement can be precise and accurate, precise but not accurate, and accurate but not precise. If one considers lidar profiling as a measurement process, one should draw the conclusion that the results of this process generally are neither accurate nor precise. Taking into consideration the large number of hidden systematic errors in lidar signals and the set of implicit premises and assumptions in the lidar-equation solutions one should conclude that the lidar signal inversion result may yield extremely large distortions in the derived optical profiles of the atmosphere.

As discussed in the previous sections, profiling of the atmosphere includes three independent operations: (i) recording the lidar signal $P_{\Sigma}(r)$, which is the sum of the backscattered light and background, (ii) separating and square-range-correcting the backscattered signal $P(r)$ and (iii) inverting this signal into a profile of the atmospheric parameter of interest. As pointed out in Section 1.1, there are no grounds to expect that one can transform light energy from the lidar telescope into an electrical signal and then separate the range-dependent backscatter component from the constant offset without introducing systematic distortions into the backscattering signal.

Eventually, the uncertainty in the lidar signal $P(r)$ within the properly established operative range, can be estimated. Therefore, the process of determining the total lidar signal may be considered as a *measurement*. The situation dramatically changes when the square-range-corrected backscatter signal is separated and inverted into an optical parameter, such as the optical depth or the extinction coefficient. As shown in Section 1.1, such an operation is based on the use of assumptions and/or implicit premises. The use of any *a priori* assumption invalidates quantifying the exact uncertainty of the atmospheric profile of interest.

To clarify this observation, one should recall the definition of the term "measurement." Unfortunately, there is no standard definition of this term. Probably, the most general definition of this term was given by Stevens (1968), who defined "measurement" as "the assignment of numbers to aspects of objects or events according to one or another rule or convention." As pointed out in the book by Pedhazur and Shmelkin (1991), numbers are assigned to aspects of objects, not to the object themselves. Any measurement should provide three values: the level of the measurement, the dimensions, and the uncertainty of the measured value. Thus, if the measurement of the extinction coefficient could be done, this procedure ought to yield not only its numerical value, but also its uncertainty, both absolute and relative. Meanwhile, when considering the extremely complicated structure of the real lidar signal, which is to be inverted into the atmospheric profile of interest, the task is unsolvable in principle; the inversion cannot yield a measurement result. In other words, the term "measurement" is not applicable when the lidar signal is inverted into optical properties of the atmosphere using one or another lidar-equation solution model, for example, such as described in Section 1.2. As shown in Chapter 2, this statement is valid not only for the elastic-lidar signals, but also for Raman lidar, HSRL, and DIAL.

The key problem of remote sensing of the atmosphere with lidar is that the inversion of the lidar signal into the optical profile is always an ill-posed problem. Moreover, the principle of maximum likelihood generally does not work here; the lidar inversion technique is based on other principles. The application of the well-known methods of statistical regularization is actually a quasi-rational solution to the problem of lidar profiling of the atmosphere. The statistical methods are rational only when the involved data obey relatively simple statistical laws. Then the determination of the most likely mean profile and its uncertainty is a mathematical task with known standard solutions.

A comprehensive analysis of this issue shows that for any type of atmospheric profiling, including profiling with splitting lidars considered in Chapter 2, the most significant output distortions appear during signal inversion, when the optical parameters of the atmosphere are extracted from the lidar signal. One can state with absolute certainty that when inverting real lidar signals into optical parameters, the researcher actually makes some *simulation*, which is based on past lidar observations. This simulation is based on some explicit assumptions and implicit premises, some statistics, and finally, on the researcher's intuition and common sense. For such a simulation based on previous lidar observations, one can use the shorter term, *a posteriori simulation*. Obviously, after performing this simulation, the researcher obtains only some approximate estimate of the actual atmospheric profile of interest with an uncertainty that cannot be reliably quantified. Actually, lidar profiling of the atmosphere is more art than science.

The related question is whether the commonly used lidar-equation solutions are rigid analytical solutions or just computational models. According to the web-site http://en.wikipedia.org/wiki/Analytical_solution, the term *analytical solution* should be restricted to a closed-form expression, which can be expressed analytically in terms of a finite number of certain well-known functions, such as constants, variables, elementary operations, exponent and logarithm, etc. Unfortunately, the lidar-equation solutions cannot be generally expressed analytically using only well-known functions. Therefore, these lidar-equation solutions should be treated as computational *solution models*. In other words, the term "solution model" looks more correct than the term "lidar equation solution." Therefore, the concept of the solution model in this book is confined to a mathematical representation of the atmospheric profile of interest, such as the optical depth and extinction and backscatter coefficient, based on the measured lidar signal and some assumptions and premises. Other models, which extrapolate the possible behavior of such profiles outside the lidar operative range, are discussed in Section 2.5.3. Such extrapolated modeling can be made for the zones where lidar data points either are absent or cannot be directly used to get an acceptable inversion result.

Let us continue the consideration of the terms related to the subject under discussion. In literature, the process of creating and analyzing a digital prototype of a physical model to predict its performance in the real world is sometimes defined as simulation modeling. Such a term may cause readers confusion because the term "modeling" is generally used for the process of creating and validating a model. The alternative term is "simulated experiment" (Winsberg, 2003). In our analyses, for the notion of computer simulation not based on a previous lidar observation, the term *numerical simulation* will be used. This term will describe mathematically the process of determining the backscattered signal from a synthetic atmosphere and the inversion of this signal into an atmospheric profile using a certain solution model. Such a numerical simulation allows imitating the atmospheric conditions of interest (for example, smoke layering in the vicinity of a wildfire) and investigating these theoretically. This is done by assessing the behavior of the backscatter signal "measured" by a virtual lidar in the investigated synthetic atmosphere, next retrieving the atmospheric profile using this signal, and comparing it with the initial profile used for the simulation.

Some comments may still be required to clarify the difference between the terms "modeling" and "simulation," which are often used interchangeably. As stressed by Cook and Skinner (2005) and stated in the "Online M&S Glossary" of the Department of Defense, modeling and simulation should be considered as different operations. A *model* is defined as "a physical, mathematical, or otherwise logical representation of a system, entity, phenomenon, or process." *Modeling* is the "application of a standard, rigorous, structured methodology to create and validate [this model];" [see also *DoD Modeling and Simulation (M&S) Glossary* (1998), Defense Modeling and Simulation Office, Department of Defense, DoD 5000.59-M (cited from www.dtic.mil/whs/directives/corres/pdf/500059m.pdf)]. The implementation of a model is considered to be static, whereas producing the model output is a simulation. As clarified in studies by Cook (2001) and Cook and Skinner (2005), simulation should be defined as a process, as a method for implementing a model. When developing analytical tools, separating the definition of modeling from simulation is extremely practical.

Finally, it is worthwhile to list the uncertainties when using a combination of experimental measurements and mathematical models. The study by Kennedy and O'Hagan (2001) nicely endorses the statement that profiling of the atmospheric optical parameters with lidar is nothing but a simulation based on past observations. According to the aforementioned authors, there are six sources of uncertainty. One can clearly see that all of these are inherent to the lidar profiling methodology [cited using the terms in Wikipedia (http://en.wikipedia.org/wiki/Uncertainty_quantification)]:

- *Parameter uncertainty*, which stems from the used mathematical model, in our case, from the solutions derived from the lidar equation. The exact input values are unknown to experimentalists and cannot be controlled when making atmospheric profiling.
- *Model inadequacy* originating from the lack of exact knowledge of the underlying physics. The same as any mathematical model, the lidar equation is only an approximation of reality. This approximation is inaccurate because, in real atmospheres, uncountable parameters may exist, which are impossible to accurately estimate.
- *Algorithmic uncertainty* originating in numerical errors and numerical approximations. Most lidar-equation solutions are too complicated to be solved uniquely. For example, the results of numerical differentiation, which is required in most inversion algorithms, will depend significantly on the particular differentiation techniques used for such an operation.
- *Parametric variability* originating, for example, in the variability of input variables to the lidar-equation solution. For example, the commonly used Mie theory is only valid for spherical particulates. Meanwhile, the presence of non-spherical particulates such as dust and smoke particles in the searched atmosphere may introduce additional uncertainty due to the differences between the solution model and real backscattering.
- *Experimental uncertainty*, which comes from the variability of experimental conditions. Experimental systematic uncertainties are inevitable and are rarely noticed in time.
- *Interpolation or extrapolation uncertainty*, which is caused by a lack of available data collected from experimental measurements. For example, the interpolation uncertainty in multiangle measurements may take place when the digitally recorded backscatter signals measured in a slope direction are recalculated for fixed altitudes.

1.3.2 Random Systematic Errors in the Derived Atmospheric Profiles: Origin and Examples

Measurement uncertainty can also be classified into two general categories: statistical aleatoric uncertainty and systematic epistemic uncertainty (Kiureghiana and Ditlevsen, 2009; Matthies, 2007). The first, statistical aleatoric uncertainty, occurs because the experimental researcher can obtain and then operate only with some average values. Questions immediately arise as to the physical sense of these averages because most atmospheric elements are not normally distributed. The second, systematic epistemic uncertainty, occurs because of effects which practically cannot be exactly estimated, such as imperfectness of the model used. Both uncertainties emerge when performing lidar profiling of the atmosphere, therefore, both types of uncertainty should be kept in mind when estimating the reliability of the output results of such profiling. In practice, it is not possible to separate their influence on the distortions in the derived atmospheric profile. It is sensible, therefore, to consider these uncertainties as components of the overall uncertainty that causes *random systematic error* in the inverted atmospheric profiles. This term, as strange as it seems, has good grounds for implementation and will be used in our next analyses. It stresses the uncontrolled variability of random and systematic errors in inverted optical profiles.

A. Experimental Uncertainty in the Nonstationary Atmosphere As mentioned in Section 1.1, the selection of lidar-signal averaging parameters is inevitably related to the selection of some *a priori* assumptions. The selection of the data averaging time is related to the assumption that the parameter measured during this time obeys some known distribution, generally, the normal distribution. A similar implicit premise is used when selecting the signal-averaging range to decrease random high-frequency noise; it is assumed that after averaging, no more significant noise components remain in the signal.

To clarify the general problem, let us consider the experimental uncertainty which appears when lidar is profiling a nonstationary atmosphere. As is known, the stationary process is a stochastic process whose joint probability distribution does not change when shifted in time or space (see http://en.wikipedia.org/wiki/Stationary_process). In a stationary atmosphere, the value of the optical parameter of interest does not vary regardless of the lidar-signal measurement time, nor do its mean and variance. On the contrary, the process in which the statistical properties change in time is nonstationary. Such a nonstationary process occurs in real atmospheres, where the optical properties may vary over time in unpredictable ways while the lidar performs sounding the atmosphere. Simply speaking, the real atmosphere does not follow the "rules of the game" invented by humans for their convenience. Moreover, there is no reliable way to address the ranges where the atmosphere is nonstationary. Often it may be quite difficult, if not impossible, to distinguish the spatial variations of the atmospheric properties and the signal low-frequency noise when its period is comparable or larger than the signal measurement time.

As follows from Eq. (1.2), in the nonstationary atmosphere, random systematic error may be introduced into the profile of the measured backscatter signal, $P(r)$, obtained through the summation of single backscatter signals $P_j(r)$ during data acquisition. In the nonstationary atmosphere, such an unpredictable error is unavoidable for any atmospheric parameter derived during a finite time interval.

To clarify this issue, let us consider typical changes of the optical parameters in a real atmosphere. Let a vertically directed virtual lidar be profiling a synthetic clear-sky atmosphere, recording the backscattering signals, $P_j(h)$, during some time interval $\Delta t^{(\mathrm{signal})}$. While the signal is recorded, a small aerosol cloud intersects the lidar line of sight at the height h (Fig. 1.6); this intersection occurs during the time interval $\Delta t^{(\mathrm{cloud})}$, lasting less than the measurement time $\Delta t^{(\mathrm{signal})}$. Accordingly, during the intersection time, the lidar measures additional backscattering from the cloud. Obviously, the resulting backscatter signal $P(h)$, determined by the summation of the single backscatter signals $P_j(h)$, and the corresponding extinction coefficient retrieved from $P(h)$ in such an optical situation will depend on the difference in time intervals, $\Delta t^{(\mathrm{signal})}$ and $\Delta t^{(\mathrm{cloud})}$. Suppose that at the fixed height h,

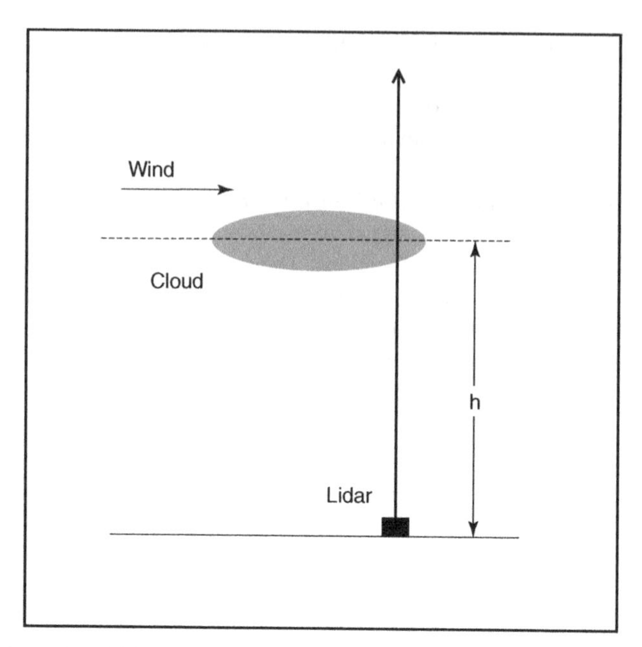

Fig. 1.6 Schematic of lidar profiling of the atmospheric optical properties in a real, "non-frozen" atmosphere.

the background extinction coefficient for the clear sky is $\kappa^{(bgr)}(h) = 0.1\ \text{km}^{-1}$, the extinction coefficient of the cloud is $\kappa^{(cloud)}(h) = 1\,\text{km}^{-1}$, and the signal recording time $\Delta t^{(signal)}$ is significantly larger than the time period $\Delta t^{(cloud)}$. Even in an ideal case, when there are no distortions to spoil the measured backscattered signal, the retrieved extinction coefficient at the height h will differ from the extinction coefficient of the clear sky, and this difference will be systematic. The derived extinction coefficient will overestimate the relative background value $\kappa^{(bgr)}(h)$ and the level of this overestimation will depend both on the extinction coefficient of the cloud and on the difference between the lidar-signal recording time and the time during which the cloud crosses the lidar field of view. If, for example, the ratio $\Delta t^{(signal)}/\Delta t^{(cloud)} = 3$, then the average value of the derived extinction coefficient at the height h will be equal 0.43 km^{-1}; if $\Delta t^{(signal)}/\Delta t^{(cloud)} = 10$, then the retrieved extinction coefficient is 0.2 km^{-1}, significantly closer to $\kappa^{(bgr)}(h)$. If the time ratio is much larger, for example, $\Delta t^{(signal)}/\Delta t^{(cloud)} = 1000$, then the retrieved extinction coefficient is equal to 0.101 km^{-1}, that is, the systematic difference between the measured and background extinction coefficient is only 1%. Note that in such a case, the atmosphere cannot be considered as horizontally homogeneous even in the statistical sense, because the variations of the extinction-coefficient profile do not obey any predictable statistical distribution.

The required time during which the atmosphere should remain "frozen" depends on the type of the used lidar; it may change from some seconds for conventional elastic lidar up to a half an hour or more for Raman lidar. For example, during the comparison of the water vapor profiles obtained with a Raman lidar and the concurrent routine radiosondes by the German Meteorological Service, lidar raw data were integrated for 20 min (Reichardt et al., 2012); Raman lidar profiling of the aerosol extinction and backscattering, discussed in the study by Ansmann et al. (1992), required averaging times up to 26 min; a study by Larchevêque et al. (2002) required averaging times of more than 30 min; the measurement time for the Raman lidar used for monitoring extinction and backscattering in African dust layers was up to 1 h (De Tomasi et al., 2003).

To illustrate the results of quantifying the random systematic error in such an atmospheric situation, let us perform a more sophisticated numerical simulation. Let the lidar signal at the wavelength 532 nm be recorded in a synthetic atmosphere in which the extinction coefficient continuously decreases with height from 0.14 km at ground level to 0.006 km at the height of 10 km. While the lidar signal is being recorded, two local aerosol layers intercept the lidar line of sight. The first one at the height 2500–2800 m remains within the lidar field of view during the interval, $\Delta t_1{}^{(cloud)} = 0.1\Delta t^{(signal)}$. Then it disappears and shortly thereafter, another local aerosol layer appears, this time at the height of 6000–6500 m. That layer remains within the lidar field of view during a larger time interval, $\Delta t_2{}^{(cloud)} = 0.2\Delta t^{(signal)}$. Clear-sky conditions take place during the remaining time interval of $0.7\Delta t^{(signal)}$.

The corresponding square-height-corrected signals of the above virtual zenith-directed lidar, measured during the absence and the presence of the aerosol layers, are shown in Fig. 1.7 as Curves 1, 2, and 3. Curve 1 is the signal measured while no cloud crosses the lidar field of view. Curves 2 and 3 show the signals while the upper and lower local clouds intersect the lidar field of view, respectively. The thick curve 4 shows the resulting signal $P(h)h^2$ averaged over the total measurement time $\Delta t^{(signal)}$. Presumably, the virtual lidar provides an ideally precise and accurate backscatter signal measurement.

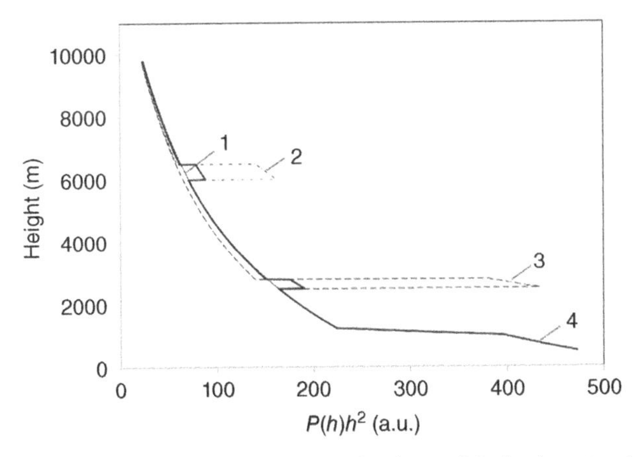

Fig. 1.7 Influence of the local aerosol layers on the shape of the backscatter signal averaged over different time intervals.

The particulate extinction-coefficient profiles, extracted from the signals of the virtual lidar are shown in Fig. 1.8. The profile of the background extinction coefficient, which monotonically decreases with height, is shown as Curve 1. Curves 2 and 3 show the extinction-coefficient profiles that correspond to the temporal intervals $\Delta t_1^{(\mathrm{cloud})}$ and $\Delta t_2^{(\mathrm{cloud})}$, when the local clouds intersect the lidar telescope field of view; Curve 4 shows the profile $\kappa_{p,\mathrm{calc}}(h)$, retrieved from the lidar signal acquired during the measurement time interval $\Delta t^{(\mathrm{signal})}$.

In Fig. 1.9, the particulate extinction coefficient $\kappa_{p,\mathrm{calc}}(h)$, retrieved from the backscatter signal obtained in the same atmospheric conditions, is shown. The horizontal extension of the local aerosol layers, and accordingly, the time intervals, $\Delta t_1^{(\mathrm{cloud})}$ and $\Delta t_2^{(\mathrm{cloud})}$, are the same as in the previous case. However, the signal is

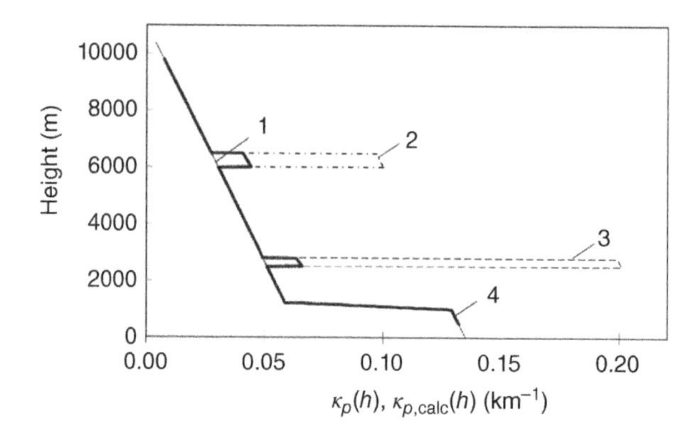

Fig. 1.8 Curves 1, 2, and 3 are the true profiles of the particulate extinction coefficient $\kappa_p(h)$ during different time intervals. The profile $\kappa_{p,\mathrm{calc}}(h)$ derived from the lidar signal accumulated during the total measurement time is shown as Curve 4.

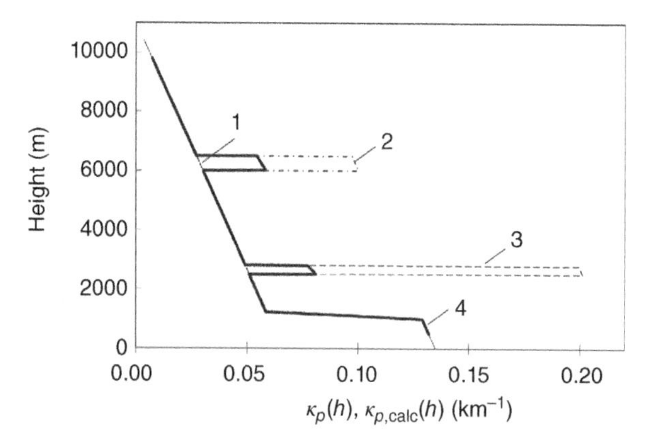

Fig. 1.9 Same as in Fig. 1.8 but where the backscatter signal is accumulated during a shorter time interval.

accumulated during a shorter time interval $\Delta t^{(\text{signal})}$, which is 40% less than in the previous case. The retrieved extinction coefficient, $\kappa_{p,\text{calc}}(h)$, has increased values of the extinction coefficient at the heights of local cloudiness, however, these are noticeably larger than in the previous case. Thus, if the temporal intervals $\Delta t_1{}^{(\text{cloud})}$ and $\Delta t_2{}^{(\text{cloud})}$ during which the cloud layering structures intersect the lidar telescope field of view are less than the time of signal acquisition, the magnitude of the corresponding bumps in the profile, $\kappa_{p,\text{calc}}(h)$, depends on the total acquisition time. Reduction in acquisition time will increase the extinction coefficient $\kappa_{p,\text{calc}}(h)$ at the heights of cloud layering structures.

The above numerical simulation makes clear the notion of random systematic error. When the condition of frozen atmosphere is not valid, the profiles of the extinction coefficient extracted from the signals acquired during a randomly selected measurement time will have systematic distortions. At heights where local layering periodically takes place, the extinction coefficient will always be overestimated as compared with the background clear-sky level, which is generally the profile of interest. The derived results depend on the specifics of local layering and the time of the lidar signal acquisition. In other words, extinction coefficients retrieved at these heights are not true values even in a statistical sense. Therefore, the use of common standard statistical estimates such as the mean value and standard deviation, have no mathematical basis.

Let us summarize. The methodology based on the accumulation of the backscatter signals during a prolonged time in real atmospheres has no grounded theoretical basis. The assumption of random fluctuations that obey common statistical laws is not valid when working in such atmospheres. Meanwhile, the tropospheric profiles derived from the averaged lidar data generally show relatively stable background structures and look sufficiently reliable. This explains why the assumption of a frozen atmosphere is still so popular among researchers. They just have no other option.

B. Parameter Uncertainty in the Boundary-Point Solution The most significant issue in lidar profiling of the atmosphere is that the systematic distortions in the derived optical profiles are generally range dependent. These distortions depend both on the distortions in the inverted signal and on the shape of the aerosol profiles along the searching laser beam. While the range-independent additive offset ΔB in the backscatter signal may be either positive or negative, but always having the same sign over the whole lidar-signal measurement range, the distortions in the optical profiles obtained from such a signal may not be systematic. The distortion caused by the offset $\Delta B \neq 0$, can result in an underestimated optical profile within one range interval and an overestimated one within the other. A similar effect can be caused by the multiplicative distortion component $[1 + \delta_P(r)]$. It may create systematic distortions in the derived optical profiles, which are positive within one interval and negative within another interval. This feature makes the resulting distortions in the inverted lidar data unpredictable even if the signal distortions are addressed. The question whether random systematic errors are present in derived atmospheric profiles is generally hard to answer.

The imperfection of the solution model may be another source of random systematic error. To illustrate how unpredictable inversion results are when, at least one *a priori* assumption is used, let us consider the random systematic error that appears when using the Fernald solution. The test profile of the particulate extinction coefficient at 532 nm used for this numerical simulation is shown in Fig. 1.10 as the bold curve. The profile imitates a clear atmosphere in which the particulate extinction coefficient decreases with height from $\kappa_p \approx 0.1\ \text{km}^{-1}$ at ground level to zero at the height $\sim 7\ \text{km}$. In addition, a local particulate layer with increased backscattering is present between the altitudes 2.5 and 4.5 km. Such layering is often produced by smoke plumes in the vicinity of wildfires due to atmospheric inversion (Kovalev et al., 2008, 2009a, 2009b, 2011). The "true" particulate backscatter-to-extinction ratio in this synthetic atmosphere is $\Pi_p = 0.05\ \text{sr}^{-1}$ in clear-air areas and $\Pi_p = 0.015\ \text{sr}^{-1}$ in the area of the particulate layer with increased backscattering.

To extract the extinction coefficient from the backscatter signal with the Fernald far-end solution [Eq. (1.22)], an *a priori* assumption about the numerical value of the vertical profile of the backscatter-to-extinction ratio is required (Section 1.2.2). As no additional information is generally available, some range-independent profile $\Pi_p = \text{const.}$ is selected. In Fig. 1.10, three retrieved profiles of the extinction coefficients are shown; these have been obtained by inverting the same signal of a virtual zenith-directed elastic lidar. The retrieval is made with *a priori* selected height-independent backscatter-to-extinction ratios, equal to $0.015\ \text{sr}^{-1}$, $0.05\ \text{sr}^{-1}$, and $0.08\ \text{sr}^{-1}$. The simulation is made under the condition that no additional errors influence the inversion result; that is, no distortion in the recorded signal takes place, no signal noise is present, and the precise boundary value is known. However, as can be seen in the figure, even in such an ideal case, the distortions in the retrieved

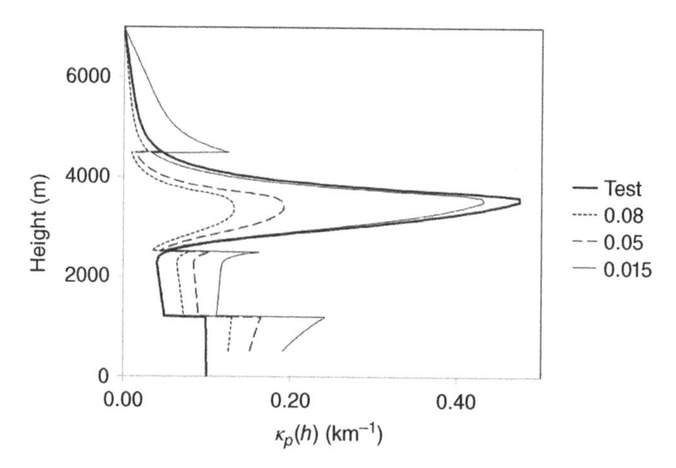

Fig. 1.10 Extinction-coefficient profiles retrieved with different backscatter-to-extinction ratios, $\Pi_p = \text{const}$; the values of the assumed ratios used for the retrieval are shown in the legend. The actual (test) extinction-coefficient profile is shown as the thick solid curve.

profiles $\kappa_p(h)$ caused by assumed constant backscatter-to-extinction ratios Π_p prove to be large.

This simulation nicely highlights the notion of random systematic error. The retrieved extinction-coefficient profile has systematic distortions of different sign over two extended height intervals, from $h_{\min} = 500$ m to $h = \sim 2.5$ km and from 2.5 to ~ 4.5 km. In the area of the polluted particulate layer, the extracted $\kappa_p(h)$ is always smaller than the "true" values, whereas below this area, the extracted $\kappa_p(h)$ is always larger than these values. Over heights greater than 4.5 km, the extracted profiles are incorrect except for the profile extracted with $\Pi_p = 0.05$ sr^{-1}, which is equal to the true value. In real lidar searching with real zenith-directed elastic lidar, there would be no way to determine which extracted profile of $\kappa_p(h)$ is closer to the actual profile.

Numerous experimental data confirm that in the Fernald solution, the uncertainty in the backscatter-to-extinction ratio has a most destructive influence on the extracted extinction coefficient, and this issue has no acceptable solution. This was nicely demonstrated in the experimental study by Ferrare et al. (1998), where comparisons of the inversion results obtained by elastic and Raman lidar were made. The profiles of the extinction coefficient obtained with Raman lidar were compared with the inversion results extracted from elastic-lidar signals using Fernald's method with three height-independent values of Π_p (0.0125 sr^{-1}, 0.0167 sr^{-1} and 0.025 sr^{-1}). As with our numerical experiment, the largest discrepancies between the Raman and elastic lidar were fixed in the areas of maximal gradient in the extinction coefficient.

Let us briefly summarize the basic specifics of random systematic error in lidar profiling of the atmosphere. As with the random error, the random systematic error is range dependent. However, unlike ordinary random error, it can have the same sign over extremely extended altitude ranges, causing the extracted profile of the extinction coefficient to be systematically shifted relative to the true profile over these areas. The sign and the value of the shift and the length of the corresponding altitude intervals will depend on specifics of the searched aerosol profile, on the adequacy of quantities selected *a priori* to actual ones, and on hidden systematic distortions in the backscatter signal.

1.4 ERROR FACTOR IN LIDAR DATA INVERSION

As stated in Section 1.3, instead of acquiring the result of the lidar measurement, the lidar researcher actually gets a result of some *a posteriori* simulation. The result of such a simulation depends heavily on the assumptions and premises inherent to the selected algorithm and on how these are compatible with the atmospheric situation in which the lidar signals were measured. The principal reason for such a situation lies in the fact that the inversion of the lidar signal into the optical profile is a typical ill-posed problem (Tikhonov and Arsenin, 1977; Lavrent'ev et al., 1980).

Unfortunately, there is one more aggravating factor, which can significantly distort the result of lidar profiling of the atmosphere: the solution of the lidar equation may be ill conditioned, or simply speaking, it can be extremely unstable. The solution is considered as being ill conditioned if a minor error in the input data results in a significantly larger error in the output value. Such an effect is typical for many lidar algorithms, especially, when profiling is made in clear atmospheres.

To address this issue in the mathematical science, the notion of the condition number is used (Thefethen and Bau, 1997). The condition number (CN) defines the relative change of the output function originating from a change in an input element of this function. The condition number for the inverted function $f(x)$ at the point x is defined as the ratio of the fractional change in $f(x)$ to a fractional change in x in the limit as Δx becomes infinitesimally small, that is,

$$CN = \lim_{\Delta x \to 0} \left[\frac{\frac{\|f(x+\Delta x)-f(x)\|}{\|f(x)\|}}{\frac{\|f(x)\|}{\|x\|}} \right]. \tag{1.30}$$

The condition number expresses how sensitive the output function $f(x)$ is to small changes (or small errors) in its argument x. The ill-conditioned solution is indicated by a large condition number. In contrast, the well-conditioned solution has a small condition number.

The direct application of the condition number defined in Eq. (1.30) for the analysis of lidar-signal-inversion algorithms does not allow one to perform general analysis of the specifics of these algorithms. When processing discrete lidar signals, numerical data rather than the analytical dependencies are used. The specific of the lidar signal inversion is that any output parameter is extracted from a number of data points and represents some mean output value within a range interval $(r, r + \Delta r)$. For example, when using the sliding numerical derivative, the range interval from r to $(r + s)$ is selected where s is the range resolution, length of which can reach several hundred meters. The selected value of the range resolution significantly influences the shape of the derived optical profile, and this feature needs to be taken into account when implementing a parameter like the condition number for lidar data analysis. First, such a parameter, termed below as the *error factor* (EF), should be range dependent; second, it should permit obtaining the analytical relationship between the error in the input parameter and the corresponding error in the output parameter. In a general form, the error factor $EF[F(x), x]$ can be defined as the absolute value of the ratio of the fractional error in the output function $F(x)$ to the fractional error of the corresponding input element x, that is,

$$EF[\delta F(x), \delta x] = \left| \frac{\delta F(x)}{\delta x} \right|. \tag{1.31}$$

The estimate of the error factor allows one to obtain a general understanding of whether the solution is ill- or well-conditioned. Sensibly, one can consider the lidar-equation solution as well-conditioned when the error factor does not exceed unity. If the error exceeds unity, but still is not more than 3–5, the solution can be considered as normally conditioned. Accordingly, the solution can be treated as ill-conditioned if the error factor exceeds 5.

To clarify the notion of the error factor, let us consider specifics of the inversion of a lidar signal into the extinction coefficient when using the conventional slope method (Kunz and Leeuw, 1993; Kovalev and Eichinger, 2004). This method is used in rare instances - either when determining the overlap function or when estimating the mean extinction coefficient over an extended area (Hughes et al., 1985; Elouragini and Flamant, 1996; Powell et al., 2000). The theoretical basis of the slope method is based on simple lidar-equation formulas for homogeneous atmospheres. In such an atmosphere, the square-range-corrected backscatter signal of the elastic lidar in the complete overlap zone is written in the form

$$P(r)r^2 = C\beta_\pi \ \exp(-2kr), \tag{1.32}$$

where κ is the range-independent total (molecular and particulate) extinction coefficient. If the logarithms of the square-range-corrected backscatter signals at distances r_1 and $r_2 = r_1 + \Delta r$ are precisely determined, the unbiased extinction coefficient κ can be found by determining the slope of the line which passes through the points bounded by r_1 and r_2 (Fig. 1.11), that is,

$$\kappa = \frac{\ln\lfloor P(r_1)r_1^2\rfloor - \ln\lfloor P(r_2)r_2^2\rfloor}{2\Delta r}. \tag{1.33}$$

Now let us assume that the signal at the range r_1 was measured precisely, whereas the signal at the range r_2 was measured with some relative error δP_2. In this case, instead of the true value $P(r_2)$ a distorted signal $\langle P(r_2)\rangle$ is recorded:

$$\langle P(r_2)\rangle = P(r_2)[1 + \delta P_2], \tag{1.34}$$

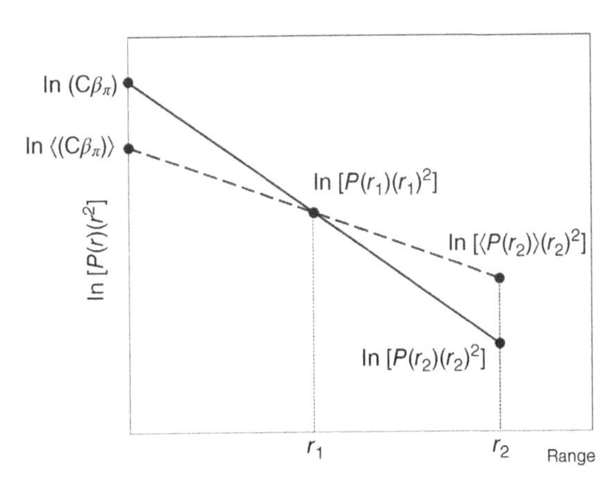

Fig. 1.11 The solid line shows the slope of the function $\ln[P(r)r^2]$ when the signals $P(r_1)$ and $P(r_2)$ are precisely measured. The dashed line shows the slope of this function when signal $\langle P(r_2)\rangle$ is overestimated.

When this distorted signal is used instead the correct $P(r_2)$ in Eq. (1.33), the slope of the line changes, accordingly, a shifted value of the extinction coefficient $\kappa + \Delta\kappa$ is obtained. The absolute error of the shifted extinction coefficient is

$$\Delta k = \frac{-\ln[1 + \delta P_2]}{2\Delta r}. \tag{1.35}$$

If the error is small, that is, $\delta P_2 \ll 1$, then $\ln[1 + \delta P_2] \approx \delta P_2$, and the relative error of the extinction coefficient $\delta\kappa = \Delta\kappa/\kappa$ can be found as

$$\delta k = \frac{-\delta P_2}{2\kappa\Delta r} = \left[\frac{-1}{2\tau\,(\Delta r)}\right]\delta P_2, \tag{1.36}$$

where $\tau(\Delta r)$ is the total optical depth over the range interval from r_1 to r_2. Accordingly, the ratio of the output relative error of the extinction coefficient, to the error of the input signal δP_2 is

$$\frac{\delta\kappa}{\delta P_2} = \frac{-1}{2\tau(\Delta r)}, \tag{1.37}$$

The absolute value of the term on the right side of Eq. (1.37) is the error factor, it can be written as

$$EF(\delta\kappa,\,\delta P_2) = \frac{1}{2\tau(\Delta r)}. \tag{1.38}$$

The error factor in Eq. (1.38) determines the proportion between the relative error in the measured signal P_2 and the corresponding error $\delta\kappa$ when the extinction coefficient is calculated using the slope of the dashed line in Fig. 1.11. In other words, the error of the retrieved extinction coefficient will be proportional to the error factor, that is, to the inverse value of the doubled optical depth $\tau(\Delta r)$. The error will depend on how clear the searched atmosphere is. If the atmosphere is moderately polluted, and the corresponding optical depth $\tau(\Delta r) = 0.5$, the condition factor $EF(\delta\kappa,\,\delta P_2) = 1$; that is, the relative error of the derived extinction coefficient is the same as the error in the measured signal P_2. However, in a clearer atmosphere, the situation will be worse. If $\tau(\Delta r) = 0.1$, then $EF(\delta\kappa,\,\delta P_2) = 5$, that is, the accuracy in the obtained extinction coefficient will be five times worse than that in the measured signal P_2; if the optical depth $\tau(\Delta r) = 0.01$ or less, $EF(\delta\kappa,\,\delta P_2) \geq 50$, and obtaining an accurate extinction coefficient could be difficult to achieve without increasing the range $(r_1,\,r_2)$, and accordingly, $\tau(\Delta r)$. One can conclude that the solution Eq. (1.33) is well conditioned for polluted atmospheres, where $\tau(\Delta r) \geq 0.5$, and ill-conditioned in clear atmospheres where $\tau(\Delta r) \ll 0.5$.

In the above analysis, the two-point variant of the slope method is used. Such a simple variant facilitates understanding the notion of the error factor when it is applied to the slope method, including that used for numerical differentiation. It will be shown in Chapter 2 that, in most cases, the error factor is proportional the inverse of the optical depth of the interval used as a range resolution.

Note that unlike the relative error, the absolute error $\Delta\kappa$ of the retrieved extinction coefficient does not depend on the optical situation. As follows from Eq. (1.35), for the relatively small signal error, when $\delta P_2 \ll 1$, the absolute error in the derived extinction coefficient is $\Delta\kappa = -0.5\delta P_2/\Delta r$. In other words, the absolute error caused by the error in the recorded signal will remain the same in all atmospheric conditions as long as the range resolution is not changed.

Commonly, the profile of the extinction coefficient is the main parameter of researchers' interest. However, in some cases, the backscatter term β_π or even the relative backscatter term $C\beta_\pi$ becomes the subject of interest. In a homogeneous atmosphere, the latter term can be found by extrapolation of the slope line that intersects the data points at the ranges r_1 and r_2 down to the range $r = 0$, with the next determination of the exponent of this term (Kunz and Leeuw, 1993; Kovalev et al., 2011a, 2011b, 2011c). Obviously, for the case when the error $\delta P_2 = 0$, the true slope (the solid line in Fig. 1.11) will yield the correct value of ln $(C\beta_\pi)$, whereas the slope of the dashed line will yield the shifted value $\ln\langle(C\beta_\pi)\rangle$. Let us consider how the accuracy of the retrieved product $C\beta_\pi$ will depend on the signal error δP_2 and under what conditions the solution for $C\beta_\pi$ is well-conditioned. At the range r_1, both lines in Fig. 1.41 intercept, accordingly,

$$\ln\langle(C\beta_\pi)\rangle - 2\langle\kappa\rangle r_1 = \ln(C\beta_\pi) - 2\kappa r_1. \tag{1.39}$$

Simple transformations of Eq. (1.39) yield the relationship between the error δP_2 and the relative error of the backscatter coefficient in the form

$$\ln[1 + \delta\beta_\pi] = \frac{r_1}{\Delta r}[-\ln(1 + \delta P_2)], \tag{1.40}$$

where $\delta\beta_\pi = \frac{\Delta\beta_\pi}{\beta_\pi}$. If the error $\delta P_2 \ll 1$ and $\delta\beta_\pi \ll 1$, Eq. (1.40) can be rewritten in the simpler form

$$\delta\beta_\pi = -\delta P_2\frac{r_1}{\Delta r} = -\delta P_2\frac{\tau(0, r_1)}{\tau(\Delta r)}. \tag{1.41}$$

Accordingly, the corresponding error factor $EF(\delta\beta_\pi, \delta P_2)$ is

$$EF(\delta\beta_\pi, \delta P_2) = \frac{\tau(0, r_1)}{\tau(\Delta r)}. \tag{1.42}$$

It follows from this formula that to reduce the error factor down to unity or less, the condition $\tau(0, r_1) \leq \tau(\Delta r)$ should be satisfied.

The relationships between the relative errors δP_2 and the corresponding errors, $\delta\kappa$ and $\delta\beta_\pi$, when $r_1 = 1000$ m and $r_2 = 2000$ m, are shown in Table 1.3. The parameters were calculated for a homogeneous atmosphere with the extinction coefficient ranging from 0.05 to 1 km^{-1}. The corresponding error factors, $EF(\delta\kappa, \delta P_2)$ and $EF(\delta\beta_\pi, \delta P_2)$, are shown within parentheses.

It follows from the table that the relative errors, $\delta\kappa$ and $\delta\beta_\pi$, and both error factors, $EF(\delta\kappa, \delta P_2)$ and $EF(\delta\beta_\pi, \delta P_2)$, are equal when $\kappa = 0.5$ km^{-1}, and accordingly, $\tau(\Delta r) = \tau(0, r_1) = 0.5$. In clear atmospheres, when the optical depth $\tau(\Delta r)$ is small,

TABLE 1.3 Relationships Between the Errors $\delta\kappa$, $\delta\beta_\pi$, and δP_2 (%)

κ, km^{-1}	δP_2, %	$\delta\kappa$, % and $EF(\delta\kappa, \delta P_2)$	$\delta\beta_\pi$, % and $EF(\delta\beta_\pi, \delta P_2)$	$\delta\kappa/\delta\beta_\pi$,
0.05	1.0	−10 (10)	−1 (1)	10
	3.0	−30 (10)	−3 (1)	
	5.0	−50 (10)	−5 (1)	
0.1	1.0	−5 (5)	−1 (1)	5
	3.0	−15 (5)	−3 (1)	
	5.0	−25 (5)	−5 (1)	
0.2	1.0	−2.5 (2.5)	−1 (1)	2.5
	3.0	−12.5 (2.5)	−3 (1)	
	5.0	−8.3 (2.5)	−5 (1)	
0.5	1.0	−1 (1)	−1 (1)	1
	3.0	−3 (1)	−3 (1)	
	5.0	−5 (1)	−5 (1)	
1.0	1.0	−0.5 (0.5)	−1 (1)	0.5
	3.0	−1.5 (0.5)	−3 (1)	
	5.0	−2.5 (0.5)	−5 (1)	

$EF(\delta\beta_\pi, \delta P_2)$ will be significantly less than $EF(\delta\kappa, \delta P_2)$. Accordingly, the formula for the extraction of the extinction coefficient with the slope method in such an atmosphere may be an ill-conditioned solution, whereas the formula for the extraction of the backscatter term will remain a well-conditioned solution. Similar results were obtained in the study by Kunz and Leeuw (1993), who compared the accuracy of the extinction and backscatter coefficients derived from the noisy signals using the slope method. The most important conclusion that follows from these analyses is that in a clear atmosphere, the backscatter coefficient can be extracted with better accuracy than the extinction coefficient. As will be shown in Chapter 3, this observation can be used to improve the inversion results when profiling the atmosphere in the multiangle mode (Kovalev et al., 2011a, 2011b, 2011c).

One side note is required. As is known, the use of the least squares method when a large number of data points within the range interval Δr are available provides better accuracy in determining the slope as compared to the two-point variant (Tailor, 1997; Bevington and Robinson, 2003). However, this is true only under the condition that relatively stringent requirements for the inverted data points are satisfied. First, the regressed data points should not have systematic distortions. Second, the discrete data points, which are inverted, should be randomly distributed relative to the true slope and their distribution should obey common statistics. Meanwhile, the validity of these requirements for real lidar data is, generally, an implicit premise rather than fact. In such a lidar signal, and accordingly, in its logarithm, the deviations of the discrete data points are not necessarily statistically distributed relative to the true slope. Therefore, the statistical estimates cannot be taken as the final word without a comprehensive analysis of the possible systematic errors in the lidar signal.

Let us clarify this issue considering some details in the case when a virtual elastic lidar operates in a synthetic homogeneous atmosphere. The ideal backscatter lidar signal $P(r)$ which would be obtained in this atmosphere is shown in Fig. 1.12 as the dashed curve. Such a signal shape would be obtained under the condition that neither noise nor other distortions are present in the recorded signal. However, in any real lidar signal, the data points will be influenced by signal noise and the local heterogeneity of the particulates, which always takes place even in a homogeneous atmosphere. In other words, in a real measurement, the recorded data points will be scattered relative to the dashed curve. However, they may be nonrandomly scattered (in the statistical sense) but obey much more complicated laws. In the figure, such signal data points are shown as filled circles. It was assumed that no systematic distortions in the lidar signal were present, so that the data points are not corrupted systematically.

The logarithms of the ideal and "real" square-range-corrected signals are shown in Fig. 1.13. The data points of the logarithm of the "real" square-range-corrected signal $\ln[P(r)r^2]$ are shown as the filled circles, and their linear fit $Y(r)$ is shown as the solid line; the logarithm of the ideal square-range-corrected signal is shown as the dashed line. One can see that the slope line, which corresponds to the true extinction coefficient used for the simulation, does not coincide with the linear fit $Y(r)$. The slope of the linear fit, determined over the interval $\Delta r = 600$ m yields the shifted extinction coefficient $\langle \kappa \rangle = 0.554\ \text{km}^{-1}$, which differs from the true extinction coefficient $\kappa = 0.5\ \text{km}^{-1}$. Thus, the extinction coefficient extracted from the slope of the linear fit $Y(r)$ using the least square method has the absolute error $0.054\ \text{km}^{-1}$. Meanwhile, the standard deviation of the linear fit $Y(r)$ is equal to $0.33\ \text{km}^{-1}$. Obviously, it has nothing in common with the real inversion uncertainty. First, it determines the random fluctuations of the data points relative to the solid slope, not relative to the true

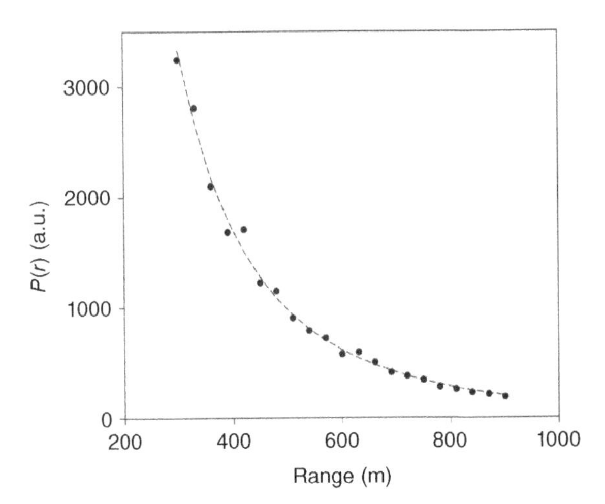

Fig. 1.12 Backscatter signal, $P(r)$, calculated for a synthetic homogeneous atmosphere with $\kappa = 0.5\ \text{km}^{-1}$ and $\beta_\pi(r) = \beta_\pi = \text{const}$. The dashed curve shows the ideal signal with no fluctuations and the scattered filled circles show a "real" lidar signal, respectively.

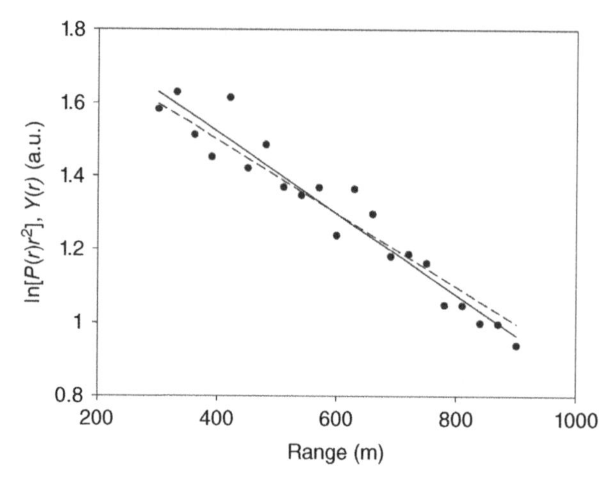

Fig. 1.13 Filled circles are the data points of the logarithm of the "real" square-range-corrected signal shown in previous figure as similar filled circles. The solid line is the linear fit $Y(r)$ obtained for these data points using the least squares method, and the dashed line is the profile $\ln[P(r)r^2]$ calculated using the ideal profile of $P(r)$ where the data points are not scattered.

dashed slope. Second, there is no evidence that the scattered data points obey a normal distribution. Therefore, the standard deviation cannot be considered as a trustworthy estimate of the uncertainty in the derived extinction coefficient.

Let us briefly summarize. The solution for the extinction coefficient obtained with the least square method is valid only for the statistically random error in the processed data points, which is not necessarily true in real conditions. When determining the extinction coefficient through the examination of the slope of the linear fit, the normal distribution of the data points relative to the "true" slope line is the implicit premise. Without an additional analysis, there is no evidence that it is valid. It should also be mentioned that for the same data points, one would obtain a different slope for the linear fit $Y(r)$ and accordingly, different extinction coefficient if differently weighted data points for the regression were used (Tailor, 1997). In other words, any lidar-data inversion results yield atmospheric profiles which are valid only for the solution selected for the inversion and its implicit premises. One can maintain that it is a mistake to invert lidar data restricting ourselves to mathematically rigid statistical models. The real atmosphere obeys much more complicated laws than our simplified statistics, and any sensible model, even if it has no rigid mathematical basis, should not be groundlessly rejected. Finally, it is worthwhile to point out that the slope method is the most common procedure for data processing in many types of the lidars, for example, Raman lidar, DIAL, and HSRL. All the restrictions inherent to the slope method should be taken into consideration while processing lidar data recorded in real atmospheres.

In the above simulations, the output errors in the retrieved extinction coefficient implemented during the inversion procedure, were produced by errors in the measured lidar signals; no explicit *a priori* assumption was involved. When such an *a priori* assumption is required, the inversion accuracy depends not only on the error factor, but also on the validity of this assumption. Accordingly, the output error quantification becomes more complicated, if not impossible.

Let us illustrate such an issue in the boundary-point solution, considered in Section 1.2. In the most general form, the solution in Eq. (1.22) can be rewritten in the form

$$\kappa_w(h) = \frac{Z_w(h)}{\frac{Z_w(h_b)}{\kappa_w(h_b)} - 2\int_{h_b}^{h} Z_w(h')dh'}, \tag{1.43}$$

where

$$\kappa_w(h) = \kappa_p(h) + \alpha_\pi(h)\kappa_m(h),$$

and accordingly $\kappa_w(h_b)$ is the "weighted" extinction coefficient at the selected reference point h_b. If this reference value is estimated with the relative error $\delta\kappa_{w,b}$, that is, instead of its true value, the estimate, $\langle\kappa_w(h_b)\rangle = \kappa_w(h_b)\lfloor 1 + \delta\kappa_{w,b}\rfloor$ is used, then the derived function $\kappa_w(h)$ will also be distorted. In such a case, Eq. (1.43) transforms into the formula

$$\kappa_w(h)[1 + \delta\kappa_w(h)] = \frac{Z_w(h)}{\frac{Z_w(h_b)}{\kappa_w(h_b)[1+\delta\kappa_{w,b}]} - 2\int_{h_b}^{h} Z_w(h')dh'} \tag{1.44}$$

Eqs. (1.43) can be rewritten in the form

$$\frac{Z_w(h)}{\kappa_w(h)} = \frac{Z_w(h_b)}{\kappa_w(h_b)} - 2\int_{h_b}^{h} Z_w(h')dh', \tag{1.45}$$

accordingly

$$\frac{Z_w(h)}{\kappa_w(h)[1 + \delta\kappa_w(h)]} = \frac{Z_w(h_b)}{\kappa_w(h_b)[1 + \delta\kappa_{w,b}]} - 2\int_{h_b}^{h} Z_w(h')dh'. \tag{1.46}$$

After subtracting Eq. (1.45) from Eq. (1.46) and simple transformations, one obtains the formula

$$\frac{\delta\kappa_w(h)}{\delta\kappa_{w,b}} = \frac{Z_w(h_b)}{Z_w(h)}\frac{\kappa_w(h)[1 + \delta\kappa_w(h)]}{\kappa_w(h_b)[1 + \delta\kappa_{w,b}]}. \tag{1.47}$$

Note that here the errors $\delta\kappa_w(h)$ and $\delta\kappa_{w,b}$ are present both in the left and the right side of the formula, so that the analytical formula for the error factor would be extremely complicated and would not allow comprehensive understanding of the dependence between the errors $\delta\kappa_w(h)$ and $\delta\kappa_{w,b}$. Therefore, instead of analyzing the error factor, the analytical dependence between $\delta\kappa_w(h)$ and $\delta\kappa_{w,b}$ should be examined. This dependence, initially given in the study by Kovalev and Moosmüller (1994), is extremely simple:

$$\delta\kappa_w(h) = \frac{\delta\kappa_{w,b}}{V_w^2(h_b, h)[1 + \delta\kappa_{w,b}] - \delta\kappa_{w,b}}, \tag{1.48}$$

where

$$V_w^2(h_b, h) = \exp\left[-2\int_{h_b}^{h} \kappa_w\left(h'\right) dh'\right] = \exp[-2\tau_w(h_b, h)]; \tag{1.49}$$

and $\tau_w(h_b, h)$ is the "weighted" optical depth of the altitude interval between the heights h_b and h. The relationship between $\delta\kappa(h)$, $\delta\kappa(h_b)$, and the optical depth, $\tau_w(h_b, h)$, is shown in Fig. 1.14. The left side of the plot, where the values of $\tau_W(h_b, h)$ are negative (that is, $h_b > h$), shows the errors in the derived profile $\kappa_w(h)$ when the far-end solution is used. The right side of the plot shows the much larger errors inherent to the near-end solution.

To determine the profile of the particulate extinction coefficient $\kappa_p(h)$ from $\kappa_w(h)$, the components $\alpha_\pi(h)$ and $\kappa_m(h)$ within the profiling range, including the reference point h_b should be known. If these components are precisely known, the relative error of the particulate extinction coefficient is equal to that of the weighted extinction coefficient, that is, $\delta\kappa_p(h) = \delta\kappa_w(h)$. However in practice, the possible error in *a priori* selected $\alpha_\pi(h)$ may be unacceptably large even when the far-end solution is used.

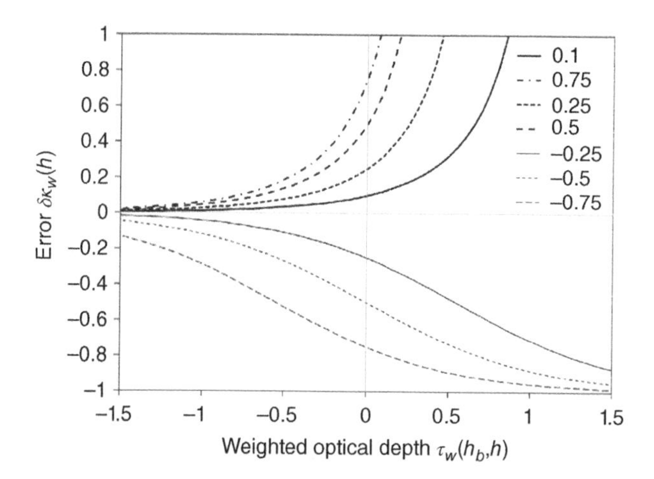

Fig. 1.14 The error $\delta\kappa_w(h)$ as a function of the optical depth $\tau_w(h_b, h)$ and the error $\delta\kappa_w(h_b)$. The value of the latter is shown in the legend. Adapted from Kovalev and Moosmüller, 1994.

1.5 BACKSCATTER SIGNAL DISTORTIONS AND CORRESPONDING ERRORS IN THE INVERTED ATMOSPHERIC PROFILES

Before making a comprehensive analysis of the lidar-signal inversion results, let us rewrite the distorted backscatter signal $\langle P(r)\rangle$ in Eq. (1.7) as the sum of the undistorted backscatter signal $P(r)$ and three accompanying distortion components, $P(r)\delta_P(r)$, $\sum w_{j,\mathrm{low}}(r)$, and the range-independent component ΔB, that is,

$$\langle P(r)\rangle = P(r) + P(r)\delta_P(r) + \sum w_{j,\mathrm{low}}(r) + \Delta B. \tag{1.50}$$

Eq. (1.50) can be considered as the most general form for the backscatter signal distorted by the additive and the multiplicative distortion components. As explained in Section 1.2, the low-frequency distortion component, $\sum w_{j,\mathrm{low}}(r)$, can originate from low-frequency induced noise in the receiver channel. The nonzero range-dependent distortion component $\delta_P(r)$ may be the result of a nonideal linear transformation of the backscattered light into the electrical signal, the presence of multiple scattering, etc.; even the incorrectly determined length of the incomplete overlap zone creates the component $\delta_P(r) \neq 0$ (see Section 1.8).

The determination of the backscatter signal over distant ranges is a significant issue compounded by the necessity for precise determination and removal of the constant shift B from the recorded lidar signal, $P_\Sigma(r)$. In practice, the researcher can only minimize the remaining offset ΔB in Eq. (1.50), rather than provide the condition $\Delta B = 0$. To clarify this issue, let us consider the remaining offset ΔB in the backscatter signal obtained by a virtual ground-based and zenith-directed lidar. To display details clearly, the vertical scale in Fig. 1.15 is reduced and the total signal is shown only

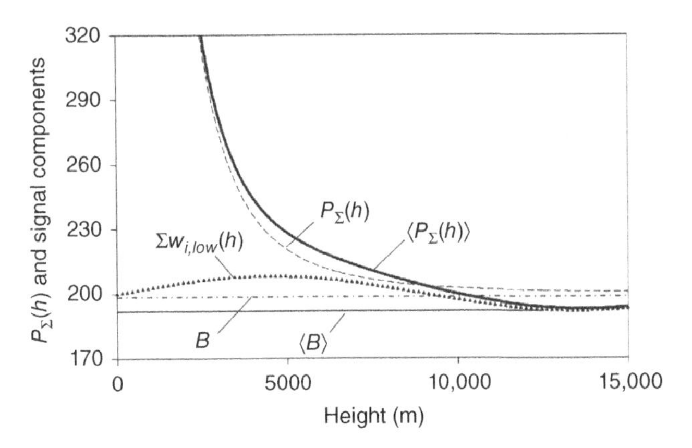

Fig. 1.15 Components of the ideal signal $P_\Sigma(h)$ and the actual distorted signal $\langle P_\Sigma(h)\rangle$ recorded with a vertically pointed virtual lidar in a simulated atmosphere. The actual shift B is shown as the dotted-dashed horizontal line, and the estimated zero-line $\langle B\rangle$ is shown as the solid horizontal line. The triangles show the induced low-frequency noise component $\sum w_{j,\mathrm{low}}(h)$ superimposed on the range-independent level B. Adapted from Kovalev et al., 2004.

over distant ranges where the backscatter signal becomes comparable with the background B. The total undistorted signal $P_{\Sigma}(h)$ versus height h is shown as the dashed curve, and the retrieved distorted signal $\langle P_{\Sigma}(h)\rangle$ as the thick solid curve. The actual additive component, $B = 200$ counts, is shown as the dashed-dotted line, whereas the estimated component $\langle B\rangle$ is shown as the solid horizontal line. The low-frequency noise component $\sum w_{j,\mathrm{low}}(h)$, centered at the level of 200 counts, is shown by the triangles as a sinusoidal curve.

There are two common ways to estimate the component B which should be subtracted from the recorded signal before the inversion of the backscatter signal is made. [The method proposed in the study by Kovalev et al. (2009b) is considered separately in Section 1.6]. First, one can record the photoreceiver signal prior to triggering the laser pulse and use this signal as a measure of the background component. Such a variant was employed, for example, in the study by Johnson et al. (2013), where background component removal was achieved by subtracting the average of the first 71 time bins from each return signal, which corresponded to recordings before the laser pulse was triggered. Then the signals were averaged and adjusted so that the range of 0 km corresponded to the laser pulse leaving the transmitter. However, the laser discharge produces electromagnetic noise over a wide spectrum, which cannot be completely suppressed. The remaining noise is superimposed on the constant offset, B, restricting the maximum operative range of the lidar. The backscatter signal $\langle P(h)\rangle$ obtained after the subtraction of even the precise background value, may have an erroneous variable shift over distant ranges comparable with the backscatter signal. In the signal shown in Fig. 1.15, the subtraction of the true offset $B = 200$ counts will result in significant shift over the ranges from, approximately, 4000 to 9000 m and negative data points at ranges greater than 10 km.

The alternative way to estimate B is by determining the recorded signal at the far end of the measured range, where the backscatter component presumably vanishes. In the case shown in Fig. 1.15, the offset B can be found by estimating the recorded signal $\langle P_{\Sigma}(h)\rangle$ over the heights 12,000–15,000 m; at these heights, the backscatter component is presumably so minor that it can be ignored. In the simulated case, such a far-end determination of the offset yielded the estimated value $\langle B\rangle = 192$ counts, so that the offset $\Delta B = 8$ counts was added to the true backscatter signal $P(h)$ forming the distorted backscatter signal $\langle P(h)\rangle$. In practical measurements, the remaining offset generally is much smaller, usually within the range ± 1 count or less; the large offset here is taken only to make all the signal distortion components in the figure discernible; such large offsets ΔB are not used in the following numerical simulations.

Thus, neither of the two methods considered for determining the constant offset B is perfect. In many cases, estimating the constant component B over the far end of the recorded signal, may be preferable as compared to estimating the pre-triggered recording. However, in this case, any remaining nonzero backscatter component due to either the molecular or the particulate scattering will result in an additional nonzero systematic shift in the estimated offset $\langle B\rangle$. This effect does not take place when a lidar signal is recorded in the stratosphere at altitudes of $70 - 100$ km, as for example, in the study by Uchino and Tabata (1991). However, when working in the lower troposphere, such an effect may occur even if only molecular loading exists over the far-end ranges used for determining the offset. To illustrate this effect, in Fig. 1.16, the simulated lidar signals at different wavelengths versus height are shown for a vertically pointed virtual lidar that operates in a synthetic molecular atmosphere with no aerosol loading at the heights $h > 5000$ m. However, the nonzero molecular scattering will create some nonzero backscatter signal at these heights. Accordingly, while determining the constant B over the far end, the estimated constant will be a little larger than the actual value of $B = 200$ counts. This will take place even if the estimate $\langle B \rangle$ is determined at the altitudes higher than 10,000 m.

The offset ΔB remaining in the backscatter signal $\langle P(h) \rangle$ will be different for different wavelengths. For example, if the constant B is estimated as the mean of $P_{\Sigma}(h)$ over the altitude range 10,000–12,000 m, and the lidar operates at the wavelength 355 nm (curve 2), the negative offset ΔB will be equal ~ 1 count; for the wavelength 532 nm (curve 3), the offset is equal ~ 2 counts. As compared with the maximum value of the lidar signal used in the numerical simulations (~ 4000 counts), these offsets seem to be insignificant (0.025–0.05%). However, in clear atmospheres, even such minor offsets can cause significant distortions in the retrieved optical profiles.

The profiles in Fig. 1.16 are calculated for a purely molecular atmosphere. In the atmospheres where aerosol loading is also present, the level of the backscatter signal at distant ranges, and accordingly, the remaining offset may be larger. The real signals over distant ranges may be corrupted with the low-frequency noise and that corruption can significantly impede the determination of an accurate estimate $\langle B \rangle$ in the recorded lidar signal. To illustrate the statement that even a seemingly insignificant offset, remaining in the backscatter signal, can significantly distort profiles of the derived extinction coefficient, let us perform simple numerical simulations. In these simulations, the situation is analyzed in which the virtual zenith-directed elastic lidar operates in a clear, but not a pure, molecular atmosphere. For such numerical

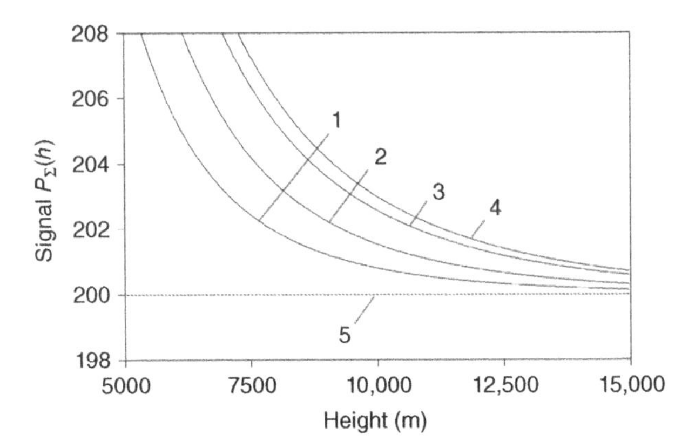

Fig. 1.16 Synthetic signals $P_{\Sigma}(h)$ at four wavelengths, 308, 355, 532, and 1064 nm (Curves 1-4, respectively) recorded in a pure molecular atmosphere with a virtual vertically pointed lidar. Curve 5 shows the actual level of the background component B for these signals. (Kovalev et al., 2004).

simulation, the background model of atmospheric aerosol loading for the wavelength 532 nm, given in the study by Zuev and Krekov (1986), is used. This model, which is in good agreement with the experimental results, obtained by Spinhirne (1976), is shown in Fig. 1.17. The inversion of the signal of the artificial lidar is made with the boundary-point solution using both forward and backward inversion algorithms, that is, the near- and far-end solutions (Section 1.2.2). For the inversion, the corresponding reference points at $h_{\rm min} = 500\,{\rm m}$ and at $h_{\rm max} = 8000\ m$ are selected. It is assumed that both boundary values are precisely known, and the distortion in the retrieved extinction-coefficient profiles occurs only because of the nonzero offset ΔB and random noise.

In Figs. 1.18 and 1.19, the inversion results are shown for the case when the remaining offset ΔB in the inverted backscatter signals is equal to -1 and $+1$ count, respectively. The maximal signal, presumably recorded by a 12-bit digitizer is approximately 4000 counts. In both figures, the model profile of the extinction coefficient is shown as a thick dashed curve. The thick solid curves 1 and 2 show the profiles of the extinction coefficient obtained with the near-end and far-end solutions when the lidar signals are not corrupted by random noise; the thin dashed curves show the extinction coefficients retrieved with the signals corrupted with random noise. For the signals corrupted by noise, the reference signal for the far-end solution was found as an average of 11 consecutive data points, that is, as the average over the height interval of 200 m. As the boundary values are precisely known, the near-end solution provides a more accurate inversion result than the far-end solution in the near zone from 500 to $\sim$ 3000 m. Within this zone, the near-end solution results in the systematic shifts in the derived extinction coefficient, $\sim 10-15\%$, whereas the far-end solution yields significantly larger distortions. Note also that for the near-end

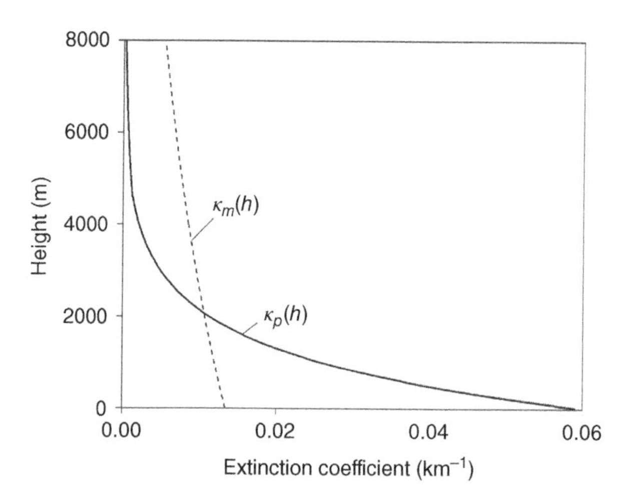

Fig. 1.17 The extinction-coefficient profile $\kappa_p(h)$ at the wavelength 532 nm, taken from the study by Zuev and Krekov (1986) and the molecular extinction profile $\kappa_m(h)$ used in the simulation.

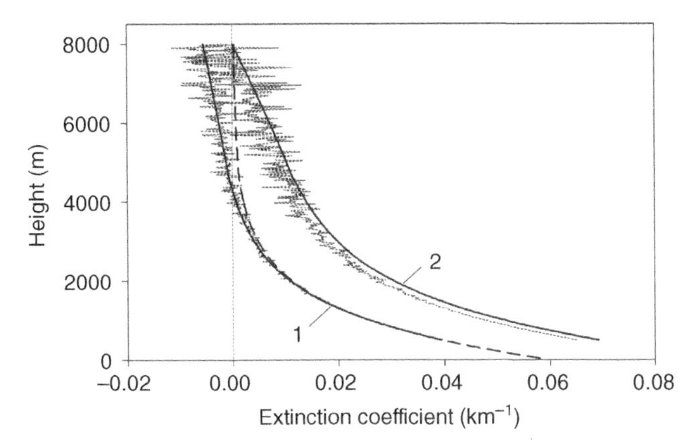

Fig. 1.18 Inversion results obtained from the signals of the artificial zenith-directed lidar in the atmospheric conditions shown in Fig. 1.17 when the remaining offset in the inverted backscatter signal $\Delta B = -1$ count. The thick solid Curves 1 and 2 show the extinction-coefficient profiles retrieved from the signals with no random noise with the near- and far-end solutions, respectively. The thin dotted curves show the same profiles retrieved from the signals corrupted with random noise. The thick dashed curve is the "true" profile $\kappa_p(h)$, used for the simulation.

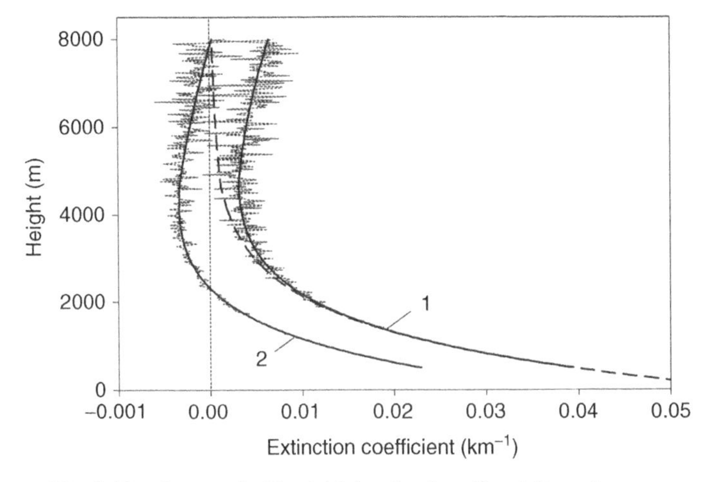

Fig. 1.19 Same as in Fig. 1.18, but for the offset $\Delta B = +1$ count.

solution, the zones of minimum systematic and minimum random errors coincide, so that the total uncertainty is generally less than that for the far-end solution. Thus, in clear atmospheres, the near-end solution may be preferable as compared to the far-end solution.

As stated in Section 1.1, there are different sources of both the variable and range-independent shifts in the recorded lidar signal, which cannot be ignored while profiling of the atmosphere. However, the offset ΔB that remains in the backscatter signal after the subtraction of the estimated offset $\langle B \rangle$ from the recorded signal is among the most influential sources of corruption of the derived optical profiles. This is one of the more aggravating factors in lidar signal inversion and must be given high priority along with the signal-to-noise ratio (SNR). In this respect, two alternative factors can be analyzed: either the signal-to-background radiation ratio or the signal-to-offset ratio (SOR). The notion of the signal-to-background ratio (SBR), similar to the SNR, was implemented by Agishev and Comeron (2002). The authors analyzed the signal/background radiation ratio and concluded that a low level of this ratio may be responsible for the poor accuracy of the derived optical profiles. For comprehensive analyses, it is sensible to use a more general function, the SOR, as the offset originates in the combination of the background sky radiation and the electric offset of the digitizer. The remaining offset ΔB can be either positive or negative, therefore, it is sensible to define the SOR ignoring the sign of the offset; that is, SOR should be defined as the ratio of the backscatter signal $P(h)$ to the absolute value of the offset, ΔB. Similar to SNR, the signal-to-offset ratio should be examined as a function of range or height. For the zenith-directed lidar, it can be defined as

$$\mathrm{SOR}(h) = \frac{P(h)}{|\Delta B|}. \tag{1.51}$$

Unfortunately, there is no simple and reliable method that would allow direct estimating of the remaining offset ΔB in real signals. For numerical simulations, an approximate estimate of $\mathrm{SOR}(h)$ can be made by relating ΔB with the standard deviation (STD) of the signal $P_{\Sigma}(h)$ over distant ranges where the backscatter signal $P(h) \to 0$ and accordingly, $P_{\Sigma}(h) \to B$. To perform such a simulation, one should determine the mean and STD of the signal $P_{\Sigma}(h)$ over the appropriate distant ranges. The standard deviation can be taken as a basis for determining the offset ΔB. The assumed offset may be found using the simple formula

$$|\Delta B| = n^{(\Delta B)} \mathrm{STD}\left[P_{\Sigma}(h)\right], \tag{1.52}$$

where the mean $P_{\Sigma}(h)$ is determined over distant ranges, and $n^{(\Delta B)} \geq 1$ is a constant selected for the simulation. Accordingly,

$$\mathrm{SOR}(h) = \frac{1}{n^{(\Delta B)}} \left\{ \frac{P(h)}{\mathrm{STD}\left[P_{\Sigma}(h)\right]} \right\}. \tag{1.53}$$

In principle, such a method can be applied not only for the numerical simulation, but also for simulations that use real lidar signals. The STD should be found within the restricted interval over distant ranges where the backscatter term presumably vanishes. For zenith-directed lidar, such an estimate can be made by selecting the clear-sky height intervals, free of noticeable aerosol loading.

Thus, when examining the results of lidar profiling, both the SNR(h) and the SOR(h) should be taken into account. As is known, there is no standard level of the minimum SNR that should be satisfied at the maximum operative height h_{max}. In practice, such a minimal level, SNR_{min}, is established arbitrarily by the researcher. The same principle can be applied when selecting an acceptable minimum for the SOR_{min}.

Let us illustrate this idea using the simulated lidar signal at 532 nm obtained in a pure molecular atmosphere (Curve 3 in Fig. 1.16), considering the case when the signal is corrupted by both quasi-random noise and unknown nonzero offset ΔB. In Fig. 1.20, the actual SOR(h) is shown as the thick solid curve. The profiles of SOR(h) determined with $n^{(\Delta B)} = 1$ and $n^{(\Delta B)} = 2$ are shown as the solid and dotted thin curves. The standard deviation of the signal $P_{\Sigma}(h)$ is determined over the height interval from 12,000 to 15,000 m. Suppose, after the standard calculation of the SNR, it was established that selected SNR_{min} is reached at $h = 9000$ m. Now, having two profiles of SOR(h), shown in Fig. 1.20, one should decide the maximum height that can be chosen to satisfy the selected minimum SOR, for example, $SOR_{min} = 10$; this level is shown in the figure as the horizontal line. At the height 9000 m, the level of SOR(h) determined with $n^{(\Delta B)} = 1$ satisfies the selected level of SOR_{min}. However, SOR(h) determined with $n^{(\Delta B)} = 2$ is significantly below this SOR_{min}. To satisfy it, the researcher should decrease the maximum operative height h_{max} down to at least 7000 m. This maximum height will reliably satisfy both requirements.

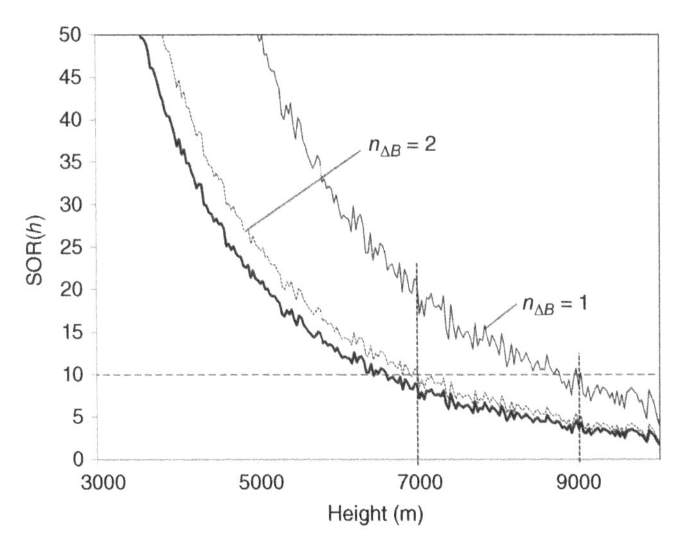

Fig. 1.20 The true signal-to-offset ratio (the thick solid curve) for the signal of the virtual lidar and the SOR(h), calculated using Eq. (1.53) when $n^{(\Delta B)} = 1$ and $n^{(\Delta B)} = 2$.

Let us summarize. In order to avoid ungrounded conclusions about the optimal maximum range of atmospheric profiling and the accuracy of the derived extinction coefficient, based on pure statistics, systematic additive distortions of the lidar signal also need to be taken into consideration. Even minor systematic distortions of the lidar signal may significantly distort the inversion result, yielding significant distortions in the derived extinction-coefficient profiles, especially, over distant ranges. The analysis of signal systematic distortions, especially those caused by the nonzero offset ΔB can produce a more accurate estimation of the validity of the lidar profiling results. When simulations are based on the results of real lidar searching, this can be done by estimating $\langle B \rangle$ through the determination of the STD of the signal $P_{\Sigma}(h)$ over distant ranges. Taking into consideration both the functions SNR(h) and $\widetilde{\text{SOR}}(h)$, one can better determine the lidar maximal operative range and the optical profile of interest within it. Anyway, one can maintain that in lidar profiling of the atmosphere, an estimation of the possible errors and distortions that do not obey normal statistics should be considered a must.

1.6 DETERMINATION OF THE CONSTANT OFFSET IN THE RECORDED LIDAR SIGNAL USING THE SLOPE METHOD

As discussed in the previous section, there are two common methods for determining the additive component B created by a daytime background illumination and the electrical or digital offset. First, one can record the output photoreceiver signal before the laser pulse is emitted to the atmosphere, and use this pre-triggered signal as a measure of the component B. The second way is to use the recorded signal over distant ranges, where the backscatter signal decreases to an ignorable small value. It has been shown that neither approach is perfect, so the backscatter signal extracted from the total signal $P_{\Sigma}(r)$ may have some constant shift ΔB remaining after the subtraction of the non-precise offset estimate $\langle B \rangle$. In practice, this remaining shift is generally small; however, even a minor shift can yield large distortions in the inverted atmospheric profile over distant ranges, especially in clear atmospheres (Kovalev, 2004; Kovalev et al., 2007a; Adam et al., 2007).

In this section, an alternative way for determining the constant offset in the total signal is discussed. In this method, the component B is also determined at the far end of the recorded signal. However, unlike the technique discussed in Section 1.5, initially, the signal $P_{\Sigma}(r)$ recorded by lidar is square-range-corrected. This operation is done before estimation and subtraction of the offset B. While performing this procedure, the monotonic change in the signal, due to the molecular component of the atmosphere may also be compensated. After these operations are made, the total offset B is estimated by determining the slope of the square-range-corrected signal $P_{\Sigma}(r)$ versus a special function. In a general case, this new variable is defined as a ratio of the squared range r^2 to the product of two molecular components, the molecular backscatter coefficient and the molecular two-way transmittance.

The slope method of determining the constant offset in a lidar signal can be used both for zenith-directed and scanning lidars. In this section, the technique will be discussed for zenith-directed lidar. The details of such a technique for multiangle searching will be discussed in Chapter 3.

1.6.1 Algorithm and Solution Uncertainty

The total signal $P_\Sigma(r)$ recorded by lidar at the range r can be written in the form [Eq. (1.6)]

$$P_\Sigma(r) = P(r) + \delta_P(r)P(r) + \sum w_j(r) + B, \tag{1.54}$$

where $P(r)$ is the undistorted backscatter signal over a zone of complete overlap,

$$P(r) = \frac{1}{r^2} C\beta_\pi(r)T_\Sigma^2(0, r).$$

As two methods of determining the constant offset B, discussed in Section 1.5, the slope method works properly if the effects caused by the components $\delta_P(r)$ and $\sum w_{j,\mathrm{low}}(r)$ are suppressed or compensated, so that the signal recorded at the far end includes only two terms, that is,

$$P_\Sigma(r) = P(r) + B. \tag{1.55}$$

To determine and remove the constant component B, the square-range correction of the total signal $P_\Sigma(r)$ combined with the removal of the monotonic change in the backscatter signal, due to the molecular component of the atmosphere is made. This operation transforms the signal $P_\Sigma(r)$ into the function $y^*(r)$ determined as

$$y^*(r) = P_\Sigma(r)z(r), \tag{1.56}$$

where the new variable $z(r)$ is defined as

$$z(r) = \frac{r^2}{\beta_{\pi,m}(r)T_m^2(0,r)}. \tag{1.57}$$

According to Eqs. (1.55) and (1.57), the function $y^*(r)$ can be written in the form

$$y^*(r) = C[1 + R_\beta(r)]T_p^2(0, r) + Bz(r), \tag{1.58}$$

where $R_\beta(r) = \beta_{\pi,p}(r)/\beta_{\pi,m}(r)$.

To determine the component B using Eq. (1.58), a restricted range interval from r_1 to r_2 over the far end of the recorded signal is selected and the linear fit of the data points $y^*(r)$ is found. The basic requirement for determining the constant B is

$$C[1 + R_\beta(r)]T_p^2(0, r) = \text{const}, \tag{1.59}$$

that is, the first term in the right side of Eq. (1.58) should be range independent within the selected interval (r_1, r_2). Under such a condition, Eq. (1.58) transforms into a linear equation with the variable $z(r)$, that is,

$$y^*(z) = Y_0 + Bz, \tag{1.60}$$

where $Y_0 = C[1 + R_\beta(r)]T_p^2(0, r)$ and the variable r in the symbol $z(r)$ is omitted for simplicity. The offset B can now be found as the slope of the linear fit determined for the data points $y^*(z)$ within the selected restricted range interval (r_1, r_2).

To satisfy the condition in Eq. (1.59) for zenith-directed lidar, the corresponding height interval (h_1, h_2) should be chosen in the zone where pure molecular scattering takes place; that is, where $R_\beta(h) = 0$, and accordingly, $T_p^2(0, h) = \text{const}$. The alternative assumption is that over the selected height interval (h_1, h_2), the ratio $\beta_{\pi,p}(h)/\beta_{\pi,m}(h) \approx \text{const}$, and the extinction coefficient $\kappa_p(h)$ is small, so that its influence on the slope of $T_p^2(0, h)$ within this interval is negligible. Simple numerical simulations show that the presence of moderate aerosol loading within a restricted altitude range (h_1, h_2) is not critical; Y_0 in Eq. (1.60) will remain practically constant if no large gradient in $R_\beta(h)$ takes place and the particulate extinction coefficient, $\kappa_p(h)$ is negligible, so that the marginal transmission terms, $T_p^2(0, h_1)$ and $T_p^2(0, h_2)$, differ insignificantly. When determining the offset $\langle B\rangle$ the minor changes in Y_0 will only slightly influence the slope of the linear fit.

After the estimate of the offset $\langle B\rangle$ is found, the backscatter signal of interest $\langle P(h)\rangle$ is determined in the conventional way, that is, as the difference between the recorded signal $P_\Sigma(h)$ and $\langle B\rangle$. Owing to random and systematic distortions, unavoidable in any real signal, the estimate $\langle B\rangle$ will not be precisely equal to the actual offset B. However, if the requirement in Eq. (1.59) is met, the remaining offset ΔB is minor, generally significantly less than when using the far-end method of determining the offset discussed in Section 1.5.

There are two important advantages of the slope method as compared to the far-end method of determining the constant offset considered in Section 1.5. The first one is that the molecular loading of the atmosphere is compensated so that its monotonic change with height is removed from the function $y^*(r)$. Accordingly, the corresponding error, discussed in Section 1.5, does not influence the estimated value of $\langle B\rangle$ even when the shorter laser wavelength is used. The second advantage of the slope method is that the presence of a moderate aerosol loading over the far end does not dramatically increase the error in the estimate of the lidar-signal offset. The error in $\langle B\rangle$ may only be caused by monotonic change of aerosol loading over the altitude range $h_1 - h_2$ rather than its nonzero level.

To make clear the influence of the nonzero aerosol loading over the assumed aerosol-free altitudes, let us consider the theoretical error in the estimated offset $\langle B\rangle$ when the ratio $R_\beta(h) > 0$. For determining the offset, the simplest two-point solution rather than the common least square method is used. The use of such a solution yields a relatively simple formula for the dependence between the resulting error in the estimated offset $\langle B\rangle$ and the corresponding marginal ratios $R_\beta(h_1)$ and $R_\beta(h_2)$. Using the data points of the function $y^*(z)$ at points z_1 and z_2, corresponding to the heights, h_1 and h_2, one obtains the formula

$$\langle B\rangle = \frac{y^*(z_2) - y^*(z_1)}{z_2 - z_1} = B + \frac{C\{[1 + R_\beta(h_2)]T_p^2(0, h_2) - [1 + R_\beta(h_1)]T_p^2(0, h_1)\}}{z_2 - z_1}. \tag{1.61}$$

Simple transformation of Eq. (1.61) yields the following formula for the error $\Delta B = B - \langle B\rangle$:

$$\Delta B = \left[\frac{z_2}{z_2 - z_1}\right] P(h_2) - \left[\frac{z_1}{z_2 - z_1}\right] P(h_1), \tag{1.62}$$

where $P(h_1)$ and $P(h_2)$ are the backscatter signals at h_1 and h_2. To clarify the essence of Eq. (1.62), it can be rewritten as

$$\Delta B = P(h_1)\left(\frac{z_1}{z_2 - z_1}\right)\left\{\left(\frac{1 + R_\beta(h_2)}{1 + R_\beta(h_1)}\right)\exp[-2\tau_p(h_1, h_2)] - 1\right\}, \tag{1.63}$$

where $\tau_p(h_1, h_2)$ is the particulate optical depth over the selected altitude interval (h_1, h_2). As follows from Eq. (1.63), the remaining offset ΔB is small under the following conditions:

1. The altitude h_1 is selected over a distant range, so that the backscatter signal $P(h_1)$ is small.
2. No noticeable gradients in $R_\beta(h)$ are present within the selected interval (h_1, h_2).
3. The particulate optical depth $\tau_p(h_1, h_2)$ within this height interval is small.

In principle, determining the slope of the linear fit of $y^*(z)$ using the standard least square method instead of the two-point solution may reduce the error ΔB. However, one should keep in mind that the presence of unavoidable systematic errors in the signal may significantly restrict such an opportunity to improve the accuracy of determining ΔB.

1.6.2 Numerical Simulations and Experimental Data

To clarify the essence of the method, let us consider a virtual ground-based and vertically pointed lidar at the wavelength 355 nm that operates in a relatively clear atmosphere. In this simulated atmosphere, the particulate extinction coefficient monotonically decreases with the height from $\kappa_p(h) = 0.17\,\text{km}^{-1}$ at ground level down to $\kappa_p(h) = 0.026\,\text{km}^{-1}$ at the height $h = 12{,}000$ m. Two turbid layers are present in the atmosphere. One takes place at the heights from 900 to 1200 m, and the second is located between the heights 4000 and 4350 m. The maximal extinction coefficients within these polluted layers are $\kappa_p(h) = 0.18\,\text{km}^{-1}$ and $\kappa_p(h) = 0.2\,\text{km}^{-1}$, respectively. In Fig. 1.21, this profile of $\kappa_p(h)$ is shown as the

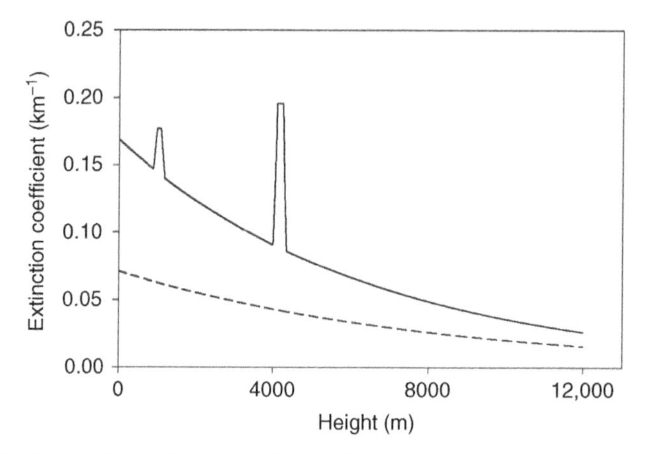

Fig. 1.21 Profiles of the vertical particulate extinction coefficient (the solid curve) and the molecular extinction coefficient (the dashed curve) used for the numerical simulation. (Kovalev et al., 2009a).

solid curve, and the corresponding molecular extinction-coefficient profile $\kappa_m(h)$ as the dashed curve. The simulated total signal $P_\Sigma(h)$ used for the inversion is the sum of the noisy backscatter signal $P(h)$ and the offset $B = 300$ counts (Fig. 1.22).

The estimate of the offset $\langle B \rangle$ was found using two alternative methods. First, the common method for determining the offset as a mean value of the recorded signal $P_\Sigma(h)$ over distant ranges was used. The mean value is determined for the altitude range 9000–11,000 m and results in the offset $\langle B(P_\Sigma) \rangle = 300.25$ counts. Second, the offset $\langle B \rangle$ was found by determining the slope of the linear fit for the function $y^*(z)$ in Eq. (1.60). As stated above, the derivative $dy^*(z)/dz$ should be found at distant ranges,

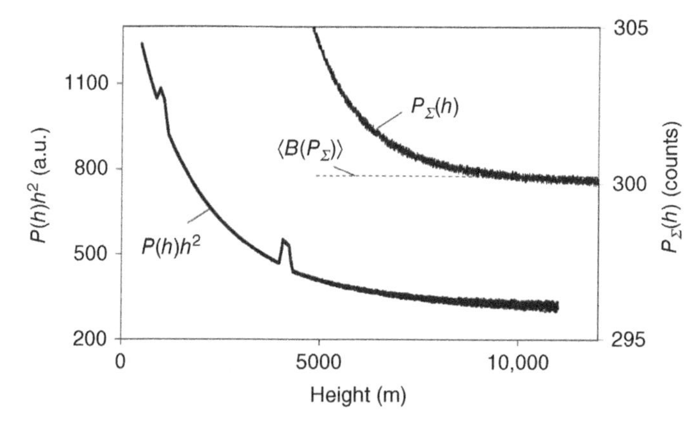

Fig. 1.22 Profile of the recorded noise-corrupted signal $P_\Sigma(h)$ and the corresponding square-range-corrected backscatter signal $P(h)h^2$ calculated for the virtual ground-based zenith-directed lidar in the simulated atmosphere.

close to h_{max}, where the fulfillment of the condition given in Eq. (1.59) is presumably most likely. However, to provide the reader with a general notion of the variability of the derivative over the total lidar operative range, in Fig. 1.23, the derivative profiles are shown over the extended height interval, from $\sim$ 2400 to 12,000 m.

When determining the derivative $dy^*(z)/dz$ through numerical differentiation, two different sliding range resolutions $s_1 = 200$ m and $s_2 = 2000$ m were used. Note that the range resolutions s_1 and s_2 are range independent, accordingly, the corresponding Δz is variable.

As can be expected, the numerical derivative calculated with the smaller range resolution, shown as gray cross symbols, is significantly noisier than that obtained with the larger range resolution, shown as the thick solid curve. In the numerical simulation, it is assumed that only high-frequency random noise influences the recorded lidar signal. Under such conditions, the mean values of the estimated offset, averaged within the height interval 9000 – 11,000 m, are practically the same, $\langle B(s_1)\rangle =$ 299.93 counts and $\langle B(s_2)\rangle = 299.95$ counts, respectively. They practically coincide with the actual value, $B = 300$ counts. Note that both $\langle B(s_1)\rangle$ and $\langle B(s_2)\rangle$ are less than the true offset, $B = 300$ counts, whereas the estimate $\langle B(P_\Sigma)\rangle$ is larger than B. Such a systematic difference between $\langle B(s)\rangle$ and $\langle B(P_\Sigma)\rangle$ may be a useful specific of the slope method. In practical sense, the opposite signs in the remaining shifts, $\Delta B(s) =$ $B - \langle B(s)\rangle$ and $\Delta B(P_\Sigma) = B - B(P_\Sigma)$, can allow some estimation of the upper and lower uncertainty limits in the calculated $P(h)$. This observation gives the opportunity to estimate the possible uncertainty level in the inverted extinction-coefficient profile.

In Fig. 1.24, the square-range-corrected backscatter signals $P(h)h^2$ obtained after removal of the offsets $\langle B(s_1)\rangle$, $\langle B(s_2)\rangle$, and $\langle B(P_\Sigma)\rangle$ are shown over distant

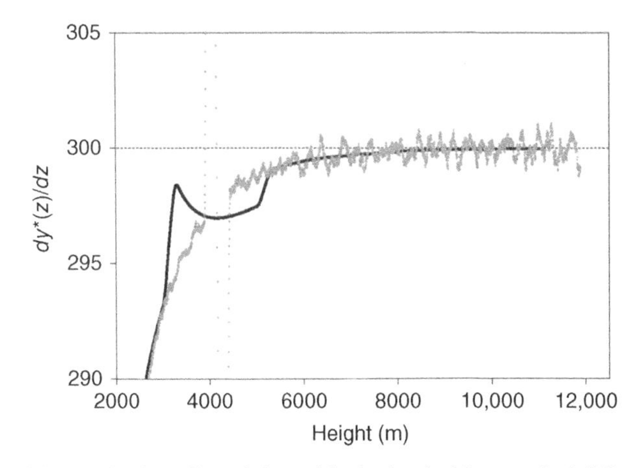

Fig. 1.23 Height-resolved profiles of $dy^*(z)/dz$ obtained with numerical differentiation. The gray cross symbols show the derivative profile versus height obtained with the sliding range resolution $s_1 = 200$ m and the thick curve shows that obtained with $s_2 = 2000$ m. The dotted horizontal line is the actual signal offset $B = 300$ counts.(Kovalev et al., 2009a).

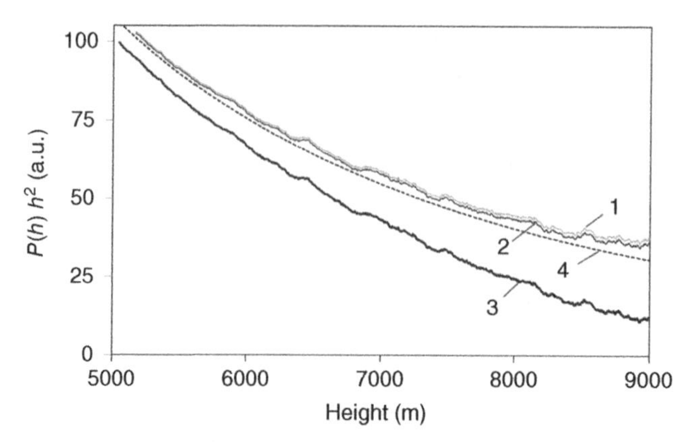

Fig. 1.24 Square-range-corrected backscatter signals determined from the simulated noise-corrupted signal, $P_{\Sigma}(h)$ in Fig. 1.22, using the estimates $\langle B(s_1)\rangle$, $\langle B(s_2)\rangle$, and $\langle B(P_{\Sigma})\rangle$ (Curves 1, 2, and 3, respectively). The simulated signal, not corrupted by noise, is shown as Curve 4. (Kovalev et al., 2009a).

ranges, for the heights $h > 5000$ m (Curves 1, 2, and 3, respectively). The simulated signal, not corrupted by the noise, is shown as Curve 4. One can see that the square-range-corrected signals calculated with estimates $\langle B(s_1)\rangle$ and $\langle B(s_2)\rangle$ are larger than the actual simulated signal, whereas the signal calculated with the estimate $\langle B(P_{\Sigma})\rangle$ is smaller than the actual signal. Moreover, when using the estimate $\langle B(P_{\Sigma})\rangle$, the backscatter signal becomes negative over heights greater than 10,000 m. Such a distortion can occur even when the remaining offset $\Delta B(P_{\Sigma})$ in the backscatter signal is minor; in our simulation, it is only −0.25 counts. In some cases, the backscatter signal obtained with the estimate $\langle B(s)\rangle$ may also be negative over distant ranges. A simple algebraic transformation of Eq. (1.62) shows that the shift $\Delta B(s)$ remaining in the backscatter signal is positive when the following simple inequality is true:

$$\frac{P(h_2)}{P(h_1)} < \frac{z_1}{z_2}. \tag{1.64}$$

This inequality may be invalid in the areas where the heterogeneous atmospheric layers with increased backscattering take place. In such areas, the numerical derivatives determined with different step sizes may significantly differ from each other. This effect, clearly seen in Fig. 1.23 between the heights ~3000 and 5000 m, puts an additional constraint for the heights acceptable for estimating the offset $\langle B(s)\rangle$.

Fig. 1.25 shows the vertical profiles of the particulate extinction coefficient retrieved from the backscatter signals. For the retrieval, the far-end solution was used. It was assumed that the lidar ratio was constant over all heights and its precise value was known, the same as the exact boundary conditions at the reference point $h_b = 8000$ m. The uncertainty in the retrieved extinction-coefficient profiles is

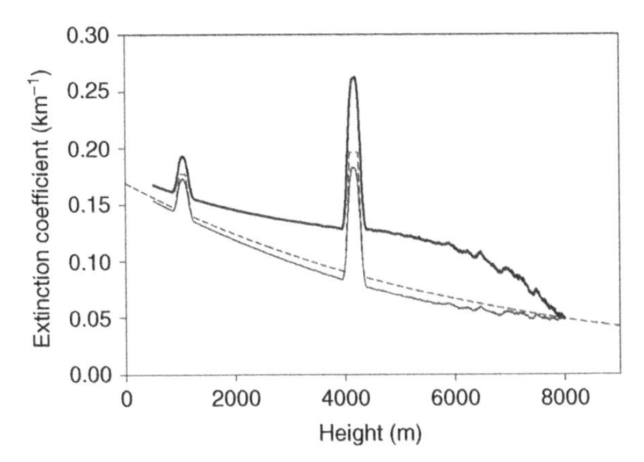

Fig. 1.25 Vertical profiles of the particulate extinction coefficient retrieved with the far-end solution using the estimates $\langle B(s_1)\rangle$ and $\langle B(P_\Sigma)\rangle$ (the thin and thick solid curves, respectively). The dotted curve shows the test profile used in the numerical experiment, same as in Fig. 1.21. (Kovalev et al., 2009a).

introduced solely due to the nonzero shift remaining in the backscatter signal after subtracting the estimated offset $\langle B\rangle$. The thin solid curve shows the profile derived with the estimate $\langle B(s_1)\rangle = 299.93$ counts. This profile, the same as that obtained with $\langle B(s_2)\rangle$ (not shown in the figure), is very close to the test profile of the extinction coefficient shown as the dotted curve. The thick curve is the extinction-coefficient profile obtained when the estimate $\langle B(P_\Sigma)\rangle = 300.25$ counts was used. Note that even such a small signal shift $\Delta B(P_\Sigma) = -0.25$ count may cause an error equal to 50% or more in the retrieved extinction coefficient.

Let us summarize. The use of the slope technique for determining the offset in the total signal $P_\Sigma(h)$ allows a more accurate estimation of the offset in the lidar signal, and accordingly, yields a more accurate lidar-signal inversion result. The simultaneous use of the slope and the conventional techniques for determining the offset can allow a more accurate estimation of the level of distortions in the derived extinction-coefficient profiles. Ignoring the additive distortion in the lidar signal may result in a significantly overestimated accuracy of the derived extinction-coefficient profiles.

1.7 EXAMINATION OF THE REMAINING OFFSET IN THE BACKSCATTER SIGNAL BY ANALYZING THE SHAPE OF THE INTEGRATED SIGNAL

When performing laboratory inspection of the lidar instrumentation before searching the atmosphere, the researchers' attention is focused mainly on improving hardware operation, minimizing instrumental distortions, and obtaining good test results. Later, when analyzing the data of atmospheric profiling, analysis of random noise becomes the task of highest priority, whereas the influence of a systematic offset is assumed negligible and ignored, sometimes without proper grounds. Meanwhile, as shown in the previous sections, even a seemingly negligible remaining offset in the inverted backscatter signal can significantly spoil the inversion result. When the boundary point in the far-end solution is chosen in the aerosol-free area, the influence of such a systematic offset in the area close the boundary point may be reduced, but some additional distortion in the derived extinction coefficient will be present over lower heights (e.g., see Figs. 1.18,1.19, and 1.25).

To address this and other issues related to the presence of the nonzero offset in the inverted backscatter signal, it is extremely important to estimate the maximum operative range up to which the distortions in the backscatter signal will not significantly spoil the inverted optical profiles. To achieve this goal, a relatively simple method can be used. To clarify this method, let us consider a typical square-range-corrected backscatter signal, which is corrupted by high-frequency random noise, contains masked low-frequency noise components, and the remaining systematic offset ΔB. As follows from Eq. (1.50) in Section 1.5, such a distorted square-range-corrected signal versus range can be written as

$$\langle P(r)\rangle r^2 = P(r)r^2 + P(r)\delta_P(r)r^2 + r^2 \sum w_j(r) + \Delta B r^2, \tag{1.65}$$

where the first term in the right side of the formula $P(r)r^2$ is the undistorted square-range-corrected backscatter component. Note that in the formula, an unsmoothed signal is assumed, therefore, the wide range spectra component $\sum w_j(r)$ which is the sum of the low- and high-frequency components, $\sum w_{j,\mathrm{low}}(r)$ and $\sum w_{j,\mathrm{high}}(r)$, is used.

The integral of the square-range-corrected signal $\langle P(r)\rangle r^2$ accumulated within the range interval from some starting point r_a to r can be written as

$$I(r_a, r) = \int_{r_a}^{r} \langle P(r')\rangle (r')^2 dr' = \int_{r_a}^{r} P(r')(r')^2 dr' + I_{\delta,w}(r_a, r) + \frac{1}{3}\Delta B(r^3 - r_a^3). \tag{1.66}$$

The first integral in the right side of the equation is the undistorted square-range-corrected backscatter signal $P(r)r^2$ accumulated within the range (r_a, r). The integral $I_{\delta,w}(r_a, r)$, is defined as the sum of the integrated products, $P(r)\delta_P(r)r^2$ and $r^2 \sum w_j(r)$, that is,

$$I_{\delta,w}(r_a, r) = \int_{r_a}^{r} \left[P\left(r'\right) \delta_P(r') + \sum w_j(r')\right] (r')^2 dr'. \tag{1.67}$$

The third term in Eq. (1.66) determines the accumulated distortion caused by nonzero difference between the true offset B and its estimated value $\langle B\rangle$.

Let us consider how the components in Eq. (1.65) influence the shape of the integral $I(r_a, r)$. The undistorted backscatter signal $P(r)$ is always a positive quantity, therefore, the integral $\int_{r_a}^{r} P(r')(r')^2 dr'$ in Eq. (1.66) can have only positive increments within the lidar operative range when the upper bound of the integral increases.

These increments asymptotically tend to zero over the ranges where the backscatter signal vanishes. The profile of the second term $I_{\delta,w}(r_a, r)$ depends on the random distortions of the lidar signal and may create some random fluctuations within the lidar operative range. The behavior of the third term $\frac{1}{3}\Delta B(r^3 - r_a^3)$, depends on the sign the offset ΔB remaining in the backscatter signal; this term can either increase or decrease the total integral $I(r_a, r)$. Under the condition of absence of any distortion component in the backscatter signal, that is, when $I_{\delta,w}(r_a, r) = 0$ and $\Delta B = 0$, the integral $I(r_a, r)$ will have only a systematic positive increment with increase of the range. These increments of the backscatter signal vanish at distant ranges. It means that starting at some distant range, the undistorted integral $I(r_a, r)$ will asymptotically approach a constant level. However, if distortion components are present in the backscatter signal, the integral over distant ranges can either decrease or even sharply increase with range, instead of being approximately constant. The remaining nonzero offset ΔB may have the most destructive influence on the shape of the integral, $I(r_a, r)$.

Let us illustrate the above statements using both real and simulated lidar data. In Fig. 1.26, an example of a real square-range-corrected backscatter signal, corrupted by random noise, and its integral $I(r_a, r)$ is presented. One can see that after reaching a maximum value at the range close to $r = 6000$ m, the integral slightly decreases. As stated above, the decrease of the integral can be caused by two reasons. First, its decrease can be caused by a slow decrease in the integral $I_{\delta,w}(r_a, r)$; second, such a decrease may be caused by a nonzero negative offset ΔB. Most likely for the case under consideration, the latter assumption is wrong. Numerical simulations show that the presence of even a small nonzero offset has much more destructive consequences, and will cause much more significant changes in the integral $I(r_a, r)$ than in Fig. 1.26. In any case, the sensible maximum range, r_{max}, for this signal should be selected

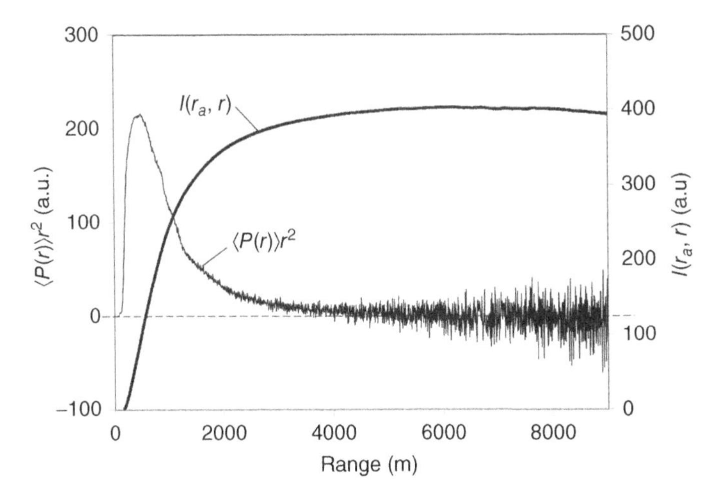

Fig. 1.26 A real square-range-corrected backscatter signal at the wavelength 355 nm, corrupted by random noise, and its integral.

within the range where no erroneous reduction in the integral $I(r_a, r)$ versus range takes place. In the case under consideration, the selected maximal operative range for the signal should not exceed ~ 6000 m.

A side note is required here. The obvious question may arise, why does not simple smoothing of a noisy signal allow for reliable determination of the optimal maximal range r_{max}? The answer to this question is that any smoothing requires the initial selection of the smoothing level. Depending on the level of the noise in the signal, different smoothing levels can be required; meanwhile, there are no standard principles for how such a smoothing level should be selected. This is illustrated by Fig. 1.27, which shows the signals obtained after applying different levels of smoothing to the noisy signal in Fig. 1.26. The gray dots represent the data points of the signal smoothed over the range 300 m, whereas the solid curve shows the same signal smoothed over the range 1000 m. When the smaller smoothing range is used, the first negative values in the smoothed signal appear at the range 5340 m, whereas the larger smoothing level shifts such negative points to the range 6100 m. The latter is close to the range established through analysis of the integral in Fig.1.26. However, analysis of the integral $I(r_a, r)$ permits the establishment of r_{max} without any *a priori* selection of the smoothing level. Moreover, such an analysis based on consideration of the integrated signal may provide additional advantages. Analyzing the integral $I(r_a, r)$, one can not only determine the maximal operative range but also detect the presence and level of signal distortions. In some cases, one can even correct the initially estimated inaccurate offset $\langle B \rangle$.

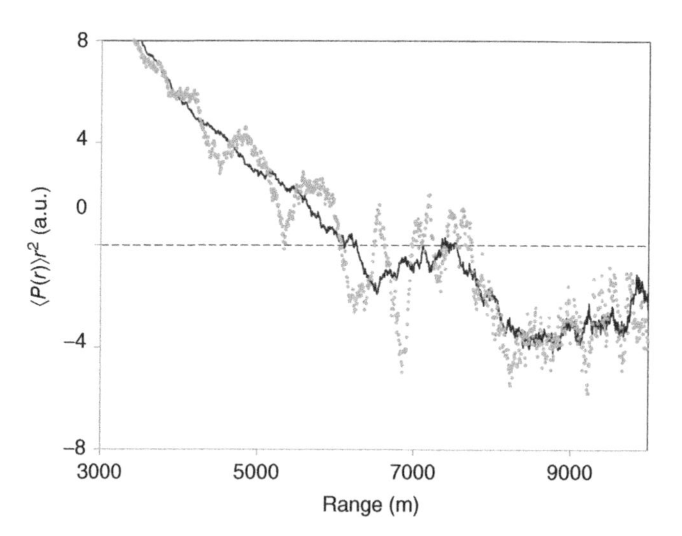

Fig. 1.27 The smoothed square-range-corrected signal $\langle P(r) \rangle r^2$ given in Fig. 1.26. The gray dots represent the data points of the signal smoothed over the range 300 m and the solid curve shows the signal smoothed over the range 1000 m. For better visualization, the signal is shown only over distant ranges.

The method for discriminating and reducing the systematic signal distortions by analyzing the shape of the integral $I(r_a, r)$ is clarified by numerical simulations, the results of which are given in Figs. 1.28–1.32. For these simulations, a synthetic atmosphere is selected in which the particulate extinction coefficient at the wavelength of 532 nm decreases with height from $\kappa_p(h) = 0, 17\,\text{km}^{-1}$ at $h = 2000$ m to $\kappa_p(h) = 0.025\,\text{km}^{-1}$ at $h = 10,000$ m. Additionally, a thin turbid layer with maximal $\kappa_p(h) = 1.0\,\text{km}^{-1}$ exists at the height ~ 6000 m (Fig. 1.28).

The signal $P_\Sigma(h)$ recorded in this atmosphere by the virtual zenith-directed lidar is the sum of the noisy backscatter signal and the constant offset $B = 200$ counts. This signal, shown in Fig. 1.29, is distorted by quasi-random noise, whose level is relatively insignificant; its magnitude does not exceed ~ 1.0 count.

The first operation, which has to be made before the lidar signal can be inverted into the optical profile of interest, is the estimation and subtraction of the estimated constant $\langle B \rangle$ from the total signal $P_\Sigma(h)$. In the simulation, the constant determined by recording the lidar signal before triggering the laser pulse yielded $\langle B \rangle = 199.6$ counts, whereas when determined over the distant ranges, it yielded $\langle B \rangle = 200.2$ counts. The latter was obtained by averaging the signals $P_\Sigma(h)$ over the heights from 8000 to 10,000 m, assuming that the backscatter component at these heights is small enough to be ignored. The obvious question immediately arises, which value of $\langle B \rangle$, 199.6 or 200.2 counts, will yield a more accurate profile of the backscatter signal, and accordingly, more accurate inversion result?

The comparison of the two alternative integrated signals $I(r_a, r)$ can help answer this question. In Fig. 1.30, the square-range-corrected backscatter signal, obtained with the first estimate $\langle B \rangle = 199.6$ counts is shown as the thin dashed curve, and its integral, $I(h_a, h)$, as a thick solid curve. One can see that the integral has large increments, and accordingly, an unrealistic increase over distant ranges. Such a shape of

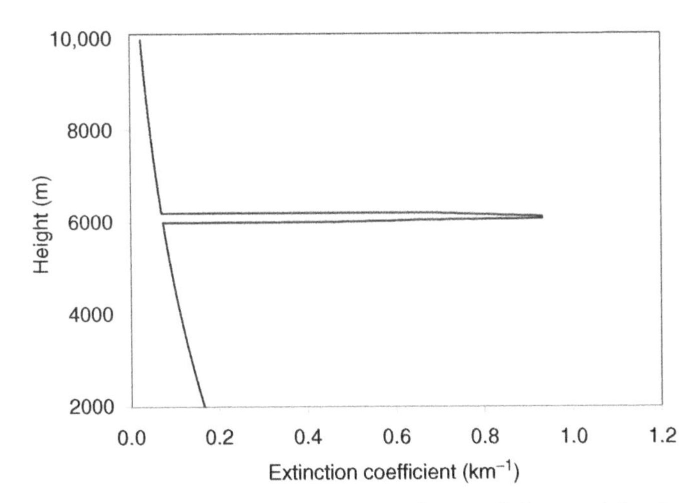

Fig. 1.28 Vertical profile of the particulate extinction coefficient used for the numerical simulation.

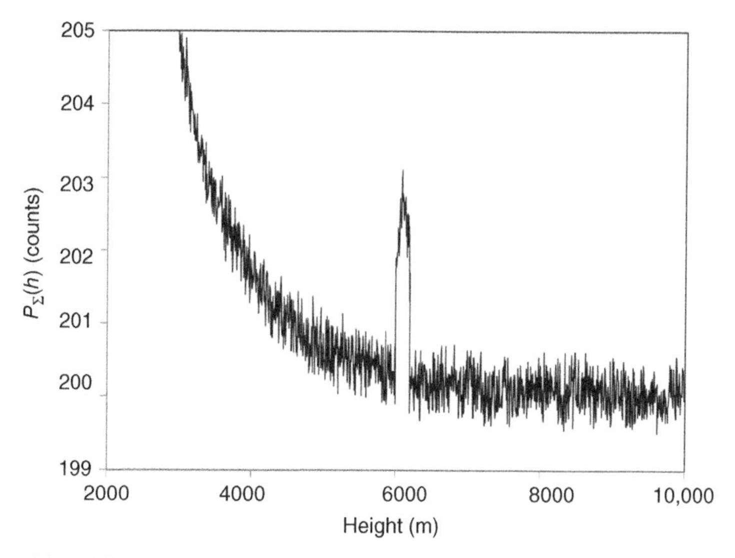

Fig. 1.29 The synthetic signal $P_\Sigma(h)$ used in the numerical simulation.

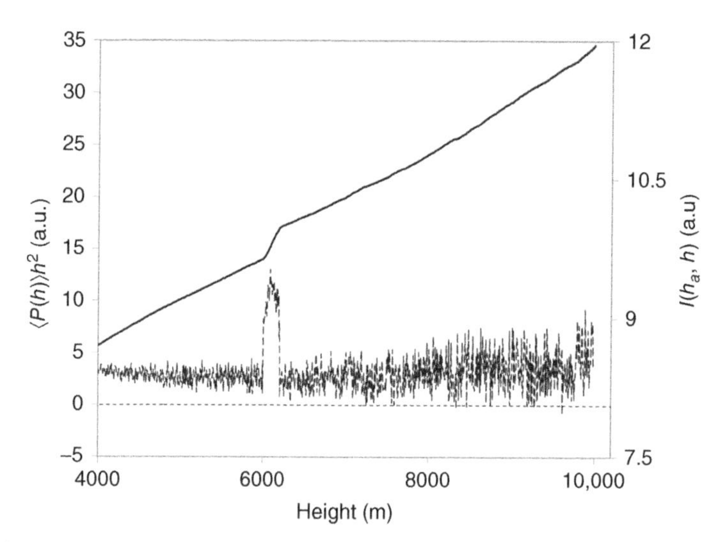

Fig. 1.30 The thin dashed curve shows the square-range-corrected backscatter signal $\langle P(h)\rangle h^2$ obtained from the synthetic signal $P_\Sigma(h)$ in Fig. 1.29 with the estimated shift $\langle B\rangle = 199.6$ count. The integral of the signal is shown as the bold curve.

the integral corresponds to particulate loading in the atmosphere, which is approximately the same over the heights from 4000 to 10,000 m. For a conventional cloudless atmosphere, such behavior of the aerosol loading looks unrealistic, so the shift $\langle B \rangle = 199.6$ counts should be rejected as not being accurate enough.

The backscatter signal $\langle P(h) \rangle h^2$ and the corresponding integral $I(h_a, h)$ determined with the estimate $\langle B \rangle = 200.2$ counts are shown in Fig. 1.31. In this case, the shape of the integral has the nonphysical decrease of the integral over distant ranges. Such a decrease indicates that that the second estimate $\langle B \rangle$, would result in erroneous negative backscatter data points over high altitudes. Thus, the estimate, $\langle B \rangle = 200.2$ counts is overestimated and it should also be rejected.

The obvious guess that emerges after analyzing the shapes of the two integrals $I(h_a, h)$ above is that the optimal estimate of $\langle B \rangle$ lies somewhere between 196.6 and 200.2 counts. The new set of calculations show that the selection of $\langle B \rangle = 200$ counts provides the most sensible shape of the integral $I(h_a, h)$ (Fig. 1.32). In spite of the presence of substantial quasi-random random noise in the square-range-corrected signal, this value coincides with the actual value of B used in the numerical simulation. The shape of the integrals $I(h_a, h)$ in Figs. 1.30–1.32 shows the high sensitivity of these integrals to even minor changes in the estimated offset $\langle B \rangle$.

In some cases, analysis of the real experimental data shows even more distorted shapes of the integral $I(r_a, r)$. Moreover, local sharp and erroneous changes in the integral may take place even within central zones of the lidar operative range. For example, the integral can reach its maximum value at relatively short distances from the lidar and then sharply decrease. Such an effect can be caused by the same two sources: it may occur when either the integral $I_{\delta,w}(r_a, r)$ in Eq. (1.76) has the nonzero value or the estimate $\langle B \rangle$ is not accurate.

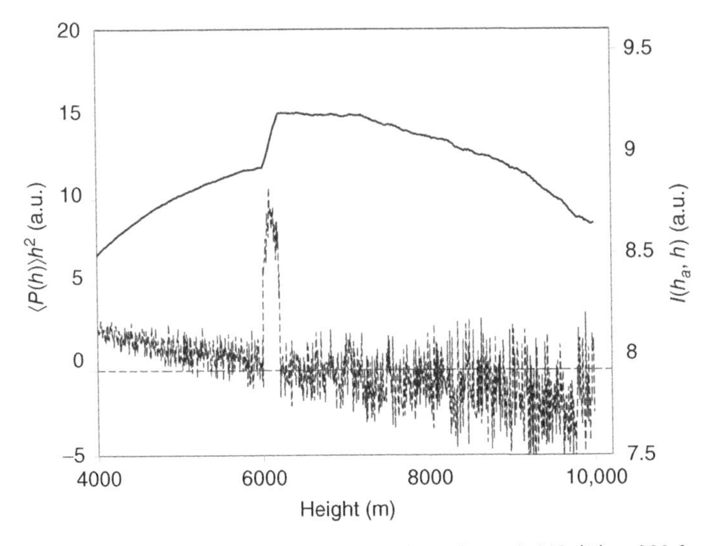

Fig. 1.31 Same as in Fig. 1.30 but when using the estimated shift $\langle B \rangle = 200.2$ counts.

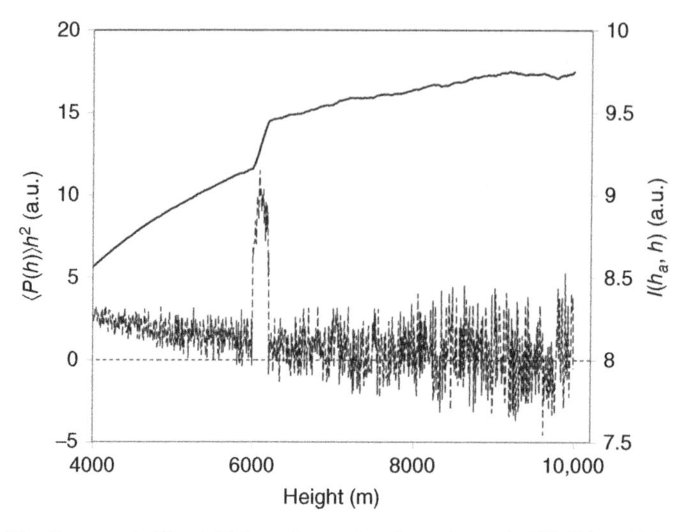

Fig. 1.32 Same as in Fig. 1.30 but when using the estimated shift $\langle B \rangle = 200$ counts.

The presence of the local particulate layers with sharp boundaries within the lidar operative range may significantly impede the estimation and removal of the background component from the recorded signal. In Figs 1.33 and 1.34, two square-range-corrected backscatter signals obtained from the same signal $P_{\Sigma}(h)$ recorded at the wavelength 355 nm, are shown as the gray filled circles. In both

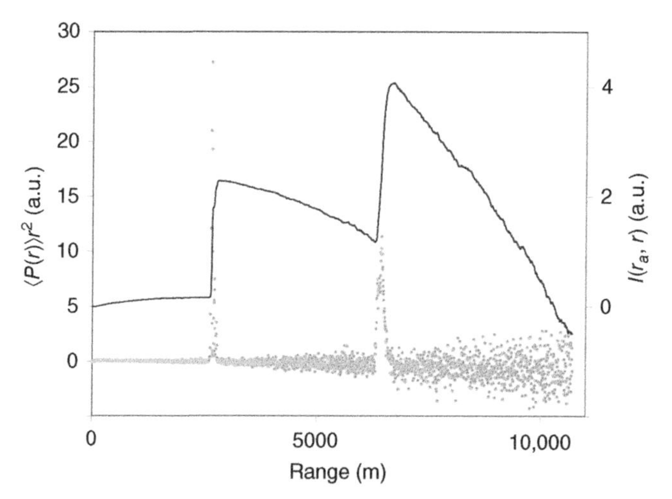

Fig. 1.33 The square-range-corrected backscatter signal obtained from the real signal $P_{\Sigma}(h)$ with the estimated offset $\langle B \rangle = 611.2$ count (the gray filled circles). The corresponding integral $I(r_a, r)$, is shown as the solid curve.

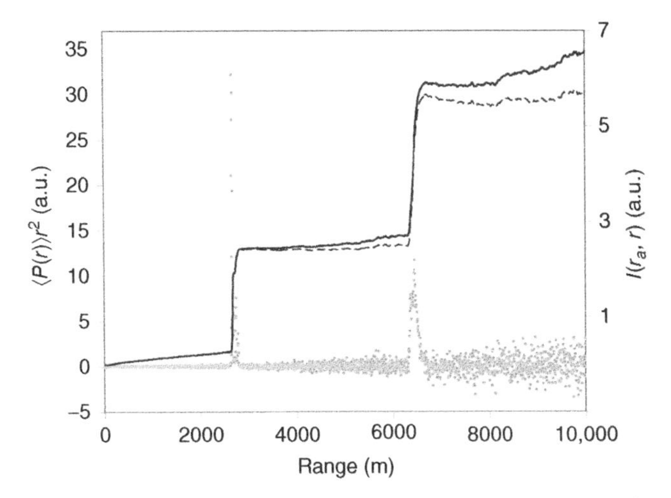

Fig. 1.34 The gray filled circles show the backscatter signal obtained from the same signal $P_{\Sigma}(h)$ with $\langle B \rangle = 608.8$ counts; the dashed curve is the corresponding integral $I(r_a, r)$. The thick solid curve shows the integral shape obtained with $\langle B \rangle = 608.4$ counts.

signals, two separated disturbances at the distances of 2700 and ~ 6500 m are clearly seen.

The backscatter signal shown in Fig. 1.33 was determined using the estimate $\langle B \rangle = 611.2$ counts derived from the recorded lidar signal over the far-end zone using the slope method. The corresponding integral, shown in the figure as the solid curve, has a nonphysical decrease between the disturbances and over the far end of the operative range. The analysis of this case showed that the more realistic shape of the integral $I(r_a, r)$ can be obtained by the minor decrease of the estimate $\langle B \rangle$. In Fig. 1.34, the backscatter signal obtained with the decreased $\langle B \rangle = 608.8$ counts and the corresponding integral are shown as the filled circles and the dashed curve, respectively. Such a minor decrease in the value of $\langle B \rangle$ down to 608.8 counts, produces an approximately constant integral in the areas outside the disturbance zones for the heights greater than 2700 m. However, taking into consideration the presence of the molecular component at the wavelength 355 nm, one can expect that a slight increase of the integral beyond the spikes at the ranges of 2700 and ~ 6500 m would be more realistic. Such behavior of the integral can be obtained by a minor reduction in the previous estimate $\langle B \rangle$ by selecting 608.4 counts. The corresponding integral, shown in Fig. 1.34 as a solid curve, reveals the realistic sharp increase of $I(r_a, r)$ at both boundaries of the increased backscattering and a very slight increase of the integral outside these spikes. This shape of the integral agrees well with the practical point that the backscattering within the two separate layers is significantly larger than that in the clear air outside.

Let us summarize:

1. The most typical shape of the integral $I(r_a, r)$ is an initial sharp monotonic increase with range over the nearest zones, which then transforms into a slow

continual increase over distant zones. When analyzing the shape of the integral, one should focus on the behavior of the integral over distant areas, keeping in mind that the saturation of the integral over distant areas is real in clear atmospheres, especially, when profiling of the atmosphere is carried out at the wavelengths in which the molecular component is negligible.

2. The shape of the integral $I(r_a, r)$ is extremely sensitive to the discrepancy between the actual offset B and its estimate $\langle B \rangle$. This property makes it easy to determine whether the value $\langle B \rangle$ should be corrected and in which direction. The selection of $\langle B \rangle$ less than the optimal value results in the systematic increase of the integral $I(r_a, r)$ over the whole measurement range. Physically, such an increase can only be interpreted as the continuous increase of particulate loading with height. Excluding extreme optical situations, for example, such as in smoky polluted atmospheres, systematic increase of the particulate loading at high altitudes in the cloudless atmosphere looks implausible.
3. The optimal shape of the backscatter signal profile can be obtained in three consecutive steps: (i) initial estimation of the offset $\langle B \rangle$ and its removal from the recorded signal; (ii) calculation of the corresponding integral $I(r_a, r)$ and analysis of its behavior over the total operative range; and (iii) adjustment of the estimated shift $\langle B \rangle$ to obtain the most sensible shape of the integral, $I(r_a, r)$. Such "tuning" produces either an insignificant monotonic increase of the integral $I(r_a, r)$ over the distant ranges inherent to clear air, or the invariability of the integral, which is sensible when both the molecular and the particulate components are minor. The first situation can be expected when operating at wavelengths close to the near UV region, where the molecular component is relatively large; the second situation is inherent to profiling of the atmosphere in the infrared spectra.

Two difficulties in the above correction technique may restrict the possibility of reducing distortion in the backscattered signal $\langle P(r) \rangle$. First, there is a general issue in determining the optimal shape of the integral $I(r_a, r)$ over the far-end range. When the integral slightly increases over distant ranges, the question arises whether this increase is caused by the presence of a real backscatter component or whether it is the result of the difference between the true constant shift B and its estimate $\langle B \rangle$. At best, the only possibility of clarifying this question is through the comparison of the slope of the integral over the near and distant zones of the measured range. The large increase of the slope over distant zones generally signals that the estimation of $\langle B \rangle$ is not accurate enough. However, if such an increase is moderate, the researcher should decide whether this increase is sensible for the atmosphere under investigation. This can be clarified *a posteriori*, that is, after deriving the particulate component from the signal over these distant ranges. However, the answer to the question cannot be found without using some assumption or assumptions about the searched atmosphere.

The second difficulty of this correction technique is the presence in Eq. (1.66) of a nonzero component $I_{\delta,w}(r_a, r)$. In the above analysis, it was assumed that $I_{\delta,w}(r_a, r) \approx 0$. If this assumption is not valid, the integral $I(r_a, r)$, may have an erroneous shape, and no corrections in the presumably incorrectly determined offset $\langle B \rangle$ can improve the situation. Such a case is illustrated in Fig. 1.35, where the integrals $I(r_a, r)$ were calculated using a real lidar signal. The value of the initially estimated offset in the recorded signal was $\langle B \rangle = 442.7$ counts.

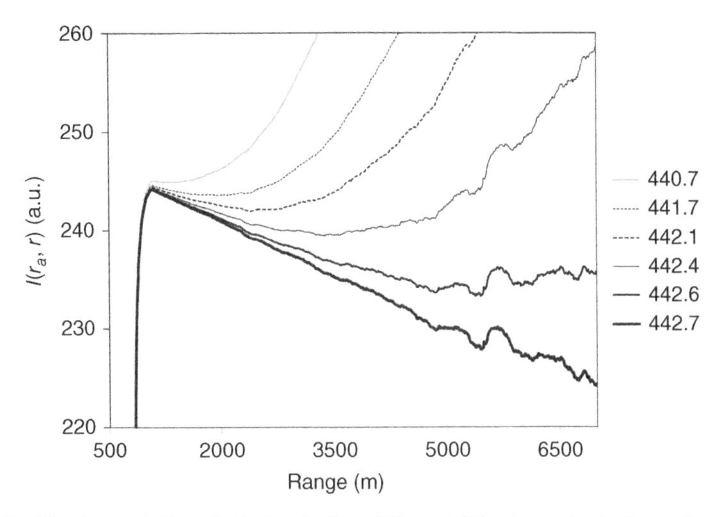

Fig. 1.35 The integral $I(r_a, r)$ obtained when different $\langle B \rangle$, shown in the legend, were used for extracting the backscatter component from the distorted lidar signal $P_\Sigma(r)$.

The corresponding integral $I(r_a, r)$ obtained after subtracting this $\langle B \rangle$ from the recorded signal $P_\Sigma(r)$ is shown in the figure as the thick solid line. For the whole operative range, starting at $r \approx 1$ km, the integral shows a nonphysical decrease with range. In accordance with the principles discussed above, a stepped reduction in $\langle B \rangle$ was performed to obtain a sensible shape of the integral. The decrease in $\langle B \rangle$ reduced the erroneous negative slope in the integral $I(r_a, r)$, but simultaneously increased the nonphysical positive slope over distant ranges. When the estimate $\langle B \rangle$ was step-by-step reduced, the lengths of the intervals where $I(r_a, r)$ decreased become shorter, but the nonphysical increase of the integral over the distant zones became more significant. Further analysis revealed the presence of a large variable component $I_{\delta,w}(r_a, r)$ in the backscatter signal. Obviously, no useful information about the optical properties of the sounded atmosphere can be extracted from such a distorted signal.

1.8 ISSUES IN THE EXAMINATION OF THE LIDAR OVERLAP FUNCTION

1.8.1 Influence of Distortions in the Lidar Signal when Determining the Overlap Function

Additive and multiplicative distortions in the recorded lidar signal influence not only the accuracy of routine lidar profiling data but also the accuracy of the lidar tests and calibration. In particular, these distortions may significantly impede experimental determination of the shape of the overlap function, $q(r)$. The related problem is that the distortion components during routine atmospheric profiling may significantly differ from those during the overlap calibration procedure.

As is known, there is no reliable method to determine analytically the shape of the incomplete overlap function of the real lidar. The analytical equation given in the book by Measures (1984) and then repeated in some other publications, as with analytical formulas proposed, for example, in the study by Stelmaszczyk et al. (2005), unfortunately, are impractical. Actually, the shape of the overlap function can be properly determined only experimentally. Only a thorough experimental examination of this function allows determining its shape within the incomplete overlap zone and establishing the range where the condition $q(r)$ = const. is true. The accurate shape of $q(r)$ within the incomplete overlap zone is required when the data points from this zone are to be used for signal inversion. This is a relatively rare case, used only for lidar systems with extremely extended range of the incomplete overlap zone. The inversion of the signals measured in the incomplete overlap zone requires special methodologies, which are different for different lidar systems (McDermid et al., 1995; Wandinger and Ansmann, 2002; Berkoff et al., 2003; Eloranta et al., 2004, Biavati et al., 2011).

Because of the importance of the overlap issue, methods for experimental determination of the length of the incomplete overlap zone, which commonly defines the minimal operative range of the lidar, have been widely discussed in literature. Before discussing these methods, it is necessary to define some terms, or more correctly, to define two similar terms, "the maximum length of the incomplete overlap zone" and "the minimum range of the complete overlap zone," which define the same range r_0 (Fig. 1.36). According to the general formulation in Section 1.1, this term defines the minimum range, where the cross section of the laser beam is completely within the field of view of the receiver's telescope.

However, the optical scheme in Fig. 1.36 is only a rough schematic of reality. First, the boundaries of a laser beam are not so nicely defined as in the figure, and second, the diameter of a laser beam cross section in the near zone does not change linearly with range. Nevertheless, the shape of the overlap function $q(r)$ is similar for all lidars: there is a slight increase in the very beginning, then the increments of $q(r)$ become larger, and then, closer to r_0, the increments becomes fewer, and finally, and at the ranges $r \geq r_0$, the overlap function reaches the constant maximum value $q(r) = 1$.

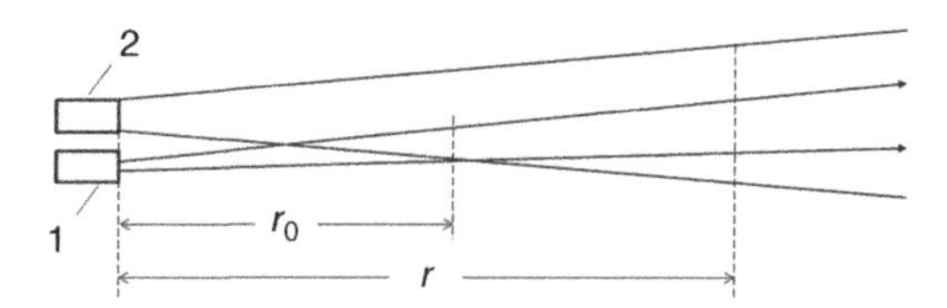

Fig. 1.36 Schematic of the monostatic biaxial lidar in which incomplete overlap takes place up to the range r_0. Here, 1: the laser; 2: the receiver's telescope.

There is some issue related to the sharp decrease in the backscatter signal within the lidar near zone. At a glance, it appears that in the incomplete overlap zone where $q(r) < 1$, the backscatter signal $P(r)$ will be significantly reduced, so its maximal value will be reached somewhere in the vicinity of the range r_0. However, this is not so. The maximal value of the backscatter signal is mostly obtained in the area of the incomplete overlap zone. An illustration of this statement is shown in Fig. 1.37 (a) and (b), where the backscatter signals, obtained with virtual lidars with different incomplete overlap zones are shown. The signals are calculated for synthetic homogeneous atmospheres with different extinction coefficients, values of which are shown in the legend.

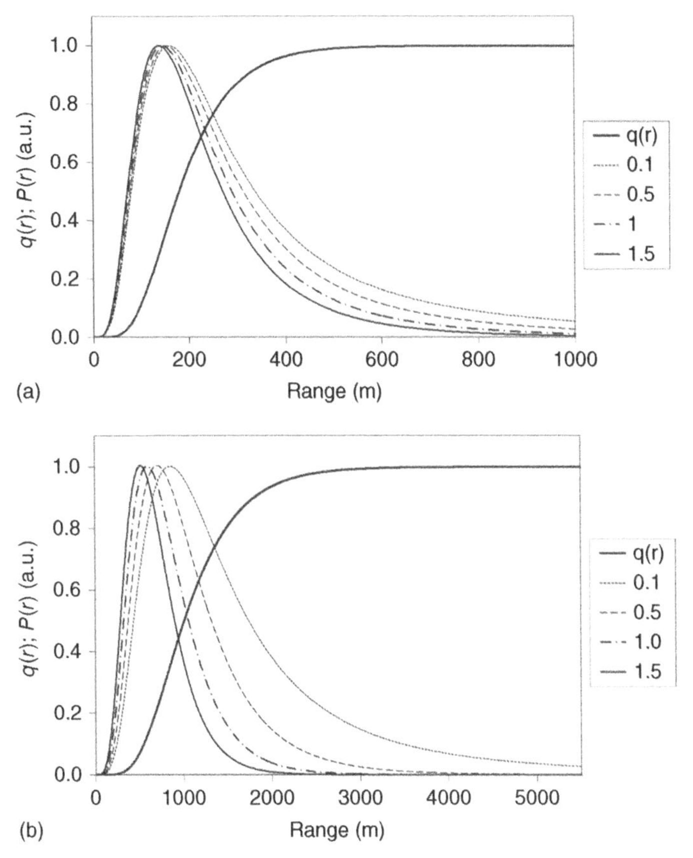

Fig. 1.37 (a) The thin dotted, dashed, dotted-dashed, and solid curves are the backscatter signals measured by the virtual lidar in synthetic homogeneous atmospheres and normalized to unity. The values of the extinction coefficient in km^{-1} are shown in the legend. The bold solid curve is the overlap function of the synthetic lidar, in which the incomplete overlap zone extends up to 500 m. (b) Same as in (a) but the incomplete overlap zone of the artificial lidar extends up to 3000 m.

The maximal length of the incomplete overlap zone for the lidars is 500 and 3000 m, respectively. These overlap functions, shown as thick solid lines in Fig. 1.37 (a) and (b), were simulated using the simple formula

$$q(r) = \left[1 - \exp\left(-a_q \frac{r}{r_0}\right)\right]^{b_q}, \tag{1.68}$$

where r_0 is the range in which the overlap function reaches the level $q(r_0) = 0.997$. The functions $q(r)$ shown in Fig. 1.37 (a) and (b) were calculated with $a_q = 7, b_q = 3$. One can see that within the complete overlap zone, $r > r_0$, the backscatter signal is significantly less than that in the near zone. Obviously, for the zenith-directed lidar, which operates in a clear cloudless atmosphere, the maximum backscatter signal may be shifted toward even smaller ranges. This means that when profiling in clear atmospheres, the signals of zenith- directed lidar with the best SNR can only be used for inversion if the shape of the overlap function in the near end is properly calibrated. As will be shown below, accurate inverting of lidar data in the incomplete overlap zone is an extremely difficult task, and, therefore, is rarely done.

Another unpleasant effect related to the shift of the maximal signal toward the near zone is the exposure of the photomultiplier detector to the intense lidar returns from the incomplete overlap zone. This can cause the so-called afterpulsing effect, which occurs in the photomultiplier tube after the abrupt termination of a short light pulse and may significantly distort the lidar signal over distant ranges (see Section 1.1). To reduce the problem, the high-intensity lidar returns from the near ranges can be shielded from the detection system, as is done in the study by McDermid et al. (1995). Here this requirement was achieved by using a small aperture in the plane of an optical chopper.

The above issue is significantly reduced when lidar is used for the investigation of distant layers with increased backscattering, such as smoke plumes or dust clouds. In such cases, the maximal signal is commonly shifted from the zone of the incomplete overlap. To illustrate such a case, in Fig. 1.38 the backscatter signal of the virtual lidar with maximal incomplete overlap range $r_0 = 500$ m is shown in a synthetic homogeneous atmosphere with $\kappa = 0.1\,\text{km}^{-1}$, in which the layer with increased backscattering is present over the ranges $500 - 700$ m. In this case, the maximum signal is shifted to the range 580 m, which is outside the incomplete overlap zone.

Now let us take a short excursion into the history of creating and modifying the methods for determining the overlap function. The simplest way for determining the shape of $q(r)$ and the corresponding range of the maximum incomplete overlap r_0 is the use of the common slope method in a homogeneous atmosphere. However, this method can only practically be used for lidars with a relatively short incomplete overlap zone. The simplest variant of this method was proposed by Sasano et al. (1979). The authors proposed to determine the shape of the overlap function in a very clear atmosphere, in which the path-transmission term can be taken as unity, so the

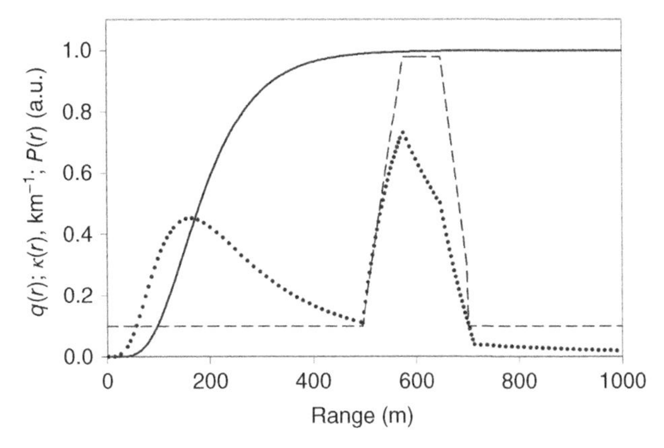

Fig. 1.38 The dashed curve shows the extinction-coefficient profile in a synthetic atmosphere, where the layer with increased backscattering exists over the ranges 500 – 700 m. The dotted curve is the synthetic backscatter signal measured by the artificial lidar in this atmosphere, and the bold solid curve is the overlap function of this lidar.

backscatter signal is directly proportional to $q(r)$. Obviously, such a method cannot be used even if the molecular component is large, that is, when the lidar operates at a wavelength shorter than the commonly used 532 nm, for example, at 355 nm. An alternative variant for determination of $q(r)$ in a moderately polluted atmosphere was proposed in the studies by Ignatenko (1985) and Tomine et al. (1989). The authors assumed that under favorable optical situations, a slightly turbid atmosphere may be treated as statistically homogeneous if a large number of lidar signals were averaged. Unlike the method of Sasano et al. (1979), they proposed to determine the mean extinction coefficient in such an atmosphere by the slope method, that is, by using the linear fit of the logarithm of the square-range-corrected signal versus range over the range interval where the logarithm forms a straight line.

A simple analysis of this idea shows that overlap calibration under reduced visibility should be avoided. Although one obtains better SNRs, the accuracy of the overlap-function determination in such an atmosphere may be insufficient for lidar searching in clearer atmospheres. The determination of the shape of the overlap function should preferably be done under high atmospheric transparency, as close as possible to the minimum extinction-coefficient range met in the next routine profiling of the atmosphere. In other words, if a lidar is assumed to operate under visibilities between ~30 and 50 km, its overlap function should be investigated at the visibility close to 50 km rather than at the visibility of 10 km. The grounds for this recommendation will be given below.

During the past decades, a number of alternative methods for experimentally determining the overlap function $q(r)$ were proposed. There is no need to discuss these in detail, and we will mention only some of these. One can mention the idea of determining the overlap function in an inhomogeneous atmosphere, considered in the study by Dho et al. (1997). To determine $q(r)$ in the incomplete overlap zone in such an atmosphere, the authors proposed to use the polynomial fit for the signal logarithm, assuming the homogeneity of the atmosphere in a statistical sense. The question that immediately arises about such a method is whether the atmosphere really obeys our statistical laws, and if yes, under what meteorological conditions. The simple method of optical transmitter-receiver adjustment in the conditions of restricted terrain by using a remote target was proposed by Sassen and Dodd (1982). An alternative method of transmitter-receiver optic alignment based on a model shape of the overlap function was proposed by Fiorani et al. (1998). The common specific of these and similar methods of determining the shape of $q(r)$ is the absence of proper analyses of the errors inherent to these methods.

To provide a general vision of the overlap issue, let us consider the most common principle for investigating the short-range overlap function and related calibration errors. The lidar return is measured in a horizontal direction in the homogeneous atmosphere, and the logarithm of the range-corrected backscatter signal $\ln[P(r)r^2]$ is the basic function for the overlap calibration. Let us initially consider the ideal case, assuming that no systematic distortions of the backscatter signal, obtained in an ideally homogeneous atmosphere, take place. The logarithm of such an ideal square-range-corrected signal obeys the formula

$$y(r) = \ln[P(r)r^2] = \ln[q(r)] + \ln[C\beta_\pi] - 2\kappa r. \tag{1.69}$$

Over ranges where the overlap is complete, that is, $q(r) = 1$, Eq. (1.69) transforms into a linear function of the range. The linear fit of the logarithm of the range-corrected signal, determined within the range interval $r_1 - r_2$, where $r_1 > r_0$ and $r_2 > r_1$, can be written as

$$Y(r) = Y_0 - 2\kappa r, \tag{1.70}$$

where $Y_0 = \ln\ [C\beta_\pi]$ is the intercept of the extrapolated linear fit with the Y-axis, and the slope of the linear fit is unequivocally related to the total extinction coefficient, $\kappa = \text{const}$. After determining these parameters through the slope of $Y(r)$ within the complete overlap area, the shape of the overlap can be determined from the formula,

$$\ln q(r) = y(r) - Y(r); \tag{1.71}$$

in the complete overlap area, $y(r) = Y(r)$ and $q(r) = 1$.

Unfortunately, this simple formula determines the overlap function when using the signals measured by an ideal lidar, which ideally transforms the input backscatter light into the electrical signal. Moreover, Eq. (1.71) is rigidly valid only under the unrealistic assumption that there is no random noise in the backscatter signal. Meanwhile, as follows from Eq.(1.65), even in an ideal homogeneous atmosphere, the square-range-corrected signal of real lidar should be written in a more complicated form:

$$\langle P(r)\rangle r^2 = Cq(r)\beta_\pi(r)\exp(-2\kappa r) + r^2[W_P(r) + \Delta B]; \tag{1.72}$$

where $\langle P(r)\rangle$ is the backscatter signal distorted by the multiplicative and additive distortion components, and the distortion function $W_P(r)$ is defined as

$$W_P(r) = P(r)\delta_P(r) + \sum w_j(r),$$

where $P(r)$ is the ideal, undistorted backscatter signal.

As was noted in Section1.1, there may be different sources of appearance of nonzero range-dependent distortion $W_P(r)$. Let us briefly recall some of these. The photomultiplier of the lidar receiving block may have variable sensitivity across the photocathode, as described in the study by Simeonov et al. (1999). This effect may be combined with another distortion investigated by Agishev and Comeron (2002), which also may occur in the signal measured in the near zone of the lidar. The fields of view for the recorded signal of the monostatic biaxial lidar and background radiation may be different; the signal field of view is generally significantly less than the field of view for the background radiation. Accordingly, the distance between the volume image and the focal plane of the receiving optics depends on how far the pulse scattering volume is from the lidar. The closer the scattering volume is, the greater the distance between the volume image and the focal plane. This means that such lidar geometry adds some variable component, which may result in a significant increase of the multiplicative distortion in the backscatter signal in the near-end area. A similar effect of nonlinear response for a photon counting system was described in the study by Donovan et al. (1993).

Taking the logarithm of the function in Eq. (1.72) and making simple transformations, one obtains

$$\langle y(r)\rangle = \ln[\langle P(r)\rangle r^2] = \ln[C\beta_\pi(r)] + \ln[q(r)] + \ln\left[1 + \frac{W_P(r) + \Delta B}{P(r)}\right] - 2\kappa r. \tag{1.73}$$

The linear fit for the distorted data points $\langle y(r)\rangle$ for the same range interval $r_1 - r_2$, within the complete overlap obeys the formula

$$\langle Y(r)\rangle = Y_0^* - 2\langle\kappa\rangle r, \tag{1.74}$$

the slope and intercept of which differs from those in the ideal linear fit in Eq. (1.70). From the distorted linear fit $\langle Y(r)\rangle$ the distorted overlap function $\langle q(r)\rangle$ is obtained, the logarithm of which is written in the form similar to that in Eq. (1.71):

$$\ln\langle q(r)\rangle = \langle y(r)\rangle - \langle Y(r)\rangle, \tag{1.75}$$

but the shape of this function may significantly differ from that of Eq. (1.71).

To demonstrate the consequences of systematic distortions in the actual backscatter signal, let us examine the overlap function derived from the virtual elastic lidar that operates in an ideally homogeneous atmosphere. The lidar operates at the wavelength 532 nm, and the maximum length of the incomplete overlap zone is $r_0 = 500$ m, similar to that shown as the thick solid curve in Fig. 1.37 (a). In the numerical simulation that follows, it is assumed for simplicity that only the remaining offset $\Delta B \neq 0$ distorts the backscatter signal shape, whereas other distortions are absent. To provide realistic results, the lidar signal was distorted by pseudorandom noise.

A comprehensive analysis, made by the author, revealed that when the backscatter signal has even a small remaining offset $\Delta B \neq 0$, the retrieved linear fit $\langle Y(r) \rangle$, and accordingly, the overlap function $\langle q(r) \rangle$, may significantly differ from the functions $Y(r)$ and $q(r)$, which would be obtained from the undistorted backscatter signal. Moreover, the level of distortion in these functions depends on how clear the atmosphere is when the overlap function is investigated. Larger distortions occur in more polluted atmospheres. To illustrate this effect, the profile of the synthetic overlap function was determined twice, under visibilities of 10 and 50 km. The simulation results shown below are obtained for the case of a minor overestimation of the constant shift B in the recorded signal. It was estimated that $\langle B \rangle = 200.4$ counts, whereas the true value $B = 200$ counts, yielding the remaining offset in the backscatter signal $\Delta B = 0.4$ count. Assuming that for recording the lidar signal, a 12-bit digitizer was used, the ratio of the maximal signal (4096 count) to the offset, ΔB, was more than 10,000.

Fig. 1.39 shows the derived function $\langle y(r) \rangle$ and its linear fit, obtained from the backscatter signal in the slightly polluted atmosphere with the visibility ~ 10 km. The linear fit $\langle Y(r) \rangle$ was determined over the range interval from $r_1 = 600$ to $r_2 = 3000$ m. This range is optimum, as the selection of $r_2 > 3000$ m would significantly increase the distortion of the linear fit in such an atmosphere. In Fig. 1.40, the ratios of the overlap function $\langle q(r) \rangle$ retrieved from the above distorted backscatter signal with the remaining offset $\Delta B = -0.4$ count, determined under visibilities at 50 and

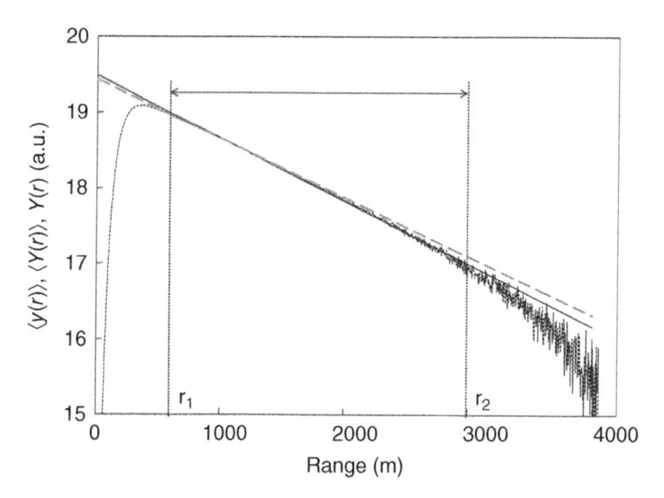

Fig. 1.39 The distorted function $\langle y(r) \rangle$ (the dotted curve) and its linear fit $\langle Y(r) \rangle$ (the solid line) obtained from the backscatter signal with the remaining offset $\Delta B = -0.4$ count, under a visibility of 10 km. The gray dashed line is the linear fit which would be obtained in the ideal case of no systematic distortion in the recorded signal.

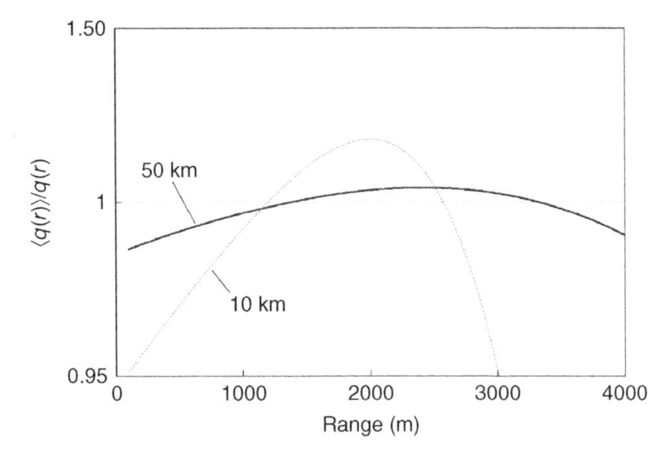

Fig. 1.40 Ratios of the distorted overlap functions $\langle q(r)\rangle$ to the true $q(r)$ retrieved under visibilities of 50 and 10 km to the actual function $q(r)$ when the remaining offset is equal −0.4 count.

10 km, to the true function $q(r)$ are shown. Note that the systematic distortions in the retrieved $\langle q(r)\rangle$ take place not only in the incomplete overlap area (500 m) but far beyond it.

Thus, when a lidar signal with nonzero distortions is used for determining the overlap function, the latter may be also distorted. The level of the distortions in the derived function $\langle q(r)\rangle$ depends on atmospheric conditions, in particular, on the visibility of the atmosphere when the function is determined. However, when the overlap function is determined in an atmosphere with high visibility, these distortions are minor.

1.8.2 Issues of Lidar Signal Inversion within the Incomplete Overlap Area

Inversion of the lidar signal in the incomplete overlap zone is generally avoided by researchers. When the data points from the area of the incomplete overlap need to be inverted into the profiles of the optical parameter of interest, specific problems appear. Practice reveals that in the near-end zone, where $q(r) < 1$, the overlap function during routine measurements may differ from that determined even recently, several days before the measurement (Wandinger and Ansmann, 2002). One can assume that the variations in the remaining offset ΔB combined with influence of the distortion factor $W_P(r)$ which takes place both during the calibration period and during routine lidar profiling of the atmosphere, may also be responsible for this issue, as with temperature effects, and so on.

It may become necessary to use the incomplete overlap area, for example, when the lidar eye safety is achieved by expanding the cross section of the laser beam through a large-diameter telescope in order to reduce the laser energy per pulse (Spinhirne, 1993, 1996; Spinhirne et al., 1995). The high SNR is achieved by the use of high-repetition laser pulses and extremely narrow bandpass filters. This idea is implemented in the MPL, where the Cassegrain-type telescope is used both for emission of the pulsed light and collection of the atmospheric backscatter. In addition to the eye safety, the extremely narrow field of view of the receiver optics (100 μrad or less) enables significant reduction in unwanted background light and multiple scattering components in the recorded lidar signal.

An unsettling feature of such types of lidars is the extremely long incomplete overlap zone; the minimum distance r_0 at which the backscatters signals are completely within the receiver field of view varies for different systems from 3 to 5 km (Berkoff et al., 2003; Campbell et al., 2002; Eloranta et al., 2004). To obtain information from the nearest zone, the shape of the variable overlap function $q(r)$ should be determined and used as a calibration curve to correct the signals recorded within this area in routine atmospheric profiling.

Two significant issues are related to such a calibration. First, the most straightforward method of determining $q(r)$, discussed above in Section 1.8.1, requires the possibility of measuring the lidar signal along the horizontal path with at least a 10-km clear line of sight, high visibility, and, what is most difficult, in a homogeneous atmosphere. To address this issue, Berkoff et al. (2003) proposed a special correction method for the MPL. To perform such a correction, a secondary, wide field-of-view receiver, the incomplete overlap zone of which was as small as 0.3 km, was co-aligned with the basic narrow-field-of view optics. The data obtained during simultaneous measurements of the backscatter signals with two receivers allowed estimation of the behavior of the overlap function for the narrow field-of-view receiver. Another way to solve this issue was used by the University of Wisconsin Lidar Group, which developed the HSRL for long-term cloud observations in the Arctic, at Barrow, Alaska (Eloranta et al., 2004). Here the geometric corrections were determined by varying the telescope focus while recording data. An alternative method for determining such corrections, considered by the authors, was based on comparison of the molecular signal measured during clear weather with the molecular lidar return predicted from a measured temperature profile.

The second issue in using the calibrated incomplete overlap function is its temporal instability. As mentioned, the overlap profile may differ from the calibrated one even when the calibration was made several days earlier. In addition to temporal instability, temperature instability of the overlap function may also take place. The investigation of this issue by Berkoff et al. (2005) revealed that the commercial-grade telescope changes its focal length when the instrument temperature changes, resulting in a change of the overlap function, and accordingly, distortion in the optical profile derived from that incomplete overlap zone. A similar issue was found by the University of Wisconsin Lidar Group. As a result, the researchers avoided extracting the particulate extinction coefficient from their HSRL data, focusing mainly on the extraction of the particulate backscatter cross section (http://lidar.ssec.wisc.edu). The optical depth profiles were analyzed only outside the incomplete overlap area.

To illustrate how slight changes in the shape of the overlap function in the incomplete overlap zone may influence the inversion result, let us perform a simple numerical simulation using the data of a virtual lidar, in which the incomplete overlap range extends up to range 3000 m. During lidar profiling, the overlap function was slightly shifted relative to the previously determined overlap calibration profile $q_{calibr}(r)$. The calibration profile, $q_{calibr}(r)$, is calculated with Eq. (1.68) using $r_0 = 3000$ m, the shifted overlap functions $\langle q(r) \rangle$ are simulated using r_0 equal to 2860, 2950, 3050, and 3147 m; these functions are shown in Fig. 1.41 (a) and (b). The maximal shifts between the actual and calibrated curves, $\Delta q(r) = \langle q(r) \rangle - q_{calibr}(r)$, are equal to ± 0.01 and ± 0.03, and occur between the ranges ~500 and ~ 1500 m (Fig. 1.42).

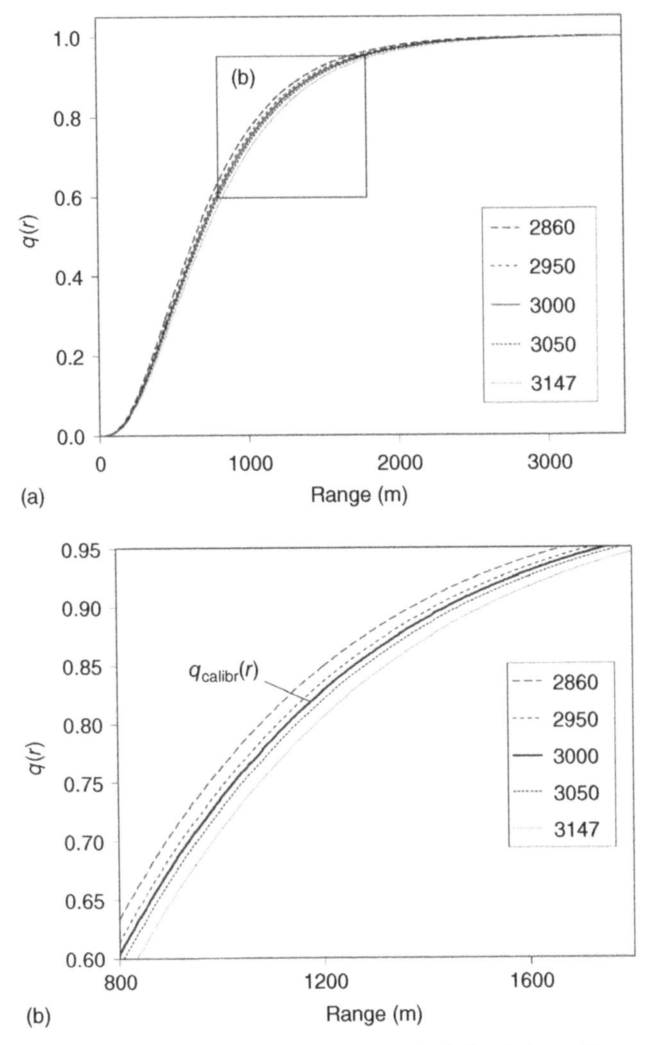

Fig. 1.41 (a) The overlap functions used in the numerical simulations. The numbers in the legend show the value of r_0 used in Eq. (1.68); (b) the same overlap functions but shown within the restricted range in order to make the shifts between the simulated functions more discernible.

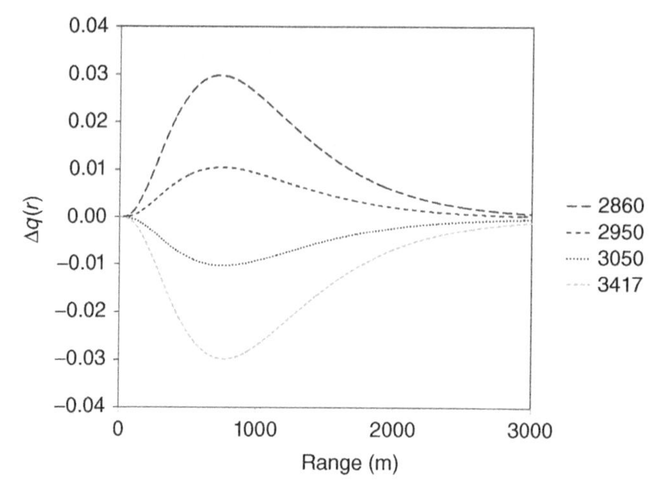

Fig. 1.42 The absolute differences $\Delta q(r)$ between the calibrated overlap function $q_{\text{calibr}}(r)$ and the shifted functions $\langle q(r) \rangle$ shown in Figs. 1.41 (a) and (b).

These shifted overlaps were used for the calculation of the shape of the synthetic signals in the area of incomplete overlap. For inverting these lidar signals into atmospheric profiles, the calibration curve $q_{\text{caliber}}(r)$ was used. The results of the numerical simulation made for such a lidar are shown in Figs. 1.43 and 1.44. In Fig. 1.43, the simplest case is considered when the virtual lidar operates in a homogeneous atmosphere with the particulate extinction coefficient $\kappa_p(r) = 0.2\ \text{km}^{-1}$. To better visualize distortions in the retrieved extinction coefficient, caused by the difference between $q_{\text{calibr}}(r)$ and $\langle q(r) \rangle$, no random noise is added to the lidar signal; the constant offset in the recorded signal is precisely determined, that is, $\Delta B = 0$; the signal

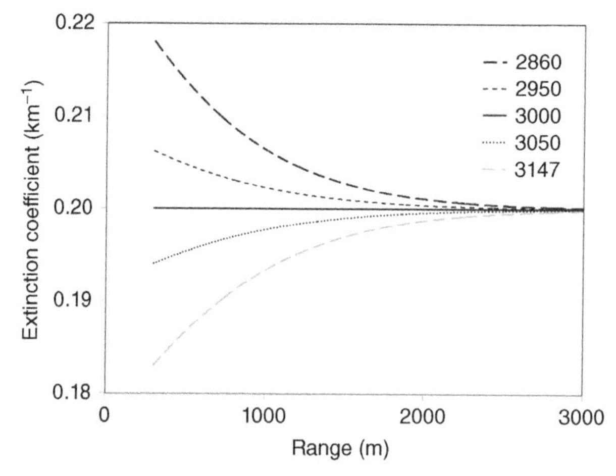

Fig. 1.43 Distortions in the particulate extinction-coefficient profiles caused by the nonzero shifts $\Delta q(r)$, shown in Figs. 1.42.

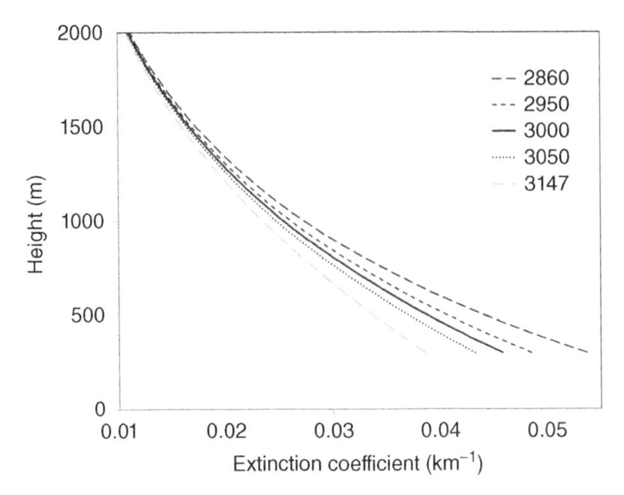

Fig. 1.44 Profiles of the particulate extinction coefficient derived from the signals of the virtual zenith-directed lidar that operates in the atmosphere with background aerosol profile given in the study by Zuev and Krekov (1986).

inversion is made using the precise boundary values of the extinction coefficient in the far-end solution. As can be seen in Fig. 1.43, even under such almost ideal conditions, small shifts in $\Delta q(r)$ can result in up to 10% distortion in the extinction coefficient in the lidar near-end zone. Obviously, the distortions would be significantly larger if other sources of error were taken into consideration.

Finally, let us consider the case, when the same virtual lidar at the wavelength 532 nm is directed in zenith and operates in the atmosphere with the background model of atmospheric aerosol given in the study by Zuev and Krekov (1986). Ideal conditions similar to those above are assumed, that is, no distortions in the lidar signal exist, except that caused by the presence of the overlap shifts $\Delta q(r) \neq 0$ and the boundary value of the extinction coefficient at the height 6000 m was precisely chosen. The profiles of the particulate extinction coefficient extracted with the far-end solution are shown in Fig. 1.44. As before, the maximum distortions in the retrieved particulate extinction coefficient $\kappa_p(h)$ occur at the near end at the height $h_{\min} = 300\,\text{m}$, where the fractional error $\delta\kappa_p(h_{\min})$ reaches $\sim 17\ \%$.

Taking into account the above results of the numerical simulation, one can conclude that inverting lidar signals from the zone of incomplete overlap may be a significant issue, caused by the temporal and temperature instability of the overlap function. One should point out that when the extinction coefficient is extracted from the incomplete overlap zone, even the far-end solution does not provide the accurate inversion result.

2

ESSENTIALS AND ISSUES IN SEPARATING THE BACKSCATTER AND TRANSMISSION TERMS IN THE LIDAR EQUATION

2.1 SEPARATION OF THE BACKSCATTER AND TRANSMISSION TERMS IN THE LIDAR EQUATION: METHODS AND INTRINSIC ASSUMPTIONS

As follows from Section 1.1, the accuracy of lidar-signal inversion cannot be rigidly estimated from single wavelength elastic backscatter data, at least, when the lidar operates in the one-directional zenith mode. The main issue is the presence of two variable functions, the backscatter coefficient $\beta_\pi(r)$ and the two-way transmission term $T_\Sigma^2(0, r)$ in the lidar equation. To overcome this issue, different methods were used that allowed separating these two functions. In this section, we will consider the most common ways to solve this problem, which are at present widely used in lidar profiling of the atmosphere.

The lidar equation for the square-range-corrected backscatter signal in the complete overlap zone can be rewritten from Eq. (1.5) as the sum of two backscatter components:

$$P(r)r^2 = C[\beta_{\pi,\,m}(r)T_\Sigma^2(0,\, r) + \beta_{\pi,\,p}(r)T_\Sigma^2(0,\, r)] \tag{2.1}$$

where the first term in the brackets is the attenuated molecular backscatter component and the second one is the attenuated particulate backscatter component. The separation of these two components is utilized in different types of lidars. The general principle for fulfilling this task is to someway remove the particulate backscatter-attenuated component, and perform the inversion procedure of the

Solutions in LIDAR Profiling of the Atmosphere, First Edition. Vladimir A. Kovalev.

remaining attenuated-backscatter molecular term $\beta_{\pi,m}(r)T_{\Sigma}^2(0, r)$. The particulate extinction coefficient, which is usually the key parameter of interest, is retrieved from the two-way transmission term $T_{\Sigma}^2(0, r)$.

To be more specific, lidars that allow separation of the molecular and particulate backscatter terms will be termed below as *splitting lidars*, and the separation operation will be termed as *splitting*. Let us consider different hardware solutions for such splitting. Direct splitting of the terms in Eq. (2.1) is performed in inelastic (Raman) lidar and in high spectral resolution lidar (HSRL). Besides these, some types of lidar, which do not do such splitting directly, allow operating in the splitting mode. The term "splitting mode" implies that the lidar searches the atmosphere in a way that allows splitting the particulate and molecular backscattering. This may be done using either special lidars or a special data processing methodology. First, the splitting mode is inherent to the differential absorption lidar (DIAL), which allows splitting the backscatter and absorption terms by determining the ratio of the backscatter signals at different wavelengths (Measures, 1984; Kovalev and Eichinger, 2004; Fujii and Fukuchi, 2005). Second, the separation of the backscatter term $\beta_{\pi}(r)$ and two-way transmittance term $T_{\Sigma}^2(0, r)$ can be achieved by the scanning elastic lidar. However, this can only be done in favorable atmospheric conditions, when the atmosphere is horizontally stratified. The essentials of data inversion of the scanning elastic lidar in horizontally stratified atmosphere are considered in Chapter 3; in this chapter, the inversion variants and their specifics are only given for the one-directional splitting lidar, generally directed toward the zenith.

Before considering the inversion methods used for processing the data of splitting lidars, one should mention two issues related to these methods. First, a statement about the presence in the lidar equation of the backscatter and extinction terms, commonly used in the scientific literature, is not formulated accurately. Unfortunately, the extinction coefficient $\kappa_p(r)$ is not present directly in the lidar equation as a separate factor, but is "packed" within the exponential term of the two-way transmittance as an unknown integrand of the integral $\int_0^r \kappa_p(r')dr'$. The exponent of the term $\exp[-2\int_0^r \kappa_p(r')dr']$ obtained from the two-way transmittance function $T_{\Sigma}^2(0, r)$ after removing the molecular component, is the basic term used to extract the profile of the particulate extinction coefficient. The latter is extracted by differentiating the optical depth profile. This operation is an issue due to the common mathematical difficulties related to numerical differentiation of a noisy function.

The second principal drawback of the commonly used splitting-lidar inversion techniques is that after splitting and removing the particulate backscatter term, only the optical depth profile retrieved from the two-way transmission term $T_{\Sigma}^2(0, r)$ is used for the extraction of the particulate extinction coefficient of interest. As mentioned above, the extraction of the extinction coefficient from the noisy optical depth profile is a significant issue and can yield distorted results. Meanwhile, valuable information about particulate loading, contained in the particulate backscatter term, generally is not used to put constraints on the derived extinction-coefficient profile.

2.1.1 Inversion Algorithm for the Signals of Raman Lidar

The inversion methodology for determining the particulate extinction coefficient with Raman lidar is based on the Raman wavelength shift of the atmospheric molecules, generally nitrogen or oxygen. The Raman signal equation for the zenith-directed lidar can be written in the form (Ansmann et al., 1990, 1992; Wandinger, 2005)

$$P_R(h) = C\frac{1}{h^2}N_R(T, p)\frac{d\sigma_{\pi, R}}{d\Omega} \times \exp\left\{-\int_0^h \left[\kappa_{p,0}\left(h'\right) + \kappa_{p,R}(h') + \kappa_{m,0}(h') + \kappa_{m,R}(h')\right]\right\} \quad (2.2)$$

where $N_R(T, p)$ is the atmospheric number density of the Raman scattering molecules, which is the temperature- and pressure-dependent factor, $\frac{d\sigma_{\pi.R}}{d\Omega}$ is the range-independent differential Raman cross section for the backward direction; $\kappa_{p,0}(h)$ and $\kappa_{m,0}(h)$ are the particulate and molecular extinction coefficients at the wavelength λ_0 of the emitted laser light and $\kappa_{p,R}(h)$ and $\kappa_{m,R}(h)$ are the corresponding particulate and molecular extinction coefficients at the shifted wavelength λ_R. The profiles $\kappa_{m,0}(h)$ and $\kappa_{m,R}(h)$ are also temperature- and pressure-dependent functions, but for simplicity, these functions in Eq. (2.2) are written as functions of height.

Let us consider the Raman lidar solution for particulate extinction coefficient using the most general form, that is, excluding the commonly used assumption about analytical dependence between $\kappa_{p,0}$ and $\kappa_{p,R}$. In such a general form, the solution for zenith-directed Raman lidar is

$$\kappa_{p,0}(h) + \kappa_{p,R}(h) = \frac{d}{dr}\left\{\ln\left[\frac{N_R(h)}{P_R(h)h^2}\right]\right\} - \kappa_{m,0,}(h) - \kappa_{m,R,}(h). \quad (2.3)$$

where the temperature-pressure-dependent function, $N_R(T, p)$ is written as a function of height.

The derivative $\frac{d}{dr}\left\{\ln\left[\frac{N_R(h)}{P_R(h)h^2}\right]\right\}$ is the key function in the Raman lidar solution from which the particulate extinction-coefficient profile of interest is extracted. Note that it depends on the derivatives of the square-range-corrected signal $P_R(h)$ and the profile $N_R(h)$ rather than on their levels. The accuracy of determining the derivative term in Eq. (2.3) is the main factor that influences the accuracy of the derived particulates components $\kappa_{p,0}$ and $\kappa_{p,R}$ in the left side of the formula.

The main issue for Raman lidar-data processing is an extremely week backscatter signal at the shifted wavelength λ_R and accordingly, a low signal-to-noise ratio in the recorded Raman signal. To facilitate atmospheric profiling, most Raman measurements are taken during nighttime; this regime allows avoiding the influence of the daylight solar background, and accordingly, improving the signal-to-noise ratio. However, even in nighttime conditions, the Raman signals are temporally and spatially averaged over extended times and extended height intervals. Accordingly, the requirement of a "frozen" atmosphere discussed in Section 1.3 should be valid for these extended times. Another issue of the Raman solution is that inversion results depend on the extinction coefficients at two wavelengths λ_0 and λ_R. To recalculate the particulate extinction-coefficient profile for a single wavelength, usually for λ_0, the so-called Angstrom exponential dependence is used, which presumably determines the relationship between the local particulate extinction coefficients at both Raman wavelengths.

To clarify this, let us take a detour into history. Initially, the Angstrom exponent was implemented in atmospheric investigations as a numeral that defines the spectral dependence of the optical thickness of the earth's atmosphere on the wavelength of the light that travels through the atmosphere (Angstrom, 1961). According to this initial study, the spectral dependence of the optical depth τ_λ on the wavelength λ can be approximated by formula $\tau_\lambda = \tau_0\, \lambda^{\alpha}$, where α is the so-called Angstrom exponent. In practice, the Angstrom exponent was computed from the optical depth measurements of the atmosphere at least at two wavelengths, and then used to find the optical depth for any wavelength of interest using the formula

$$\tau_\lambda = \tau_{\lambda_0}\left(\frac{\lambda}{\lambda_0}\right)^{-\alpha}. \tag{2.4}$$

The Angstrom exponent, which value presumably varies within the range between 0 and 4, was introduced as a simple measure of the atmospheric turbidity. It was widely used in the measurement of the radiation values, integrated over the whole atmosphere, generally utilizing pyrheliometers and sun photometers with a set of glass filters. The simple approximation in Eq. (2.4) was also widely used in computations of the particulate optical depth through the atmosphere at different wavelengths, for the characterization of particulate types in different atmospheres, in the studies of the earth radiation budget, etc. This exponent proved to be a good tracer for the origin and concentration of particulates over land and sea, for investigations of seasonal variation in atmospheric turbidity, and its dependence on the air mass, latitude, and so on. However, one should always keep in mind that the dependence in Eq. (2.4) is a surrogate. It does not follow from the Mie theory, and therefore, was initially used only for some qualitative estimates of the optical properties of the atmosphere under different conditions.

In the Raman measurements, this relationship acquired a "formal status." Now it is used under much more exacting conditions, presuming its validity for the range-resolved profiles of the particulate extinction coefficient at the wavelengths λ_0 and λ_R, that is,

$$\kappa_{p,R}(h) = \kappa_{p,0}(h)\left(\frac{\lambda_0}{\lambda_R}\right)^{\alpha}. \tag{2.5}$$

With the above height-resolved dependence, Eq. (2.3) can be rewritten in the form (Ansmann et al., 1990, 1992)

$$\kappa_{p,0}(h) = \frac{\frac{d}{dr}\left\{\ln\left[\frac{N_R(h)}{P_R(h)h^2}\right]\right\} - \kappa_{m,0}(h) - \kappa_{m,R}(h)}{1 + \left(\frac{\lambda_0}{\lambda_R}\right)^{\alpha}}. \tag{2.6}$$

Thus, in addition to the initial assumption on a frozen atmosphere, three related assumptions are used when determining the extinction coefficient $\kappa_{p,0}(h)$ with Eq. (2.6):

1. The Angstrom wavelength dependence for the atmospheric integrated characteristics, such as the optical depth, is valid at each local data point of the extinction-coefficient profile.
2. The wavelength-dependence parameter α in Eq. (2.5) is height independent, at least, within extended ranges.
3. The selected value or values of α are appropriate for the whole altitude range from h_{min} to h_{max} used for lidar profiling.

Let us summarize. Unlike the elastic lidar data, the inversion of Raman lidar signals does not require knowledge of the lidar ratio. This is a significant advantage of the Raman lidar, achieved by the absence of the particulate backscatter term in the basic Eq. (2.6). However, the removal of the backscatter term does not solve general issues of the lidar-solution uncertainty. One such issue originates from the necessity to extract the extinction-coefficient profile from the noisy and commonly systematically corrupted optical depth. The numerical differentiation procedure, which is used for this extraction, can yield unacceptably poor retrieval accuracy, especially when weak Raman lidar signals are measured in a clear atmosphere. This issue, common for all splitting lidars, will be considered in detail in the following sections.

One can also mention that in the Raman lidar instrumentation, the same as in other splitting lidars, a photon counting system is generally used. This results in additional sources of uncertainty, which include the uncertainty of photon counting statistics, photomultiplier afterpulsing uncertainty, etc. However, these are second-order issues, at least, when compared with the requirement for a "frozen" atmosphere and the inversion of noisy functions through numerical differentiation.

2.1.2 Inversion Algorithm for the Signals of High Spectral Resolution Lidar (HSRL)

Owing to Doppler broadening, the wavelengths of backscatter light from molecules and particulates in the atmosphere do not coincide precisely. In the HSRL, this effect is used to separate the backscatter returns from molecules and particulates. Splitting the particulate and molecular returns in the HRSL is achieved by using two separate receiving channels, the total channel and the molecular channel. In the total channel, the recorded photons originate in both molecular and particulate scattering, whereas in the molecular channel, the recorded photons, or at least an overwhelming number of these, originate in molecular scattering. Suppressing particulate scattering in the molecular channel is mostly achieved by using a narrowband iodine absorption filter (Piironen and Eloranta, 1994; Hair et al., 2001; Eloranta, 2005). The principle of suppressing the particulate component in the molecular channel is illustrated in Fig. 2.1. The narrow peak in the center of the dotted curve shows the spectral distribution of the

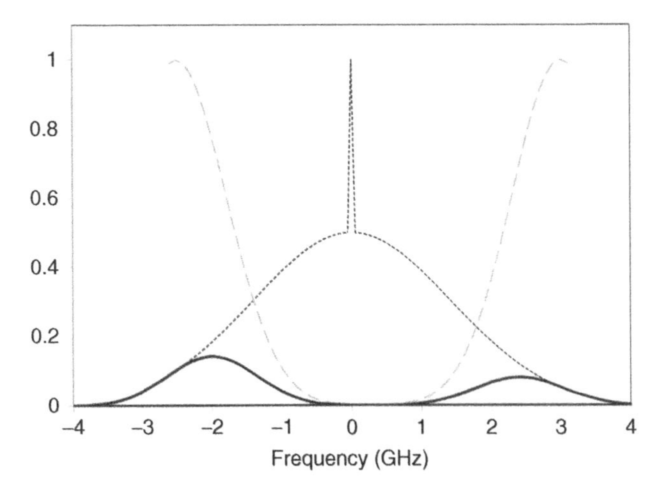

Fig. 2.1 Principle of suppression of the particulate component in the molecular channel of HSRL (Adapted from Kovalev and Eichinger, 2004).

elastically scattered light from particulates; this peak is centered relatively to the frequency of the emitted laser pulse. The wide wings on both sides of the central peak show the spectral distribution of the molecular component. The thin dashed curve shows the iodine filter transmission which blocks the particulate component and part of the molecular one.

The signal that remains after passing through the iodine filter then passes through narrow bandpass filters, which cut off the background light. The thick solid curve in the figure shows resulting molecular transmission spectra for the light that passes the iodine and narrow bandpass filters in the molecular channel. Because of such complicated filtering shapes, an internal calibration of the sensitivities of both channels is required before routine profiling of the atmosphere can be made. The molecular channel should be corrected for the amount of molecular backscatter signal blocked by the iodine filter. To extract the particulate extinction from the molecular channel data, the measured attenuated molecular backscatter-coefficient profile is compared to the molecular density profile obtained either from balloon measurements or from the selected model atmosphere (Hair et al., 2008).

The backscatter HSRL signals in the total and the molecular channels of the zenith-directed HSRL $P_{\Sigma}(h)$ and $P_m(h)$ obtained over the height interval Δh_b in the complete overlap area can be written as (Hair et al., 2001; Esselborn et al., 2008; Liu, et al., 2009)

$$P_{\Sigma}(h) = C_{\Sigma}\frac{\Delta h_b}{h^2}[\beta_{\pi,m}(h) + \beta_{\pi,p}(h)]T_{\Sigma}^2(0,h), \tag{2.7}$$

and

$$P_m(h) = C_m\frac{\Delta h_b}{h^2}[f_m(T^h, p^h)\beta_{\pi,m}(h) + f_p\beta_{\pi,p}(h)]T_{\Sigma}^2(0, h). \tag{2.8}$$

Here C_Σ and C_m are the constants for the total and molecular channels, respectively; T is the temperature, p the pressure, and the term $f_m(T^h, p^h)$ is the temperature- and pressure-dependent attenuation factor at the height h, which, for the Cabannes spectrum, passed through the iodine-vapor filter can be written as (Liu, et al., 2009)

$$f_m(T^h, p^h) = \int F(\nu) \int \Re(\nu, T^h, p^h) \quad l(\nu - \nu') \quad d\nu' \, d\nu; \tag{2.9}$$

here ν is the frequency, $F(\nu)$ is the normalized function of the iodine vapor transmission, and $\Re(\nu, T^h, p^h)$ is the normalized temperature- and pressure-dependent Cabannes scattering spectrum; $l(\nu)$ is the laser transmitting profile. For data processing, the temperature- and pressure-dependent factor $f_m(T^h, p^h)$ is converted to the height-dependent function $f_m(h)$ by using the corresponding temperature and pressure profiles obtained either from balloon measurements or from reference atmospheres.

The terms $f_m(h)$ and f_p can be considered as the weight functions for the backscatter coefficients $\beta_{\pi,\,m}(h)$ and $\beta_{\pi,\,p}(h)$, where $f_p \ll f_m(h)$. Under favorable conditions, when the particulate backscatter coefficient $\beta_{\pi,\,p}$ is not too large as compared to the molecular backscatter coefficient $\beta_{\pi,\,m}$, the component $f_p\beta_{\pi,\,p}(h)$ in Eq. (2.8) is significantly less than the component $f_m(h)\beta_{\pi,\,m}(h)$ and can be ignored. Using the estimated height-dependent function $f_m(h)$, and assuming that within the spectral range of the iodine filter $f_p\beta_{\pi,\,p}(h) \ll f_m(h)\beta_{\pi,\,m}(h)$, one can transform Eq. (2.8) into the simple form

$$P_m(h) = C_m \frac{\Delta h_b}{h^2} f_m(h)\beta_{\pi,\,m}(h)T_\Sigma^2(0,\, h). \tag{2.10}$$

The corresponding profile of the particulate extinction and backscatter coefficient versus height can be obtained with the formulas (Hair et al., 2008)

$$\kappa_p(h) = -0.5\frac{d}{dh}\left[\ln \frac{P_m(h)\,h^2}{f_m(h)\beta_{\pi,\,m}(h)}\right] - \kappa_m(h), \tag{2.11}$$

and

$$\beta_{\pi,\,p}(h) = \left[\frac{C_m}{C_\Sigma}\frac{P_\Sigma(h)}{P_m(h)} f_m(h) - 1\right]\beta_{\pi,\,m}(h). \tag{2.12}$$

The calculation of the particulate extinction coefficient with Eq. (2.11) does not require the knowledge of the constant C_Σ. To determine the backscatter coefficient with Eq. (2.12), the signals obtained in both the total and the molecular channels should be used and the ratio C_m/C_Σ should be someway estimated. This can be achieved by the use of the assumption of a particulate-free atmosphere at some reference height. As with profiling of the atmosphere with elastic lidar (Section 1.2.1), such an assumption may be an additional error source.

As follows from Eq. (2.11), the particulate extinction-coefficient profile $\kappa_p(h)$ is obtained through numerical differentiation of the square-range-corrected signal $P_m(h)$ as is done with Raman lidar data. In the HSRL inversion technique, three separate functions of the height, given within brackets in Eq. (2.11), should be differentiated, that is,

$$\kappa_p(h) = -0.5\frac{d}{dh}\{\ln[P_m(h)h^2]\} + 0.5\frac{d}{dh}\{\ln[f_m(h)]\} + 0.5\frac{d}{dh}\{\ln[\beta_{\pi,\,m}(h)]\} - \kappa_m(h), \tag{2.13}$$

Accordingly, three error sources may influence the accuracy of the inversion result. The error in the square-height-corrected signal $P_m(h)h^2$ does not depend on the accuracy of determining the vertical profiles of temperature and pressure and the Cabannes scattering spectrum, whereas the next two terms may be significantly corrupted by the inaccurately estimated profiles $T(h)$, $p(h)$, and $f_m(h)$. Analyses and experimental data show that these errors may significantly corrupt the profile of the extracted extinction coefficient $\kappa_p(h)$ (Liu et al., 2009). To determine the profile $f_m(h)$, some theoretical scattering model should be selected. The situation is exacerbated by the fact that the numerical derivatives of the functions rather than their values should be found to calculate the particulate extinction coefficient. If the terms in the equation are determined at the end points h and $h + s$ of the range resolution s, the increments of three differentiated components should be found, that is,

$$\kappa_p(h_s) = \frac{1}{2s}\left\{\ln\left[\frac{P_m(h)\,h^2}{P_m(h+s)(h+s)^2}\right] - \ln\left[\frac{f_m(h)}{f_m(h+s)}\right] - \ln\left[\frac{\beta_{\pi,\,m}(h)}{\beta_{\pi,\,m}(h+s)}\right]\right\}$$
$$-\kappa_m(h_s), \tag{2.14}$$

where $h_s = h + 0.5s$ is the middle point of the interval $(h,\ h + s)$. If the interval s is large, for example, 100 m or more, the retrieved extinction coefficient $\kappa_p(h_s)$ can be considered as some mean for this interval (for details, see Section 2.4). The error factor of the retrieved extinction coefficient is the same function of the optical depth as that in the slope method applied to elastic lidar data (Section 1.4), that is, $EF(\delta\kappa_p,\ \delta P_m) = EF(\delta\kappa_p,\ \delta f_m) = EF(\delta\kappa_p,\ \delta\beta_{\pi,m}) = [2\tau_p(s)]^{-1}$.

2.1.3 Inversion Algorithm for Signals of the Differential Absorption Lidar (DIAL)

The profiling of the gas concentration with DIAL instrumentation is based on the simultaneous measurement of the signals at two wavelengths: at the wavelength λ_{on}, within an absorption line of the gas of interest, and at the wavelength λ_{off}, outside the main absorption region of the gas (Megie and Menzies, 1980; Browell et al., 1985; Proffitt and Langford, 1997; Ehret et al., 1993). DIAL has been used in a large number of investigations of gas concentration profiles in the lower troposphere, including the gases related to the ozone depletion problem. The ozone-concentration profile is derived from the ratio of the signals $P_{on}(h)$ and $P_{off}(h)$, measured at wavelengths λ_{on} and λ_{off}. These signals for the ground-based and zenith-directed DIAL can be written as

$$P_{on}(h) = C_{on}\beta_{\pi,\,on}(h)h^{-2}\exp\left\{-2\int_0^h \left[\beta_{on}\left(h'\right) + n_{ozone}(h')\sigma_{on}\right]dh'\right\}, \tag{2.15}$$

and

$$P_{\text{off}}(h) = C_{\text{off}}\beta_{\pi,\text{ off}}(h)h^{-2}\exp\left\{-2\int_0^h \left[\beta_{\text{off}}\left(h'\right) + n_{\text{ozone}}(h')\sigma_{\text{off}}\right]dh'\right\}; \quad (2.16)$$

where $\beta_{\pi,\text{ on}}(h)$ and $\beta_{\pi,\text{ off}}(h)$ are the total (molecular and particulate) backscatter coefficients, and $\beta_{\text{on}}(h)$ and $\beta_{\text{off}}(h)$ are the total scattering coefficients at the on and off wavelengths, respectively; $n_{\text{ozone}}(h)$ is the ozone number density at the height h, and σ_{on} and σ_{off} are the ozone absorption cross sections at the wavelengths λ_{on} and λ_{off}.

The range-resolved ozone concentration $n_{\text{ozone}}(h)$ is determined with the formula (Browell et al., 1985)

$$n_{\text{ozone}}(h) = \frac{-1}{2\Delta\sigma}\frac{d}{dh}\left\{\ln\left[\frac{P_{\text{on}}(h)}{P_{\text{off}}(h)}\right]\right\} + \frac{1}{2\Delta\sigma}\frac{d}{dh}\left\{\ln\left[\frac{\beta_{\pi,\text{ on}}(h)}{\beta_{\pi,\text{ off}}(h)}\right]\right\}$$
$$-\frac{1}{\Delta\sigma}[\beta_{\text{on}}(h) - \beta_{\text{off}}(h)], \quad (2.17)$$

where $\Delta\sigma = \sigma_{\text{on}} - \sigma_{\text{off}}$ is the differential absorption cross section of ozone for the on and off wavelengths.

The first term in the right side of the formula is the basic component of the equation; the other two determine the corrections required in order to remove the distortions in the retrieved $n_{\text{ozone}}(h)$ caused by the difference between particulate and molecular scattering and backscattering at the on and off wavelengths. To obtain an accurate profile of the ozone concentration, all three components in Eq. (2.17) need to be taken into account. Accordingly, comprehensive analysis of the accuracy of the derived ozone-concentration profile should be based on the consideration of three uncertainty sources: the statistical and systematic uncertainties in the measured signals $P_{\text{on}}(h)$ and $P_{\text{off}}(h)$; a random systematic error related to the use of some model relationship that estimates the difference between the particulate backscatter terms $\beta_{\pi,\text{ on}}(h)$ and $\beta_{\pi,\text{ off}}(h)$; and a random systematic error related to the difference in the total scattering coefficients $\beta_{\text{on}}(h)$ and $\beta_{\text{off}}(h)$. Note that Eqs. (2.15) and (2.16) are valid under the condition that no other absorption component except ozone influences the signals at the on and off wavelengths. Otherwise, an additional correction to the output data should be performed based either on *a priori* assumptions or obtained by an independent measurement of that absorption component.

Accurate determination of the terms in Eq. (2.17) in a real atmosphere is not a trivial task, especially in a turbid troposphere where nonzero gradients of particulate scattering typically take place. As with the processing methods discussed above, in DIAL profiling of the ozone concentration, numerical differentiation is used. For determining $n_{\text{ozone}}(h)$, the height resolution s is selected and the required terms in Eq. (2.17) at the end points h_1 and $h_2 = h_1 + s$ are determined. In accordance with Eq. (2.17), three independent terms should be taken into account. These terms, denoted respectively as $P(s)$, $BCK(s)$, and E, can be written as (Browell, et al., 1985)

$$P(s) = \frac{1}{2s\Delta\sigma}\ln\left[\frac{P_{\text{on}}\left(h_1\right)P_{\text{off}}(h_2)}{P_{\text{off}}(h_1)P_{\text{on}}(h_2)}\right]; \quad (2.18)$$

$$BCK(s) = \frac{-1}{2s\Delta\sigma} \ln \left[\frac{\beta_{\pi,\text{on}}(h_1)\,\beta_{\pi,\text{off}}(h_2)}{\beta_{\pi,\text{off}}(h_1)\beta_{\pi,\text{on}}(h_2)} \right]; \tag{2.19}$$

and

$$E = \frac{-1}{\Delta\sigma}(\beta_{\text{on}} - \beta_{\text{off}}). \tag{2.20}$$

The estimation of the correction term E in Eq. (2.20) is based on the same inversion principle as that used in Raman lidar; in particular, the wavelength dependence similar to Eq. (2.5) and the same three assumptions as in Section 2.1.1 are used. Given that the scattering for particulates and gas molecules vary inversely with wavelength to the power of α and 4, respectively, the estimate of the correction term E within the wavelength range $(\lambda_{\text{on}} - \lambda_{\text{off}})$ can be made with the formula

$$E = \frac{\lambda_{\text{off}} - \lambda_{\text{on}}}{\Delta\sigma\lambda_{\text{off}}}[-\alpha\overline{\beta_{p,\,\text{off}}(s)} + 4\overline{\beta_{m,\,\text{off}}(s)}], \tag{2.21}$$

where $\overline{\beta_{p,\,\text{off}}(s)}$ and $\overline{\beta_{m,\,\text{off}}(s)}$ are considered the mean values of the scattering coefficient for particulates and air molecules at λ_{off} within the range-resolution interval s. If the term E is ignored, that is, the ozone concentration is not corrected for the particulate extinction, the profile of $n_{\text{ozone}}(h)$ will be slightly overestimated. According to the study cited above, the particulate extinction correction E is generally less than 10% of the ozone concentration derived from the basic term $P(s)$.

As shown in studies by Kovalev and McElroy (1994) and Kovalev and Eichinger (2004), the most significant issue in DIAL profiling of the ozone concentration is related to the uncertainty of the backscatter correction term $BCK(s)$. For determining this component, Browell et al. (1985) assumed that the particulate backscatter coefficients also obey the power law dependence on wavelength, similar to the dependence of the extinction coefficients. The basic problem is the implementation of the backscatter correction: it is not the ratio of $\beta_{\pi,\,\text{on}}(h)$ to $\beta_{\pi,\,\text{off}}(h)$, but the gradient of this ratio that should be determined. A simple analysis shows that the random systematic error in derived ozone concentration in regions with nonzero vertical gradients of particulate backscattering can be extremely large. Actually, no backscatter corrections are reliable in the troposphere in the areas of increased backscattering heterogeneity (Kovalev and McElroy, 1994).

Thus, the accuracy in determining the components $P(s)$ and $BCK(s)$ significantly influences the accuracy of profiling of the ozone concentration. For these components, the error factor depends on the magnitude of the differential absorption cross section $\Delta\sigma$ and on the range resolution s. Simple transformations show that the error factor is $EF(\delta n,\ \delta P_{\text{on/off}}) = EF(\delta n,\ \delta\beta_{\pi,\text{on/off}}) = [2\Delta\tau_{\text{abs}}(s)]^{-1}$, where $\Delta\tau_{\text{abs}} = \Delta\sigma\, n_{\text{ozone}}\, s$ is the differential absorption optical depth. The error factor is extremely sensitive to the differential absorption optical depth within the selected height resolution interval.

Table 2.1 summarizes the main assumptions and implicit premises used for the inversion of the signals measured by the splitting lidars considered in this section.

TABLE 2.1 Basic Assumptions Used for the Inversion of the Splitting Lidar Signals

Splitting Lidar Type	Data Processing Operation	Assumptions
Raman lidar	Signal inversion	(a) The wavelength dependence of the particulate extinction coefficient obeys the Angstrom exponential law (b) The exponent of the wavelength dependence is height independent over either the whole operative range of the lidar or over the selected extended zones (c) *a priori* exponent values used by researchers are valid for any local point within the lidar operative range.
HSRL	Extraction of the particulate extinction coefficient from the signal of the molecular channel	(a) The particulate component in the output of the molecular channel is minor and can be ignored (b) The Cabannes-Brillouin molecular scattering model used for system calibration provides an accurate estimation of the molecular transmission spectra after the signal passes the iodine band pass filter (c) The molecules and particulates in the atmosphere are in thermal equilibrium and have a Maxwell velocity distribution.
DIAL	Signal inversion	(a) The wavelength dependence of the particulate extinction and backscatter coefficients for the on and off wavelengths obeys the exponential law (b) The exponents in the exponential law formulas are height independent over the whole operative range of the lidar and their values are known *a priori* both for the particulate extinction and backscatter coefficients (c) Within the total operative range, the gradient of the ratio of the backscatter coefficients at the on and off wavelengths is small as compared to the gradient of the ratio of the on and off signals and can be ignored

The general assumptions cited in Table 1.1 and in Section 1.1 are not included in the Table 2.1; however, these remain valid for the splitting lidars also.

Let us summarize. The common method for splitting two unknown functions in the lidar equation is to separate and remove the particulate backscatter term $\beta_{\pi,p}(r)$ and focus the inversion procedure on the profile of the two-way transmission term $[T_\Sigma(0, r)]^2$. In other words, the extinction coefficient is extracted by processing the transmittance term, whereas the backscatter term, which also contains useful information about the particulate loading of the atmosphere, is disregarded. Such a data processing method forces the researchers to extract the profile of the extinction coefficient from the noisy optical depth profile through numerical differentiation. As will be shown in the following sections, the use of numerical differentiation of the noisy and often systematically corrupted signals or involved functions is the most significant issue for splitting lidars and lidars working in splitting mode. The accuracy of the derived particulate extinction-coefficient profile may be extremely poor; what is worse, its accuracy cannot be reliably estimated. Nevertheless, the common implicit premise is that determining the sliding numerical derivative over the selected range interval results in proper mean values of the extinction coefficient within these intervals. The standard deviation of the mean is commonly believed to be the correct estimate of the inversion uncertainty. As will be shown in the following sections, there are generally no proper grounds for such a premise.

2.2 DISTORTIONS IN THE OPTICAL DEPTH AND EXTINCTION-COEFFICIENT PROFILES DERIVED FROM RAMAN LIDAR DATA

Let us consider important specifics of processing Raman lidar data, focusing on the features of the signal inversion algorithm and on the influence of systematic distortions in the lidar signal and the selected *a priori* assumptions. The following formulas analyze the profiles of interest as functions of range and are applicable for profiling ether in zenith or in any slope direction. In the most general form, the lidar equation for the Raman signal $P_R(r)$ taken as a function of range in the complete overlap zone can be written in the form

$$P_R(r) = C\beta_{\pi,\,R}(r)r^{-2}T_0(0,\, r)T_R(0,\, r), \tag{2.22}$$

where, as in Eq. (2.2), the subscript "0" refers to the functions at the wavelength of the emitted laser pulse λ_0 and the subscript "R" refers to the functions at the shifted Raman wavelength λ_R; $T_0(0, r)$ and $T_R(0, r)$ are the one-way transmissions within the interval $(0, r)$ for the wavelengths λ_0 and λ_R, respectively. $\beta_{\pi,\,R}(r)$ is the Raman backscatter coefficient at the range r and is the product of the atmospheric number density of the Raman scatterers $N_R(r)$ and the corresponding backscatter cross section $\sigma_R(\lambda_R)$, that is,

$$\beta_{\pi,\,R}(r) = N_R(r)\sigma_R(\lambda_R).$$

After separating the particulate and molecular scattering terms, the Raman signal can be rewritten as

$$P_R(r) = C\beta_{\pi,\,R}(r) r^{-2} T_{m,\,0}(0,\,r) T_{m,\,R}(0,\,r) \times \exp\left\{-\int_0^r \kappa_{p,\,0}(r')\,dr' - \int_0^r \kappa_{p,\,R}(r')dr\right\}, \tag{2.23}$$

where $T_{m,\,0}(0,\,r)$ and $T_{m,\,R}(0,\,r)$ are the one-way molecular transmissions for the wavelengths λ_0 and λ_R, respectively; $\kappa_{p,\,0}(r)$ and $\kappa_{p,\,R}(r)$ are the particulate extinction coefficients at the same wavelengths.

2.2.1 Distortion of the Derived Extinction Coefficient Due to Uncertainty of the Angstrom Exponent

Let us start the analysis with the issue related to the presence in the Raman signal of different particulate extinction coefficients at the wavelength emitted by the laser λ_0 and at the wavelength of the Raman-shifted λ_R. As shown in Section 2.1.1, the optical profiles at these wavelengths can be related through the Angstrom exponent α [Eq. (2.5)]. Let us consider initially the simplest case when the actual exponent α in the searched atmosphere is height independent but its exact value is not known. The only option in such a situation is an *a priori* selection of the exponent, so that instead of the actual exponent α_{actual}, a shifted α_{assumed} may be used for the calculation of the extinction coefficient. When $\alpha_{\text{assumed}} \neq \alpha_{\text{actual}}$, a shifted profile of the extinction coefficient $\langle\kappa_{p,\,0}(r)\rangle$ will be obtained instead of the true profile $\kappa_{p,\,0}(r)$. As follows from Eq. (2.6), the relationship between these profiles can be written as

$$\langle\kappa_{p,\,0}(r)\rangle = \kappa_{p,\,0}(r)\frac{1 + (\lambda_0/\lambda_R)^{\alpha_{\text{actual}}}}{1 + (\lambda_0/\lambda_R)^{\alpha_{\text{assumed}}}}. \tag{2.24}$$

The corresponding fractional error of the derived extinction coefficient $\langle\kappa_{p,\,0}(r)\rangle$ is

$$\delta\kappa_{p,\,0}(r) = \frac{\langle\kappa_{p,\,0}(r)\rangle}{\kappa_{p,\,0}(r)} - 1 = \frac{(\lambda_0/\lambda_R)^{\alpha_{\text{actual}}} - (\lambda_0/\lambda_R)^{\alpha_{\text{assumed}}}}{1 + (\lambda_0/\lambda_R)^{\alpha_{\text{assumed}}}}. \tag{2.25}$$

In Table 2.2, the fractional error caused by the difference between α_{actual} and α_{assumed} is given for two wavelength pairs, 308 nm/332 nm and 351 nm/383 nm; the possible range of the actual exponent is chosen from 0 to 3, the assumed value $\alpha_{\text{assumed}} = 1$.

The selection of the proper value of the exponent α in Eq. (2.6) is generally made with guesses based on some statistical estimates. Different values of α_{assumed}, presumably appropriate to existing atmospheric conditions and possible limits of their variations, were analyzed and used in practical Raman lidar profiling (e.g., Ansmann et al., 1990, 1992; Ferrare et al, 1998; Pappalardo et al., 2004). In the studies by Ansmann et al. (1990, 1992), the authors assumed that for the elastic/inelastic channels, 308 nm and 332 nm, the most typical α in the lower atmosphere is unity; for the

TABLE 2.2 Theoretical Fractional Errors in the Extracted Extinction Coefficient Caused by the Difference Between α_{actual} and $\alpha_{assumed}$

λ_0 (nm)	λ_R (nm)	λ_0/λ_R	$\alpha_{assumed}$	α_{actual}	$\delta\kappa_{p,0}(h)$, %
308	332	0.928	1.0	0	3.75
				0.5	1.84
				1.0	0
				1.5	−1.77
				2.0	−3.48
				2.5	−5.12
				3.0	−6.71
351	383	0.916	1.0	0	4.36
				0.5	2.13
				1.0	0
				1.5	−2.04
				2.0	−4.00
				2.5	−5.87
				3.0	−7.66

cirrus clouds located between 8 and 10 km, the exponent $\alpha_{assumed} = 0$ was selected. According to the authors, the deviation of this parameter from the estimated value by 0.5 or 1 causes a relative error in the extinction coefficient of approximately 2% and 4%, respectively. According to the study by De Tomasi et al. (2003), the use of $\alpha_{assumed} = 1$ for the pair 351nm/383nm yields the uncertainty of ~ 5%, assuming that the actual exponent varies between 0 and 2. According to the study by Ferrare et al. (1998), when α_{actual} varies between 0 and 2, and $\alpha_{assumed} = 1$, the error in the derived extinction coefficient at 351 nm for the Raman signal from nitrogen at 383 nm is ±10%; the use of the Raman oxygen signal at 372 nm reduces this error down to 6%. The relative uncertainty in the differential atmospheric transmission is more significant. If α_{actual} varies between 0 and 2, the uncertainty in the differential transmission for the pair 351 nm/383 nm can increase up to 20–25%; the use of the Raman oxygen channel (372 nm) reduces the uncertainty down to 12%. The authors pointed out that the error could be reduced if the wavelength dependence were estimated using data of a sun photometer.

Thus, even in the case of the range-independent exponent, the use of some *a priori*, statistically established $\alpha_{assumed}$ can produce distortions in the derived optical profiles. Meanwhile, the simple dependence in Eq. (2.5) is a rough simplification of reality. As discussed in Section 2.1, the Angstrom coefficient was implemented in atmospheric investigations as an integrated rather than a local characteristic of the atmosphere. Therefore, the assumption that local α has the same invariable value in heterogeneous atmospheric layers, that is, in polluted boundary layers, clear air, and lofted particulate layers is actually a nonphysical assumption.

If α varies with height, the situation is much worse. To soften this issue, different constant exponents α may be used over different zones within the lidar measurement range, for example, in clear air and cloudy layers (Ansmann et al. (1990, 1992). However, in practice, this operation requires determining the locations of these zones with presumably different α before applying the solution in Eq. (2.6) or possibly, using a special iterative procedure. However, such a methodology can only be used when different layering within the lidar operative range is clearly discernible.

The assumption about a stable relationship between the local extinction coefficients at the wavelengths λ_0 and λ_R is a weak point in the Raman solution. Case studies in real atmospheres show that the actual α is not height independent. For example, the intercomparison of lidar-derived optical properties and airborne measurements during ACE-Asia revealed that the values of the Angstrom coefficient within the atmospheric column up to 7 km varied from approximately 0 to 2 (Murayama et al., 2003).

Thus, the spectral dependence with a constant Angstrom exponent given in Eq. (2.5) is a simplified approximation of reality. The real fractional error in the Raman profiling of the atmosphere can be significantly larger than that given in Table 2.2. To be closer to reality, the exponent α needs to be taken as a height-dependent quantity, that is, the quantity should be assumed as a variable. A more realistic, but unfortunately, less practical form of the Raman signal should be written as

$$P_R(r) = C\beta_{\pi,\,R}(r)r^{-2}T_{m,\,0}(0,\,r)T_{m,\,R}(0,\,r)\exp\left\{-\int_0^r \varsigma\left(r'\right)\kappa_{p,\,0}(r')dr'\right\}, \quad (2.26)$$

where

$$\varsigma(r) = 1 + (\lambda_0/\lambda_R)^{\alpha(r)}. \quad (2.27)$$

The exponent $\alpha = \alpha(r) = var.$ and, accordingly, $\zeta = \zeta(r) =$ var. The exponential term in Eq. (2.26) is determined as

$$\exp\left\{-\int_0^r \varsigma\left(r'\right)\kappa_{p,\,0}(r')dr'\right\} = \frac{P_R(r)r^2}{C\beta_{\pi,\,R}(r)T_{m,\,0}(0,\,r)T_{m,\,R}(0,\,r)}. \quad (2.28)$$

Taking the logarithm and the derivative of both sides of Eq. (2.28), and implementing the atmospheric number density of the Raman scattering molecules instead of $\beta_{\pi,\,R}(r)$, one obtains the formula

$$\varsigma(r)\kappa_{p,\,0}(0,\,r) = \frac{d}{dr}\left\{ln\left[\frac{N_R(r)}{P_R(r)r^2}\right]\right\} - \kappa_{m,\,0}(r) - \kappa_{m,\,R}(r). \quad (2.29)$$

As follows from Eqs. (2.6) and (2.29), the relationship between the extinction coefficient $\langle\kappa_{p,\,0}(r)\rangle$, estimated using the assumed constant exponent α and the true extinction coefficient $\kappa_{p,\,0}(r)$, is

$$\varsigma(r)\kappa_{p,\,0}(r) = \langle\kappa_{p,\,0}(r)\rangle\lfloor 1 + (\lambda_0/\lambda_R)^{\alpha}\rfloor,$$

and accordingly,

$$\langle \kappa_{p,0}(r) \rangle = \frac{\varsigma(r)\kappa_{p,0}(r)}{\lfloor 1 + (\lambda_0/\lambda_R)^{\alpha} \rfloor}. \tag{2.30}$$

The fractional error in the extinction coefficient obtained when the assumed $\alpha_0 = \text{const.}$ is used instead of the true $\alpha(r)$ obeys the formula

$$\delta\kappa_{p,0}(r) = \frac{\varsigma(r)}{\zeta_0} - 1, \tag{2.31}$$

where $\zeta_0 = 1 + (\lambda_0/\lambda_R)^{\alpha}$.

To get some feel of the behavior of this error, let us consider some results from a numerical simulation made for the same synthetic atmosphere as shown in Fig. 1.10. The synthetic profiles of the extinction coefficient $\kappa_{p,\,0}(h)$ and the height-dependent exponent $\alpha(h)$ used in the numerical simulation are shown in Fig. 2.2, on the left and right panel, respectively.

In Fig. 2.3, the profile of the error $\delta\kappa_{p,\,0}(h)$ determined with Eq. (2.31) for the wavelengths $\lambda_0 = 351$ nm and $\lambda_R = 383$ nm is shown, when different values of the Angstrom coefficient $\alpha = \text{const.}$, are used for the inversion. The error is range dependent, and its value ranges from approximately −8% to 9% when α is selected between 0 and 2, and from −8% to 18% when α is selected between 0 and 4. Presumably, the Angstrom exponent generally varies in a rather restricted range from approximately 0.5 to 2. However, one should keep in mind that the real local variations of $\alpha(h)$ can be significantly wider. Experimental studies show that, in some cases, its value is larger than 4 and may even be negative (Kaskaoutis et al., 2007). However, to the best of our knowledge, no comprehensive investigation of the behavior of the actual exponent $\alpha(h)$ in different atmospheres has been performed, so that one can only

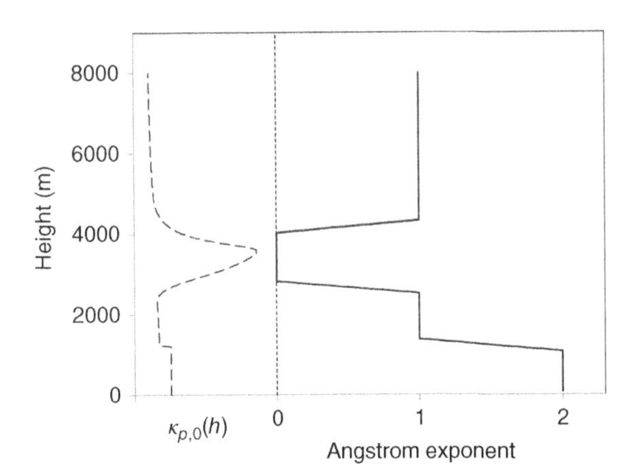

Fig. 2.2 Synthetic profiles of the extinction coefficient $\kappa_{p,\,0}(h)$ and the exponent $\alpha(h)$ used in numerical simulation. The maximum $k_{p,0}(h)$ at $h = 3560$ m is $k_{p,0}(h) = 0.47$ km^{-1}.

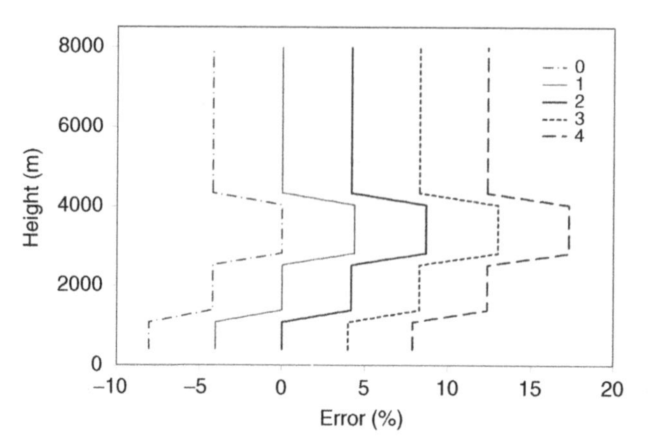

Fig. 2.3 The error $\delta\kappa_{p,\,0}(h)$, calculated using Eq. (2.31) for the pair of wavelengths $\lambda_0 = 351$ nm and $\lambda_R = 383$ nm when different constant values of α shown in the legend are used for inversion instead the true $\alpha(h)$.

speculate on what atmospheric processes might originate such " atypical" exponents. One should also keep in mind that the experimental values of the exponent cited in the literature are generally obtained from the measurements of the optical depth and represent some integrated rather than local values. As known, any local values in the atmosphere are always more scattered than the integrated ones.

Let us summarize. There is no practical way to determine the actual behavior of the exponent α versus height. Therefore, the function is generally assumed to be height independent. If the inversion of the Raman lidar data is made assuming that $\alpha =$ const., while this assumption is not valid, a shifted profile of the extinction coefficient is obtained. There is no way to estimate the true boundaries of the uncertainty in the profiles derived from the Raman lidar signal due to the uncertainty of α. Thus, from an accuracy standpoint, there is no difference when a researcher selects *a priori* a lidar ratio for the inversion of an elastic lidar signal or *a priori* assumes a constant exponent α for the inversion of a Raman lidar signal. Experimental studies, such as those published by Kaskaouts et al. (2007) and Murayama et al. (2003), raise doubts regarding the common belief that the second type of inversion is much more accurate. Possibly, instead of using the assumption $\alpha =$ const. and selecting *a priori* its value, it might be better to estimate the sum $\kappa_{p,\,0}(r) + \kappa_{p,\,R}(r)$ using the general formula in Eq. (2.3), that is, by writing this sum in the left side of Eq. (2.29) instead of the product $\zeta(r)\kappa_{p,\,0}(r)$. Then the average extinction coefficient for the two wavelengths λ_0 and λ_R,

$$\overline{\kappa_p^{(0,\,R)}}(r) = 0.5[\kappa_{p,\,0}(r) + \kappa_{p,\,R}(r)], \tag{2.32}$$

could be determined. Such a recommendation is not an ideal solution; however, the use of the average $\overline{\kappa_p^{(0,\,R)}}(r)$ would alleviate the need to select the height-independent exponent α at least, for the atmospheres with nondiscernible layering. It would allow more accurate comparison of the results of Raman profiling of the atmosphere made by different researchers, who currently use different exponents α when extracting the extinction coefficient from Raman lidar signals.

2.2.2 Errors in the Derived Optical Depth Profile Caused by Distortions in the Raman Lidar Signal

As with the elastic lidar signal, the Raman lidar signal can be distorted by a multiplicative factor $[1 + \delta_P(r)]$, and can include two additive components, the low-frequency noise component $\Sigma w_{j,\text{low}}(r)$ and the constant nonzero offset ΔB remaining after subtraction of the estimated background $\langle B \rangle$ [see Section 1.1]. Thus, instead of the ideal Raman lidar signal $P_R(r)$ in Eq. (2.23), the real Raman signal $\langle P_R(r) \rangle$ should be written in the form

$$\langle P_R(r) \rangle = P_R(r)[1 + \delta_P(r)] + \Sigma w_{j,\text{low}}(r) + \Delta B. \tag{2.33}$$

Denoting for simplicity the sum of the particulate optical depths at λ_0 and λ_R as

$$\Sigma\tau_p^{(0,\,R)}(0,\ r) = \tau_{p,\,0}(0,\ r) + \tau_{p,\,R}(0,\ r), \tag{2.34}$$

one can relate this sum, which does not depend on the assumed exponent α with the distortion quantities in Eq. (2.33) using the formula

$$\Sigma\tau_p^{(0,\,R)}(0,\ r) = \ln\left[\frac{C\beta_{\pi,\,R}(r)[1 + \delta_P(r)]T_{m,\,0}(0,r)T_{m,\,R}(0,\ r)}{[\langle P_R(r) \rangle - \Sigma w_{j,\,\text{low}}(r) - \Delta B]r^2}\right]. \tag{2.35}$$

If one could miraculously estimate all the distortion components in the distorted signal $\langle P_R(r) \rangle$, that is, the terms $[1 + \delta_P(r)]$, $\Sigma w_{j,\text{low}}(r)$, and ΔB, one would obtain with Eq. (2.35) the precise sum $\Sigma\tau_p^{(0,\,R)}(0,\ r)$. Unfortunately, there is no way to precisely determine these distortions. In practice, the distorted signal $\langle P_R(r) \rangle$ is processed assuming the absence of such distortions, that is, by assuming that the Raman signal obeys the formula in Eq. (2.23). As a result of ignoring the distortion terms, the inverted signal, $\langle P_R(r) \rangle$, will yield the incorrect sum $\langle \Sigma\tau_p^{(0,\,R)}(0,\ r) \rangle$ rather than the true $\Sigma\tau_p^{(0,\,R)}(0,\ r)$. Taking this observation into account, Eq. (2.23) should be rewritten in the form, which relates the distorted $\langle P_R(r) \rangle$ and the distorted sum $\langle \Sigma\tau_p^{(0,\,R)}(0,\ r) \rangle$, that is,

$$\langle P_R(r) \rangle = C\beta_{\pi,\,R}(r)r^{-2}T_{m,\,0}(0,\ r)T_{m,\,R}(0,\ r)\exp[-\langle \Sigma\tau_p^{(0,\,R)}(0,\ r) \rangle], \tag{2.36}$$

and accordingly,

$$\langle \Sigma\tau_p^{(0,\,R)}(0,\ r) \rangle = \ln\left[\frac{C\beta_{\pi,\,R}(r)T_{m,\,0}(0,\ r)T_{m,\,R}(0,\ r)}{\langle P_R(r) \rangle r^2}\right]. \tag{2.37}$$

Let us consider possible consequences of ignoring the signal distortion factors, starting initially with the assumption of the presence of only the nonzero offset ΔB in the processed signal, and the absence of the other two distortion terms. In this case, Eq. (2.35) reduces to the formula

$$\Sigma\tau_p^{(0,\,R)}(0,\,r) = \ln\left[\frac{C\beta_{\pi,\,R}(r)\,T_{m,\,0}(0,\,r)T_{m,\,R}(0,\,r)}{\langle P_R(r)\rangle r^2 + \Delta B r^2}\right]. \tag{2.38}$$

The difference between the true sum of the optical depths $\Sigma\tau_p^{(0,\,R)}(0,\,r)$ and its distorted estimate $\langle\Sigma\tau_p^{(0,\,R)}(0,\,r)\rangle$ can be obtained from Eqs. (2.37) and (2.38) in the form

$$\Delta[\Sigma\tau_p^{(0,\,R)}(r)] = \langle\Sigma\tau_p^{(0,\,R)}(0,\,r)\rangle - \Sigma\tau_p^{(0,\,R)}(0,\,r) = \ln\left[1 + \frac{1}{\mathrm{SOR}\,(r)}\right], \tag{2.39}$$

where the signal-to-offset ratio SOR(r) is determined as the ratio $\langle P(r)\rangle/|\Delta B|$. In Fig. 2.4, the simulated profiles of the true sum $\Sigma\tau_p^{(0,\,R)}(0,\,h)$ and the estimated $\langle\Sigma\tau_p^{(0,\,R)}(0,\,h)\rangle$, calculated for the atmospheric situation shown in Fig. 2.2 are shown as Curves 1 and 2, respectively. The difference between these, that is, the absolute error $\Delta[\Sigma\tau_p^{(0,\,R)}(h)]$ originated in the nonzero offset $\Delta B = -0.4$ counts in the signal $\langle P(r)\rangle$ is shown as Curve 3. Note that the offset ΔB is very small - its value relative to the maximal signal is 0.01%. Curve 4 shows the ratio of the absolute error $\Delta[\Sigma\tau_p^{(0,\,R)}(h)]$ to $\langle\Sigma\tau_p^{(0,\,R)}(0,\,h)\rangle$ as a percentage. One can see that the error significantly increases with height, reaching 19% at $h = 8000$ m. To have the error in the above sum that does not exceed some established maximum value, the signal-to-offset ratio SOR(r) should be restricted from the bottom. For example, if

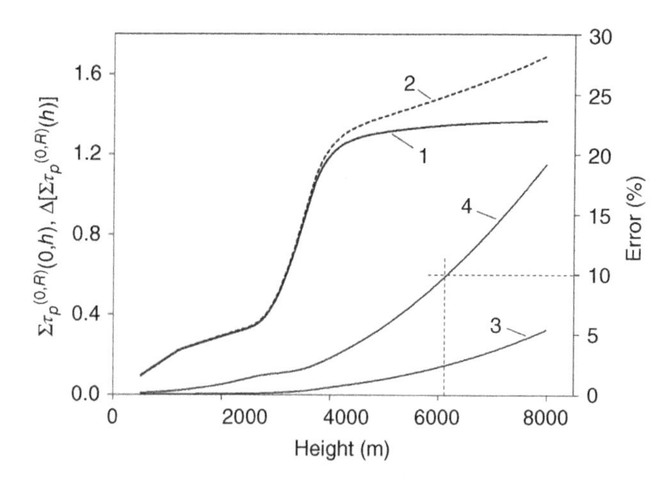

Fig. 2.4 Curves 1 and 2 are profiles of $\Sigma\tau_p^{(0,\,R)}(0,\,h)$ and $\langle\Sigma\tau_p^{(0,\,R)}(0,\,h)\rangle$ obtained for the atmospheric conditions shown in Fig. 2.2. The difference between the optical depths caused by the remaining offset in the recorded backscatter signal $\Delta B = -0.4$ counts is shown as Curve 3. Curve 4 is the corresponding fractional error as a percentage.

the established level of the maximal error is ~10%, the ratio SOR(r) should not be less than ~6.5; accordingly, the maximal height h_{max} should be less than ~6100 m.

Now let us find the relationship between the multiplicative distortion $\delta_P(r)$ and the inversion result, assuming that all other distortion components in Eq. (2.35) are absent, that is, the offset $\Delta B = 0$ and $\Sigma w_{j,\text{low}}(r) = 0$. Under these conditions, Eq. (2.35) transforms to

$$\Sigma\tau_p^{(0,\,R)}(0,\ r) = \ln\left[\frac{C\beta_{\pi,\,R}(r)\,[1+\delta_P(r)]T_{m,\,0}(0,\ r)T_{m,\,R}(0,\ r)}{\langle P_R(r)\rangle r^2}\right], \tag{2.40}$$

and accordingly, the difference between the estimated and actual sums of the optical depths is

$$\Delta[\Sigma\tau_p^{(0,\,R)}(r)] = \langle\Sigma\tau_p^{(0,\,R)}(0,\ r)\rangle - \Sigma\tau_p^{(0,\,R)}(0,\ r) = -\ln[1+\delta_P(r)], \tag{2.41}$$

that is, when the distortion factor $\delta_P(r) \ll 1$, the absolute error in the calculated sum of the optical depths is $\Delta[\Sigma\Delta\tau_p^{(0,\,R)}(r)] \approx \delta_P(r)$. As discussed in Section 1.1, the signals in the near zone, where their magnitude sharply changes over short time intervals, are most strongly impacted by this factor. Accordingly, larger errors in the derived optical depth caused by the signal multiplicative distortion can be expected in the near zone.

2.2.3 Errors in the Derived Extinction-Coefficient Profile Caused by Distortions in the Raman Lidar Signal

The error in the extinction coefficient derived from the Raman signal will primary depend on the distortions in the sum of the optical depths $\Sigma\tau_p^{(0,\,R)}(0,\ r)$ retrieved from the signal. Moreover, the difference in the slopes of the optical depth profiles, the true $\Sigma\tau_p^{(0,\,R)}(0,\ r)$ and distorted $\langle\Sigma\tau_p^{(0,\,R)}(0,\ r)\rangle$ rather than the difference in their numerical values will influence the accuracy of the derived extinction coefficient. Determining slope with numerical differentiation always produces additional errors, so that the error in the derived extinction coefficient is generally significantly larger than that in the term $\langle\Sigma\tau_p^{(0,\,R)}(0,\ h)\rangle$. A accordingly, the reasonable maximal height h_{max} for the derived extinction coefficient will be significantly less than for the inverted optical depth profile.

In Fig. 2.5, two extinction-coefficient profiles are shown, the test profile and the profile extracted from the optical depth total $\langle\Sigma\tau_p^{(0,\,R)}(0,\ h)\rangle$, shown as Curve 2 in Fig. 2.4. These profiles are close to each other up to heights ~4000 m, but then significantly diverge. As was stated above, if the maximal fractional error in the derived profile of $\langle\Sigma\tau_p^{(0,\,R)}(0,\ h)\rangle$ is restricted by 10 %, the maximal height of the optical depth profile should be limited by the height ~6100 m. Meanwhile, under the same condition, the maximal height for the derived extinction coefficient should be limited by the height ~3900 m. In real conditions, the presence of random noise in the signal can significantly reduce both heights.

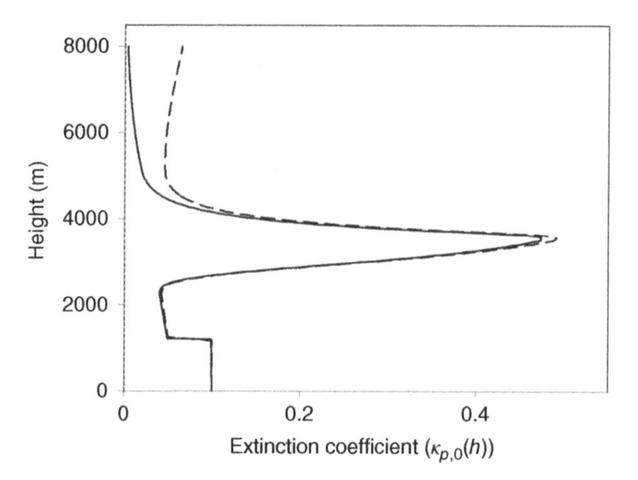

Fig. 2.5 The solid curve is the test profile of the extinction coefficient $\kappa_{p,\,0}(h)$ used for the numerical simulation and the dashed curve is the extinction coefficient extracted from the profile $\langle\Sigma\tau_p^{(0,\,R)}(0,\,r)\rangle$, shown as Curve 2 in Fig. 2.4.

Now let us perform the numerical simulation which will allow estimating the error factor when profiling the atmosphere with Raman lidar. To exclude the assumption about the behavior of the exponent α, let us use the general form of the Raman lidar solution in Eq. (2.3). Denoting the sum of the retrieved particulate extinction coefficients at λ_0 and λ_R as $\Sigma\kappa_p^{(0,\,R)}(r)$, one can rewrite Eq. (2.3) as the function of range in the form

$$\Sigma\kappa_p^{(0,\,R)}(r) = \frac{d}{dr}\left\{\ln\left[\frac{N_R(r)}{P_R(r)r^2}\right]\right\} - \Sigma\kappa_m^{(0,\,R)}(r) \tag{2.42}$$

where $\Sigma\kappa_m^{(0,\,R)}(r) = \kappa_{m,\,0}(r) + \kappa_{m,\,R}(r)$ is the sum of the molecular extinction coefficients at the wavelengths λ_0 and λ_R.

Here a side note should be given, which clarifies the method commonly used by the author in this book when extracting the extinction-coefficient profile from the signal of the splitting lidar. The conventional data processing technique of the splitting lidar data generally includes numerical differentiation, performed in most cases by using the sliding least-square linear fit. The uncertainty in the inversion result is determined through the standard deviation. However, the straightforward use of the least-square technique for the analysis does not produce a simple analytical dependence between the signal distortions and the range resolution selected for the differentiation (see Whiteman, 1999). In other words, one cannot estimate analytically the error dependencies (like error factors) when retrieving the extinction coefficient with the common least-square technique. To avoid this issue, a simpler inversion method can be used. The differential parameter of interest can be obtained by using only the end points of the range resolution s, that is, the end points of the interval from r_1 to $r_2 = r_1 + s$. Using this method, the sum of the particulate extinction coefficients $\Sigma\kappa_p^{(0,\,R)}(r)$ for the middle point $r_s = r + 0.5s$ (details see in Fig. 2.10) can be defined as

$$\Sigma\kappa_p^{(0,\,R)}(r_s) = \frac{1}{s}\ln\left[\frac{N_R(r_1)}{N_R(r_2)}\right] - \frac{1}{s}\ln\left[\frac{P_R(r_1)\,r_1^2}{P_R(r_2)r_2^2}\right] - \Sigma\kappa_m^{(0,\,R)}(r_s), \tag{2.43}$$

Now let us consider the error in the derived sum $\Sigma\kappa_p^{(0,\,R)}(r_s)$, assuming that the profile of $N_R(r)$ is precisely known, so that the error is produced only by distortions in the lidar signals at the end points r_1 and r_2. The absolute error of the sum $\Sigma\kappa_p^{(0,\,R)}(r_s)$ caused by the error of the signal at the near point $P_R(r_1)$ can be obtained from Eq. (2.43) in the form

$$\Delta[\Sigma\kappa_p^{(0,\,R)}(r_s)] = \frac{\delta\langle P_R(r_1)\rangle}{s}, \tag{2.44}$$

where $\delta\langle P_R(r_1)\rangle$ is the relative error of the recorded lidar signal at the range r_1. The corresponding relative error $\delta_1[\Sigma\kappa_p^{(0,\,R)}(r_s)]$ can be found as

$$\delta_1[\Sigma\kappa_p^{(0,\,R)}(r_s)] = \frac{1}{[\Sigma\tau_p^{(0,\,R)}(s)]}\delta\langle P_R(r_1)\rangle, \tag{2.45}$$

where

$$\Sigma\tau_p^{(0,\,r)}(s) = \tau_{p,\,0}(s) + \tau_{p,\,R}(s),$$

is the sum of the one-way optical depths of the range resolution s at the wavelengths λ_0 and λ_R. Accordingly, the corresponding error factor is

$$EF(\delta\Sigma\kappa_p^{(0,\,R)}, \delta P_R) = [\Sigma\tau_p^{(0,\,R)}(s)]^{-1}. \tag{2.46}$$

The relative error $\delta_2[\Sigma\kappa_p^{(0,\,R)}(r_s)]$ caused by the distortion in the lidar signal at the point r_2 is the function of the same error factor, that is,

$$\delta_2[\Sigma\kappa_p^{(0,\,R)}(r_s)] = \frac{1}{[\Sigma\tau_p^{(0,\,R)}(s)]}\delta\langle P_R(r_2)\rangle. \tag{2.47}$$

It follows from Eqs. (2.45) and (2.47) that the relative error in $\Sigma\kappa_p^{(0,\,R)}(r_s)$ dramatically increases when the sum of the optical depths $\Sigma\tau_p^{(0,\,R)}(s)$ is small. Therefore, to reduce the influence of the signal distortions and noise on the accuracy of the derived extinction coefficient, a larger range resolution s should be used. This requirement is inevitable when calculating the numerical derivative over distant ranges close to r_{max}. On the other hand, one should keep in mind that the length of the selected range resolution determines the effective range resolution of the derived extinction coefficient (Whiteman, 1999; Pappalardo et al., 2004).

2.3 DISTORTIONS IN THE EXTINCTION-COEFFICIENT PROFILE DERIVED FROM THE HSRL SIGNAL

As with the Raman lidar, the inversion of the HSRL signal does not require knowledge of the lidar ratio. Moreover, this instrumentation provides a much better signal-to-noise ratio than does the Raman lidar, and unlike the latter, it can reliably operate in daytime conditions. However, there are also some issues when inverting the HSRL signal, as not all errors can be reliably addressed. The estimate of the actual accuracy of the particulate extinction coefficient derived from the HSRL signals requires taking into account all the involved parameters. Accordingly, a more general formula should be analyzed, which unlike Eq. 2.14, does not ignore the term $f_p\beta_{\pi,p}(h)$, that is,

$$\kappa_p(h_s) = \frac{1}{2s}\ln\left[\frac{P_m(h_1)h_1^2}{P_m(h_2)h_2^2}\right] - \frac{1}{2s}\ln\left[\frac{f_m(h_1)\beta_{\pi,m}(h_1)+f_p\beta_{\pi,p}(h_1)}{f_m(h_2)\beta_{\pi,m}(h_2)+f_p\beta_{\pi,p}(h_2)}\right] - \kappa_m(h). \tag{2.48}$$

As in Eq. (2.14), all the quantities in the formula are determined at the end points of the extended range-resolution interval s, that is, at the points h_1 and $h_2 = h_1 + s$.

When analyzing possible errors in the extinction coefficient derived with Eq. (2.48), three possible sources of the random systematic error should be considered: (i) the error caused by the nonperfect suppression of the particulate component by the iodine filter, (ii) the error related to the uncertainty in the spectrum of Cabannes scattering, and (iii) the error caused by the distortion in the signal recorded in the molecular channel of HSRL. The error related to the uncertainty in the profile of the molecular backscatter coefficient $\beta_{\pi,m}(h)$ is relatively negligible (Eberhard, 2010; Adam, 2012), so it is not considered here.

Let us start our analysis with the first error source, caused by the presence of the iodine vapor filter in the molecular channel of HSRL. Since $f_p \neq 0$, the particulate component $\beta_{\pi,p}$ in the molecular channel is not blocked entirely, so that the molecular component is not precisely separated. In other words, the output signal in the molecular channel counts not only molecular photons but also some fraction of the particulate photons. In the atmospheric layers, where the particulate component $\beta_{\pi,p}(h)$ is significantly larger than the molecular component $\beta_{\pi,m}(h)$, the products $f_p\beta_{\pi,p}(h)$ and $f_m(h)\beta_{\pi,m}(h)$ in Eq. (2.8) may become comparable in spite of the significant inequality $f_p \ll f_m$. The level of the particulate-scattering component that penetrates to the output of the molecular channel depends on the specifics of the atmospheric profiles under investigation; the particulate component can be especially large in the areas of increased aerosol loading with sharp boundaries. Minimizing this effect is not a trivial task. Althausen et al. (2012) proposed to compensate the nonperfect suppression of the particulate backscattering in the molecular channel of HSRL by implementing an additional correction parameter. Such a parameter can be taken either *a priori*, or introduced on the basis of knowledge of the known specifics of the investigated optical profile. However, when using HSRL, the only additional information about the aerosol profiles is coded in the signals of the total channel, the inversion of which have their own sources of uncertainty. Obvious doubts emerge whether the implementation of any such correction will provide improved inversion results.

Let us consider the possible error in the derived particulate extinction coefficient caused by the nonperfect suppression of the particulate component in the iodine cell. It follows from Eq. (2.8) that the actual particulate two-way transmittance term $T_p^2(0, h)$, extracted from the molecular channel of HSRL obeys the formula

$$T_p^2(0, h) = \frac{P_m(h)h^2}{C_m T_m^2(0, h)\Delta h_b[f_m(h)\beta_{\pi, m}(h) + f_p\beta_{\pi, p}(h)]}, \tag{2.49}$$

where $T_m^2(0, h)$ is the two-way molecular transmittance within the altitude range $(0, h)$. When processing the output data of the molecular channel, the implicit premise is that the iodine vapor filter blocks the particulate component entirely, that is, the component $f_p\beta_{\pi, p}(h) = 0$. Accordingly, the estimate of the two-way particulate transmittance is made using the formula

$$\langle T_p^2(0, h)\rangle = \frac{P_m(h)h^2}{C_m T_m^2(0, h)\Delta h_b f_m(h)\beta_{\pi, m}(h)}. \tag{2.50}$$

If the above premise, $f_p\beta_{\pi, p}(h) = 0$ is not met, the estimated transmittance is distorted, that is, the estimate $\langle T_p^2(0, h)\rangle$ in Eq. (2.51) is not equal to the true two-way transmittance $T_p^2(0, h)$ defined in Eq. (2.49). Taking the logarithm of the ratio $\langle T_p^2(0, h)\rangle$ to $T_p^2(0, h)$ and defining the corresponding optical depths as $\langle \tau_p(0, h)\rangle$ and $\tau_p(0, h)$, one obtains the difference between the estimated and true optical depths in the form

$$\langle \tau_p(0, h)\rangle - \tau_p(0, h) = -0.5 \ln\lfloor 1 + R_{p, m}(h)\rfloor, \tag{2.51}$$

where $R_{p, m}(h) = \frac{f_p\beta_{\pi, p}(h)}{f_m(h)\beta_{\pi, m}(h)}$ and it is assumed that the terms C_m, $T_m^2(0, h)$, and $f_m(h)$ in Eq. (2.50) are precisely known The achieved rejection ratio of the scattered light from particulates may vary, approximately, from 1:1000 to 1:5000 (Kovalev and Eichinger, 2004), so that $R_{p, m}(h) \approx (0.001 - 0.0005)\frac{\beta_{\pi, p}(h)}{\beta_{\pi, m}(h)}$. After differentiating Eq. (2.51), one obtains the simple formula

$$\kappa_p^{(\mathrm{dif})}(h) - \kappa_p(h) = -0.5\frac{d}{dh}\{\ln[1 + R_{p, m}(h)]\}, \tag{2.52}$$

where $\kappa_p^{(\mathrm{dif})}(h)$ is the particulate extinction coefficient, derived from the optical depth $\langle \tau_p(0, h)\rangle$, that is, $\kappa_p^{(\mathrm{dif})}(h) = \frac{d}{dh}\langle \tau_p(0, h)\rangle$, and $\kappa_p(h) = \frac{d}{dh}[\tau_p(0, h)]$ is the actual extinction coefficient, which would be obtained from the nondistorted optical depth. Using the data points h_1 and $h_2 = h_1 + s$ for two-point numerical differentiation, one obtains the following formula for the error in the extinction coefficient $\kappa_p^{(\mathrm{dif})}(h)$:

$$\delta\kappa_p^{(\mathrm{dif})}(h) = \frac{\kappa_p^{(\mathrm{dif})}(h) - \kappa_p(h)}{\kappa_p(h)} = \frac{-1}{2s[\kappa_p(h)]} \ln\left[\frac{1 + R_{p, m}(h_1)}{1 + R_{p, m}(h_2)}\right], \tag{2.53}$$

where s is the range resolution, selected for numerical differentiation. As with Raman lidar, the error of the retrieved extinction coefficient $\delta\kappa_p^{(dif)}(h)$ dramatically increases when the optical depth of the range-resolution interval $\tau_p(s) = \kappa_p(h)s$ is small. One should point out that the presence of a nonzero value of f_p in Eq. (2.48) and the corresponding nonzero values $R_{p,\,m}(h_1)$ and $R_{p,\,m}(h_2)$ is not necessarily the origin of the error in the retrieved extinction coefficient. Indeed, according to Eq. (2.53), no error in $\kappa_p^{(dif)}(h)$ takes place if $R_{p,\,m}(h_1) = R_{p,\,m}(h_2)$. Obviously, the most significant distortion of the retrieved extinction coefficient will occur in the areas of sharp change of the particulate backscatter term $\beta_{\pi,\,p}(h)$ within the interval $(h_1 - h_2)$. In the multilayered atmosphere, for example, such as generally observed in the vicinity of wildfires, large errors $\delta\kappa_p^{(dif)}(h)$ would commonly be unavoidable.

Let us consider this issue using the same synthetic atmosphere as shown in Fig. 2.2. The profile of the extinction coefficient $\kappa_p(h)$ in this synthetic atmosphere is shown in Fig. 2.6 as Curve 1; Curve 2 is the profile of the corresponding ratio $R_{p,\,m}(h)$ calculated with the rejection ratio, 1:1000. The lidar ratio in this atmosphere is selected to be constant, equal to 0.05 sr^{-1} over all heights. The corresponding errors in the extinction coefficient $\kappa_p^{(dif)}(h)$ derived using numerical differentiation with range resolutions $s = 120$ m and $s = 320$ m are shown in Fig. 2.7 as the dashed and solid curves, respectively. Note that the synthetic atmosphere used in the numerical simulation is rather clear; the maximal extinction coefficient $\kappa_p(h)$ in the polluted layer is less than 0.5 km^{-1} and its gradient is relatively small. Nevertheless, the error at the boundaries of the polluted layer, at the heights ~2700 and ~3800 m, is close to 10%. Over most heights, the use of different range resolutions provides almost the same error profiles. However, the errors in the extinction coefficient retrieved with smaller range resolution $s = 120$ m become much larger in the vicinity of sharp changes of

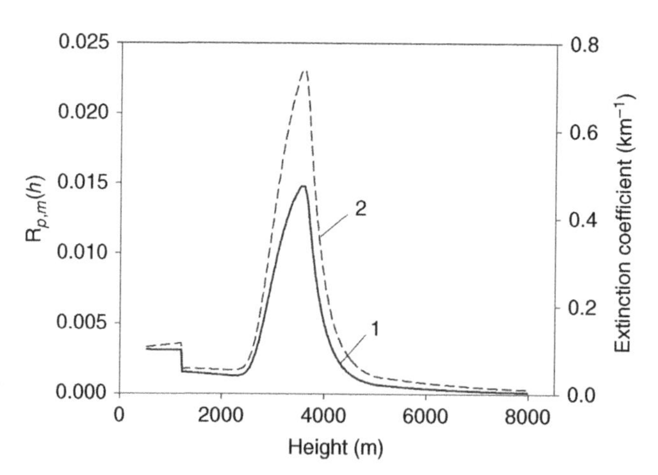

Fig. 2.6 Curve 1 is the model extinction-coefficient profile of the synthetic atmosphere used for numerical simulation. The corresponding function $R_{p,\,m}(h)$ calculated with the rejection ratio 1:1000 is shown as Curve 2.

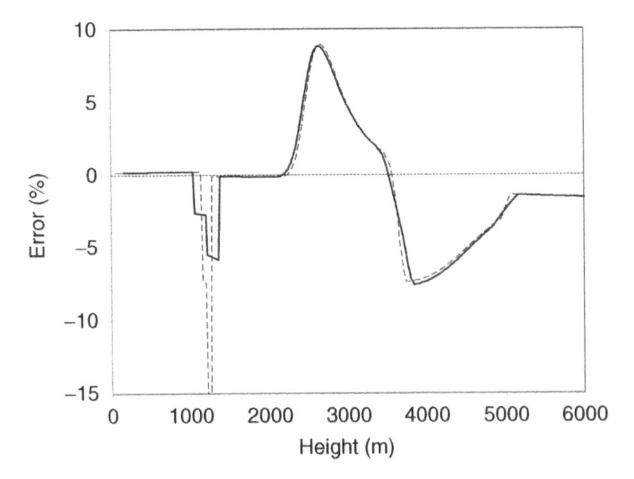

Fig. 2.7 Error profiles of the retrieved extinction coefficient $\kappa_p^{(\mathrm{dif})}(h)$ obtained with range resolutions 120 and 320 m, shown as the dashed and solid curves, respectively.

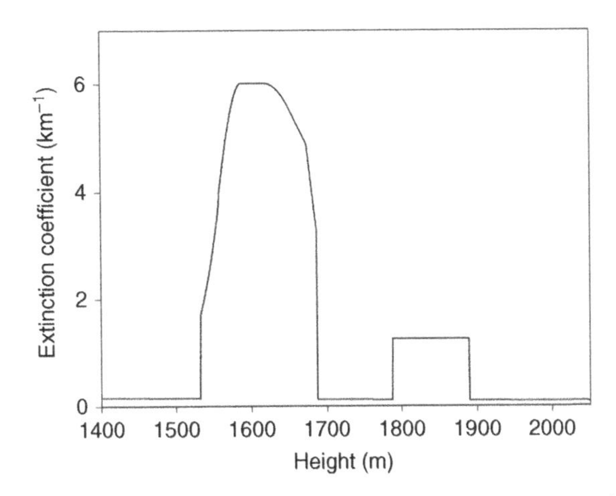

Fig. 2.8 Test profile of the particulate extinction coefficient used in the numerical experiment.

the extinction coefficient, such as at the height, ~1200 m. This effect takes place even when the absolute changes in the simulated extinction coefficient are small.

Let us consider the case when the atmosphere contains layers with increased backscattering and relatively sharp boundaries. Such a case is illustrated by Figs. 2.8 and 2.9. The shape of the extinction coefficient used in the numerical simulation is shown in Fig. 2.8. In this synthetic atmosphere, two aerosol layers with increased backscattering are located within the altitude range from 1530 to 1685 m and from 1790 to 1890 m. The maximal extinction coefficients within these layers are 6 and 1.25 km^{-1}, respectively. The relative errors in the retrieved extinction-coefficient

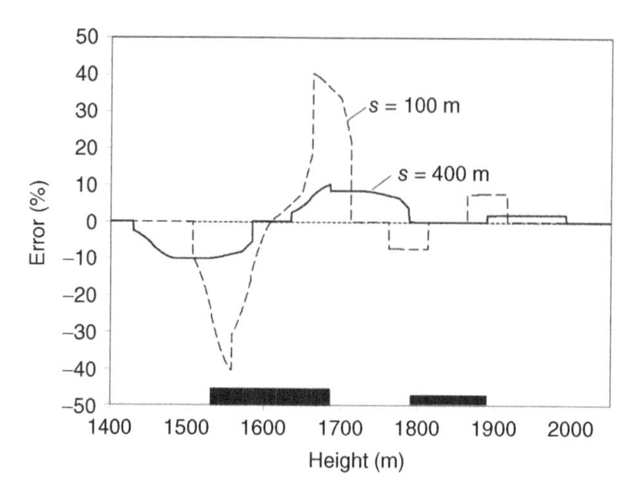

Fig. 2.9 Relative error in the extinction coefficient $\kappa_p^{(\mathrm{dif})}(h)$ derived with range resolutions 100 and 400 m. The black rectangles at the bottom show the locations of the simulated layers with increased backscattering.

profiles caused by the presence of the nonzero particulate component $f_p\beta_{\pi,\,p}(h)$ are shown in Fig. 2.9. The thin dashed curve shows the error $\delta\kappa_p^{(\mathrm{dif})}(h)$ when the extinction coefficient $\kappa_p^{(\mathrm{dif})}(h)$ is derived with the range resolution $s = 100$ m, and the thick solid curve is the error, when the range resolution $s = 400$ m, is used. The rejection ratio in both cases is equal to 1:1000, the same as in the previous case. As can be expected, when the range resolution is smaller, the error in the retrieved extinction coefficient is larger and vice versa. When $s = 100$ m, the maximum absolute error in $\kappa_p^{(\mathrm{dif})}(h)$ is close to ~40% it decreases down to 10% when $s = 400$ m. However, the decrease in maximum error when using the small range resolution is accompanied by a significant expansion of the error $\delta\kappa_p^{(\mathrm{dif})}(h)$, beyond the actual boundaries of the layer with increased backscattering.

Thus, the assumption that the product $f_p\beta_{\pi,\,p}(h) = 0$, used when processing the HSRL signal, may produce a significant error in the retrieved extinction coefficients, especially in a layered atmosphere. This result is obtained even while assuming that all the molecular components in Eqs. (2.49) and (2.50) are precisely known. Now let us consider the error in the derived extinction coefficient, related to the uncertainty in the molecular component $f_m(h)$. As stated in Section 2.1, the inversion of the HSRL signal requires estimation of the portion of the molecular scattering blocked by the iodine filter. This in turn, requires knowledge of the backscatter spectrum of the Cabannes scattering. There are different theoretical models for estimation of the Cabannes spectrum. As affirmed in the study by Liu et al. (2009), most typical models provide relatively small discrepancies in the estimated theoretical spectra. However, the aggravating factor remains that the inversion result depends on the difference in the slopes of the involved functions rather than on the differences in the numerical values of these functions. Let us consider the error caused by an inaccurately estimated term $f_m(h)$, assuming that the terms C_m and $T_m^2(0, h)$ are precisely known and $f_p(h) = 0$. If, instead of the true function $f_m(h)$, some distorted function $\langle f_m(h)\rangle = f_m(h)[1 + \delta f_m(h)]$ is used, then instead of the actual two-way transmission term, defined as

$$T_p^2(0, h) = \frac{P_m(h)h^2}{C_m T_m^2 \Delta h_b [f_m(h)\beta_{\pi, m}(h)]}, \tag{2.54}$$

some distorted two-way transmission function will be obtained, that is,

$$\langle T_p^2(0, h)\rangle = \frac{P_m(h)h^2}{C_m T_m^2 \Delta h_b \langle f_m(h)\rangle \beta_{\pi, m}(h)}. \tag{2.55}$$

Determining the ratio of the functions $\langle T_p^2(0, h)\rangle$ and $T_p^2(0, h)$ and making a simple transformation, one obtains the simple relationship between the retrieved and actual optical depths:

$$\langle \tau_p(0, h)\rangle = \tau_p(0, h) + \ln[1 + \delta f_m(h)]. \tag{2.56}$$

Applying two-point numerical differentiation to the same points h_1 and $h_2 = h_1 + s$ as in Eq. (2.53), one obtains the formula

$$\delta\kappa_p^{(\mathrm{dif})}(h) = \frac{-1}{2s[\kappa_p(h)]} \ln \left[\frac{\dfrac{f_m(h_1)}{\langle f_m(h_1)\rangle}}{\dfrac{f_m(h_2)}{\langle f_m(h_2)\rangle}} \right]. \tag{2.57}$$

Note that if $\frac{f_m(h_1)}{\langle f_m(h_1)\rangle} = \frac{f_m(h_2)}{\langle f_m(h_2)\rangle}$, the error $\delta\kappa_p^{(\mathrm{dif})}(h) = 0$. The nonzero error in Eq. (2.57) will take place only if this equality is not valid. As with the previous case, the error $\delta\kappa_p^{(\mathrm{dif})}(h)$ will be strongly influenced by the error factor, which is equal to $[2\tau_p(s)]^{-1}$, where $\tau_p(s)$ is the particulate optical depth within the range resolution s. This error factor may be extremely large when the lidar is working in a clear atmosphere where the optical depth over the commonly used range resolution s is generally much less than unity.

Finally, let us focus on the error in the derived extinction coefficient caused by distortions in the backscatter signal in the molecular channel of HSRL. The relationship between the distorted signal $\langle P_m(h)\rangle$ and nondistorted $P_m(h)$ is similar to that in the Raman signals, that is,

$$\langle P_m(h)\rangle = P_m(h)[1 + \delta_P(h)] + \Sigma w_{j,\,\mathrm{low}}(h) + \Delta B. \tag{2.58}$$

It is assumed here that preliminary signal smoothing was made, so that the high-frequency components $\Sigma w_{j,\,\mathrm{high}}(h)$ are removed from the signal. As with the above, we will consider the error in the extinction coefficient retrieved from the signals $P_m(h_1)$ and $P_m(h_2)$, determined at the end points of the selected range resolution s, that is, at the points h_1 and $h_2 = h_1 + s$. The absolute error of the extinction coefficient $\Delta\kappa_{p,1}^{(\mathrm{dif})}(h)$ caused by the distortions in the signal $P_m(h_1)$ is

$$\Delta\kappa_{p,\,1}^{(\mathrm{dif})}(h) = \frac{1}{2h}\delta\langle P_m(h_1)\rangle, \tag{2.59}$$

where $\delta\langle P_m(h_1)\rangle$ is the relative error of the signal. The extinction coefficient error $\Delta\kappa_{p,\,2}^{(\mathrm{dif})}(h)$ caused by the error of the signal at the point h_2 obeys a similar formula. When the selected height resolution s is large, the errors $\Delta\kappa_{p,\,1}^{(\mathrm{dif})}(h)$ and $\Delta\kappa_{p,\,2}^{(\mathrm{dif})}(h)$ can be considered as uncorrelated and the total fractional error can be estimated with the standard formula

$$\delta\kappa_p^{(\mathrm{dif})}(h) = \frac{1}{2\tau_p(s)}\sqrt{\{\delta\langle P_m(h_1)\rangle\}^2 + \{\delta\langle P_m(h_2)\rangle\}^2}, \tag{2.60}$$

where $\tau_p(s)$ is the particulate optical depth of the range-resolution interval, s. As follows from these formulas, the error factor for the retrieved extinction coefficient is the same as above:

$$EF[\delta\kappa_{p,}\ \delta P_m] = [2\tau_{\mathrm{p}}(s)]^{-1}. \tag{2.61}$$

As to other uncertainty sources in the inverted data of HSRL, one can mention the errors related to the inaccuracies in calibration parameters, including the overlap function $q(r)$ in the case when the signals within the incomplete overlap zone need to be inverted into optical parameters; the errors related to the photomultiplier afterpulsing effect; etc. However, in the studies by Liu et al. (2009), Rogers et al. (2009, 2011), Esselborn et al. (2008, 2009), and Hair et al. (2008), the authors maintain that most of the issues can be overcome and acceptable accuracy in the extinction coefficient retrieved from the HSRL data can be achieved. One can also expect that the use of HSRL instrumentation in the multiangle measurement mode can open up new opportunities in the application of these lidars (Kovalev et al., 2012).

Let us summarize the common features inherent to splitting lidars and add some author's comments to clarify his vision of the current state of the art. The inversion results obtained from the splitting lidar data can improve our understanding of physical processes in the atmosphere, but they still cannot be considered as measurement data. They yield the probable atmospheric profiles that obey the mathematical model (or models) selected for the inversion, the implicit premises inherent to such a model, and the assumptions used by the researcher. The solution models either ignore important systematic errors, or require *a priori* assumptions. Accordingly, the researchers have no other option except to focus basically on statistical dependencies. However, as the author has already declared in his publications, the real atmosphere obeys much more complicated laws than common statistics.

2.4 NUMERICAL DIFFERENTIATION AND THE UNCERTAINTY INHERENT IN THE INVERTED DATA

2.4.1 Basics

Numerical differentiation is the most common operation used for inversion of the lidar signals. As shown in the previous sections, numerical differentiation is an unavoidable procedure when the extinction coefficient is extracted from the signals of splitting lidars, such as the Raman or high-spectral resolution lidar. The same procedure is used when inverting signals of DIAL and for processing signals of the elastic lidar, which operates in the multiangle mode.

The harsh reality related to such a mandatory application of numerical differentiation is that no standard methods for obtaining outputs from the signals of splitting lidars exist, especially when these lidars work in a multilayered atmosphere. As shown in Chapter 1, the real lidar signals are always corrupted, and the presence of both random and systematic distortions significantly impedes the use of the numerical differentiation technique for atmospheric profiling. This issue has not been solved at present, and most likely, will not be satisfactorily solved in the near future. Numerous attempts to solve this problem explain why different retrieval techniques for performing numerical differentiation are used for atmospheric investigation.

In lidar profiling of the atmosphere, the direct differentiation technique is commonly used, which allows determining the extinction coefficient directly from the optical depth profile extracted from the lidar signal. The technique focuses on the inversion of the optical depth without taking into consideration any available auxiliary function related to the extinction coefficient of interest. Meanwhile, the alternative, rarely used indirect methods of numerical differentiation allow putting additional restrictions on the derived extinction coefficient. These methods are based on using two functions derived from the same lidar signal, the profile of the optical depth and the profile of the backscatter coefficient. Some variants of such an alternative technique are considered in the following sections; this section presents a brief survey of the common direct numerical differentiation techniques used for extracting the extinction coefficient from the optical depth profile.

The optical depth profile defined from a splitting lidar signal within the lidar operative range $r_{min} - r_{max}$ and the corresponding extinction coefficient $\kappa(r)$ within this interval are related as follows:

$$\tau(r_{min}, r) = \int_{r_{min}}^{r} \kappa(r')dr', \tag{2.62}$$

where r_{min} is a starting point from which the optical depth is accumulated. Accordingly, the extinction coefficient at the range r is determined through differentiation of the optical depth $\tau(r_{min}, r)$, that is,

$$\kappa(r) = \frac{d}{dr}[\tau(r_{min}, r)]. \tag{2.63}$$

Let us denote the total error of the inverted optical depth $\tau(r_{min}, r)$ as $\Delta\tau(r)$. When the error $\Delta\tau(r) \neq 0$, then instead of the true profile $\tau(r_{min}, r)$, the corrupted profile

$$\langle \tau(r_{min}, r) \rangle = \tau(r_{min}, r) \pm \Delta\tau(r), \tag{2.64}$$

is differentiated. Accordingly, instead of the actual profile of the extinction coefficient, a distorted profile $\kappa^{(dif)}(r)$ is found:

$$\kappa^{(dif)}(r) = \frac{d}{dr} \langle \tau(r_{min}, r) \rangle, \tag{2.65}$$

where $\kappa^{(dif)}(r)$ is an approximate solution for the extinction coefficient of interest obtained by a differential quotient.

The lidar signals and the corresponding inverted data are commonly recorded in digital form. Therefore, in practical lidar data processing, the analytical derivative at the point of interest is generally approximated by a differential quotient of the optical depth in a finite range-resolution interval. The general principles of the differentiation are based on the discretization of the interval of interest (r_1, r_2) using some differentiation step s, so that $r_i = r_1 + is$, where $i = 0, 1, \ldots, N$ and $s = (r_2 - r_1)/N$. The approximate solution of the equation for any point of the interval $r \in [r_1, r_2]$ can be written in a simplified form as

$$\kappa^{(dif)}(r_i) = \frac{1}{s}[\tau(r_1, r_{i+1}) - \tau(r_1, r_i)], \tag{2.66}$$

Theoretically, the extinction coefficient $\kappa^{(dif)}(r_i)$ in Eq. (2.66) is found more accurately when a smaller range resolution is used. Indeed, when $s \to 0$, the retrieved extinction coefficient $\kappa^{(dif)}(r_i) \to \kappa(r_i)$. When s $\neq 0$, the accuracy of the differentiation is restricted by the error $\Delta\kappa(r)$, where

$$\Delta\kappa(r) = \kappa^{(dif)}(r) - \kappa(r), \tag{2.67}$$

which depends both on the range resolution s and the absolute error $\Delta\tau(r)$.

The commonly used technique for numerical differentiation is based on Taylor's theorem. This technique estimates the derivative through the first term in Taylor's series (Wylie and Barret, 1982). The estimate of $\Delta\kappa(r)$ can be obtained using the presentation of the function $\tau(r_{min}, r)$ in the vicinity of the point r_i as the Taylor series with the Lagrange remainder term, $o(s)$, that is (see http://www.stewartcalculus.com/data/CALCULUS%20Early%20Transcendentals/upfiles/Formulas4RemainderTaylorSeries5ET.pdf),

$$\tau(r_i + s) = \tau(r_i) + \tau'(r_i)s + o(s), \tag{2.68}$$

where $\tau'(r_i) = \frac{d}{dr}[\tau(r_i)]$ and the infinitesimally small quantity $o(s)$ is a Lagrange remainder:

$$o(s) = \frac{1}{2}\tau''(\xi)s^2, \tag{2.69}$$

where $\xi \in (r_i, r_i + s)$. It is difficult to precisely determine ξ, so that the remainder $o(s)$ is commonly restricted by its assumed maximum value

$$o(s) \leq \frac{1}{2}\Big| \max_{\xi \in [r_i,\, r_i+s]} [\tau''(\xi)s^2]. \tag{2.70}$$

The second-order derivative $\tau''(r)$ at the point r_i can be presented through the finite differences as

$$\tau''(r_i) \approx \frac{1}{s^2}[\tau(r_{i+1}) - 2\tau(r_i) + \tau(r_{i-1})], \tag{2.71}$$

accordingly,

$$\kappa(r_i) = \frac{[\tau(r_{i+1}) - \tau(r_i) - o(s)]}{s}. \tag{2.72}$$

The corresponding estimate of the extinction coefficient can be written in the form

$$\kappa^{(\mathrm{dif})}(r_i) = \frac{[\langle\tau(r_{i+1})\rangle - \langle\tau(r_i)\rangle - \langle o(s)\rangle]}{s}. \tag{2.73}$$

Taking Eq. (2.64) into consideration, one obtains

$$|\Delta\kappa(r_i)| \leq |\Delta\tau(r_{i+1}) + \Delta\tau(r_i)|/s + |\langle o(s)\rangle|/s, \tag{2.74}$$

and

$$|\Delta\kappa(r_i)| \leq |\Delta\tau(r_{i+1}) + \Delta\tau(r_i)|/s + |\Delta\tau(r_{i+1}) + 2\Delta\tau(r_i) + \Delta\tau(r_{i-1})|/(2s) + |o(s)|/(2s), \tag{2.75}$$

It follows from Eq. (2.75) that when $s \to 0$, the first and the second term in the right side of the equation have no limitation and may be indefinitely large. The third term in the right side tends to zero, as here both the numerator and the denominator tend to zero when $s \to 0$. To avoid large oscillation in the extinction coefficient, special methods are used. The common method for reducing oscillations in $\kappa^{(\mathrm{dif})}(r_i)$ is to increase the range resolution s. However, increasing the range resolution increases the component $o(s)$ in Eq. (2.68) and the related uncertainty.

Thus, the requirements for the selection of the optimal range resolution are contradictory. In principle, the optimal solution can be achieved by selecting a value of s that minimizes the function $|\Delta\kappa(r_i)|$. This can be achieved by determining the approximate remainder term using $\langle o(s)\rangle$ obtained after smoothing the optical depth. In practice, such an approach significantly simplifies the estimation of the uncertainty, but this estimate is not rigorously justified, the accuracy of the derived extinction coefficient can be significantly overestimated. One should always keep in mind that the standard estimate of the random noise error ignores other possible multiplicative and additive distortion components. Therefore, when using the above formulas, a significant underestimate of the actual uncertainty in $\kappa^{(\mathrm{dif})}(r_i)$ is likely.

When processing real lidar data, the sliding range resolution can include tens or hundreds of data points, and be as long as hundreds or even thousands of meters. Therefore, it is common for the least-squares technique to be applied for numerical differentiation. However, for a better understanding of the specifics and practical issues of numerical differentiation, the simplest two-point numerical differentiation may be used. The simple analytical formulas inherent to this method of extracting the derivative allows the reader to better understand possible pitfalls in the practical differentiation procedure. When the extended range resolution s is used, it is reasonable to designate the extracted extinction coefficient to the middle point of the optical depth range-resolution interval r_s, that is,

$$\kappa^{(\mathrm{dif})}(r_s) = \frac{\tau(r_{\min}, r_s + 0.5s) - \tau(r_{\min}, r_s - 0.5s)}{s}, \tag{2.76}$$

that is, the extinction coefficient $\kappa^{(\mathrm{dif})}(r_s)$ is determined for the middle point of the interval from $r = r_s - 0.5s$ to $r = r_s + 0.5s$ (Fig. 2.10). The thin curve in the figure is a differentiated experimental optical depth profile $\tau(r_{\min}, r)$; s is the range resolution of the numerical differentiation and r_s is the middle point of this interval. The extinction coefficient $\kappa^{(\mathrm{dif})}(r_s)$ can be considered as some mean extinction coefficient over this range interval. However, one should stress that in the general case the data points within the range resolution s do not necessarily obey a normal distribution relative to the actual profile. When estimating numerical differentiation accuracy, two factors should be taken into account. The first factor is the theoretical accuracy of the differentiation technique under ideal conditions when the scattered data points of the differentiated profile obey common statistical laws. The second factor that needs to be taken into consideration is the presence in the processed optical depth profile of

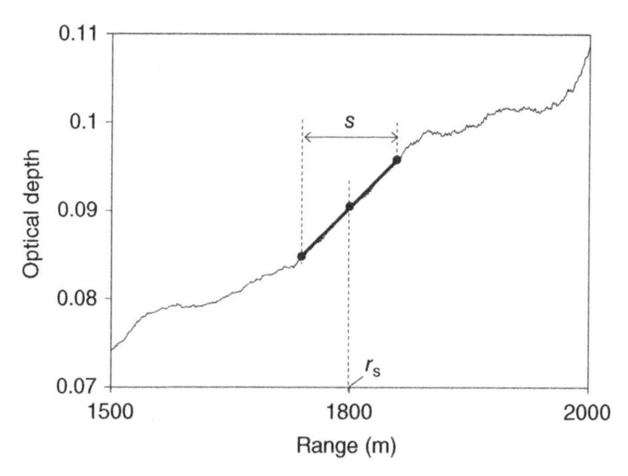

Fig. 2.10 Illustration of the numerical differentiation technique. The thin curve is the profile of the optical depth and the thick line is its linear fit within the range resolution s restricted by the end points $r_s - 0.5s$ and $r_s + 0.5s$.

the systematic and low-frequency random distortions, which do not obey the statistics. Accordingly, the mean extinction coefficient $\kappa^{(\mathrm{dif})}(r_s)$ defined in Eq. (2.76) is not necessarily the average in the statistical sense, and the corresponding standard deviation cannot be considered at the 68% confidence level of the solution (see Figs. 1.12 and 1.13 and the related text).

One related side note should be added. When using the least-square technique for numerical differentiation, the weight functions for the involved data are commonly used. The issue is that the profile of the extinction coefficient derived from the optical depth depends on the selected weight functions for the data points used for the differentiation. The use of the weight functions, such as the standard deviation or the variances, is based on the implicit premise that the data points of the differentiated function have no systematic distortions. In real conditions, such an implicit premise, taken without a thorough analysis, can mislead researcher.

2.4.2 Nonlinear Fit in the Numerical Differentiation Technique and its Issue

As stated above, the linear fit of the numerical differentiation means that only the first term in Taylor's series is used when determining the increment of the function of interest; the higher orders of the series are ignored. In other words, the application of the linear fit for the optical depth profile within the extended range resolution is based on the implicit premise that the extinction coefficient within this range is a constant value, at least, in the statistical sense. This premise is rather restricting, especially, for the large range-resolution interval, even if the linear fit is found through the least-square technique. When determining the slope of the linear fit this way, the inverted data points of the optical depth, weighted or nonweighted, are randomly distributed relative to their linear fit. Presumably, the slope of the linear fit is uniquely related to the actual mean extinction coefficient within the selected range resolution. However, the relationship between the actual extinction coefficient and the slope of the statistical linear fit can be much more complicated (see Section 1.3). A large number of studies have been published in which different, presumably more optimal, methods for determining the slope are considered. However in these, the common statistical approach to the data obtained from lidar remote sensing is used (Whiteman, 1999; Rocadenbosch et al., 2000; Volkov et al., 2002; Sicard et al., 2002; Pornsawad et al. 2008, etc.).

The use of linear approximation for numerical differentiation is the most simple and straightforward method; however, it is not the only option. Numerical differentiation can also be accomplished using a nonlinear, high-order polynomial fitting (Pelon and Megie, 1982; Kempfer et al., 1994; Fujimoto et al., 1994). The polynomial fit of a specific order can be used for analytical approximation of intermediate functions, and applied either on a local or global scale. For example, in the study of Pelon and Megie (1982), ozone concentration was found from the difference of the signal derivatives at the off and on DIAL wavelengths. Each derivative at the local range interval was determined by fitting the range-corrected signal to a second-order polynomial. McDermid et al. (1990) used a similar approach.

Let us digress to some mathematical basics to clarify such a technique. As follows from the general theory of polynomials, a polynomial function in the form

$$f(x) = a_n x^n + a_{n-1} x^{n-1} + \ldots + a_1 x + a_0, \tag{2.77}$$

where a_n, a_{n-1}, … a_0 are real numbers and n is a nonzero integer, can approximate any variable function over a restricted interval (Boyd, 200; Mason and Christopher, 2003). The optical depth $\langle \tau(0, r) \rangle$, extracted from the splitting lidar signal within the range interval from r_j to r_k can be presented as a polynomial function, $\Pi\tau(r_j, r)$, with any selected number of polynomial terms and their appropriate magnitudes, that is,

$$\Pi\tau(r_j, r) \approx a_n r^n + a_{n-1} r^{n-1} + \ldots + a_1 r + a_0. \tag{2.78}$$

The degree of the polynomial, that is, the largest exponent in variable r in Eq. (2.78) will define the level of approximation, that is, how close the polynomial fit $\Pi\tau(r_j, r)$ would approximate the actual optical depth $\tau(r_j, r)$. Obviously, one should use extreme caution when using high-order polynomial fits for noisy profiles of the optical depth extracted from lidar data. The nonlinear fit may generate false variances in the retrieved profile of interest; this overfitting effect is discussed Section 1.2. It is not possible to determine whether the oscillations in the output profile, obtained when using the polynomial fits, are false or real. Truly reliable regularization in the lidar measurements is generally impossible because of the absence of proper information that would disallow distortion effect. However, as was stated above, the simplest linear fit may also be inappropriate because of the loss of significant detail in the derived profile. In fact, all the retrieval methods based on numerical differentiation have such an issue. In order to avoid over fitting when using the polynomial fit, it is necessary to use either available information about the behavior of the aerosol loading, or at least, reliable criteria for establishing the maximum lidar measurement range. Linear approximation is commonly applied in numerical differentiation because linear fit has the simplest mathematical formulation, and there is generally no evidence that nonlinear fitting yields more accurate inversion results.

To test different differentiation techniques, special investigations were performed, such as those by Godin et al. (1999) for ozone DIAL and Pappalardo et al. (2004) for Raman lidar. In these investigations, different research groups inverted optical profiles obtained from synthetic signals of a virtual lidar using their preferred retrieval technique. In the former study, the organizers initially computed the virtual lidar signals for three synthetic vertical ozone profiles. The ozone profiles were smooth enough, but some of them contained short-range perturbations, which were located in the regions of the low and high stratosphere. These perturbations were included in order to test the vertical resolution of 10 inversion algorithms used by the participating lidar groups. Most teams used very similar data processing techniques based on linear or parabolic fit. In four algorithms, the logarithm of the on- and off-signal ratio was fitted to a straight line from which the ozone concentration was then derived. In the three other algorithms, the ozone concentration was derived from the difference in the derivative of a second-order polynomial fitted to the logarithm of each lidar signal. Only two algorithms used a higher-order polynomial to fit the logarithm of the signal ratio.

Even this unique test, in which recognized lidar specialists were involved, did not answer the question as to which algorithm can be considered optimal. The comparison showed, however, that the more simple technique generally provided better inversion results. The test revealed that all the fitting techniques, that is, the use of both linear and high-order polynomial fits, produced large bias in the inverted ozone-concentration profile at high altitudes. In fact, no technique showed sufficiently accurate results over the altitudes from 10 to 50 km, especially in areas of aerosol perturbations; over high altitudes, where the signal-to-noise ratio is relatively small, the discrepancies in the outputs proved to be extremely large.

Thus, the selection of the algorithm and the range resolution for numerical differentiation is nothing else but a tacit selection of a concrete model for determining the profile of interest. Each algorithm for determining the numerical derivative has its own smoothing parameters, that is, each algorithm is valid only for some statistical model or models. In other words, numerical differentiation is always based on some assumed behavior of the extracted characteristic over the local range of interest (Kovalev and Eichinger, 2004).

Let us summarize the general assumptions used in numerical differentiation techniques:

1. The assumed linear or nonlinear fit in the sliding derivative and the selected range resolution used for the differentiation are consistent with the actual atmospheric profile of interest.
2. Systematic and low-frequency distortions in the differentiated function are absent, and accordingly, no such distortions take place in the output profile.
3. The selected weight functions for the data points used for regression provide nonbiased inversion results both at the near- and the far-end ranges.

2.4.3 Numerical Differentiation as a Filtering Procedure

Let us consider the inversion result when the profile of the inverted optical depth $\tau(r_j, r)$ is distorted by the random noise component $\upsilon_{\text{rand}}(r)$. For simplicity, it is assumed that the noise component is just superimposed on the actual optical depth profile, that is, the distorted optical depth at the range r within the range intervals from r_j to r_k obeys the simple formula $\langle \tau(r_j, r)\rangle = \tau(r_j, r) + \upsilon_{\text{rand}}(r)$. In this case, instead of the actual extinction coefficient $\kappa(r)$ the distorted extinction coefficient $\kappa^{(\text{dif})}(r)$ is obtained through numerical differentiation, that is,

$$\kappa^{(\text{dif})}(r) = \kappa(r) + \frac{d\upsilon_{\text{rand}}}{dr} = \kappa(r) + \frac{\upsilon_{\text{rand}}(r + 0.5s) - \upsilon_{\text{rand}}(r - 0.5s)}{s}. \tag{2.79}$$

For simplicity, here and in the following, the subscript s in the range defined in Eq. (2.76) as r_s is omitted.

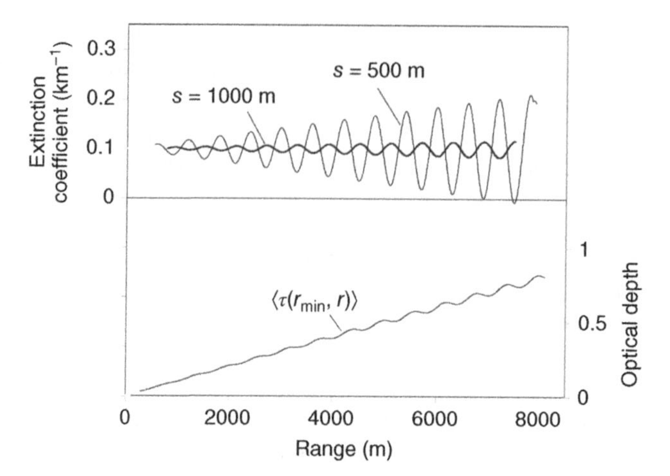

Fig. 2.11 Bottom panel: the optical depth profile $\langle\tau(r_j,\ r)\rangle$ in a synthetic homogeneous atmosphere, corrupted with noise having a spatial period of 600 m, shown in Fig. 2.12. Top panel: the profiles of the extinction coefficient retrieved from the optical depth with different range resolution s.

The influence of random noise in the optical depth on the extinction coefficient, derived with different range resolutions s, is illustrated in Figs. 2.11 and 2.13. In the bottom panel of Fig. 2.11, the distorted optical depth profile $\langle\tau(r_j,\ r)\rangle$ obtained with a virtual splitting lidar in a synthetic homogeneous atmosphere with $\kappa(r) = 0.1\ \text{km}^{-1}$ is shown. A high-frequency sinusoidal noise, $\upsilon_{\text{rand}}(r)$ having a spatial period of 600 m distorts this profile; the shape of this noise component is shown in Fig. 2.12. The

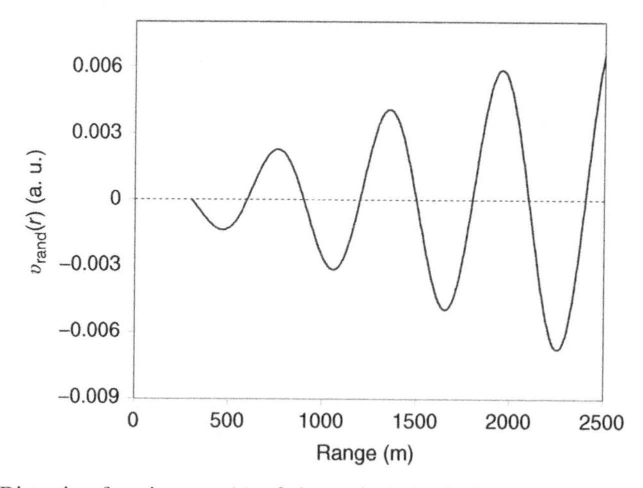

Fig. 2.12 Distortion function $\upsilon_{\text{rand}}(r)$ of the optical depth shown in the bottom panel in Fig. 2.11.

magnitude of the noise spikes increases with range, as would occur in real conditions, because of the square range correction of the noisy backscatter signal. The thin and thick solid curves in the upper panel of Fig. 2.11 show the profiles of the extinction coefficient $\kappa^{(dif)}(r)$ obtained through numerical differentiation of the optical depth with the sliding range resolutions $s = 500$ m and $s = 1000$ m, respectively. These profiles of the retrieved extinction coefficient expose the fluctuation at the same spatial frequency as in the differentiated function, $\langle\tau(r_j, r)\rangle$ in the bottom panel. As can be expected, the magnitude of fluctuation in the retrieved extinction coefficient $\kappa^{(dif)}(r)$ is less when larger s is used. Because of the absence of systematic distortions in the optical depth $\langle\tau(r_j, r)\rangle$ no systematic distortions corrupt the profile $\kappa^{(dif)}(r)$.

The same artificial atmosphere is considered in Fig. 2.13, but here the spatial period of the sinusoidal noise component $\upsilon_{rand}(r)$ is smaller than in the previous case; now it is equal to 120 m. In this case, the profile of $\kappa^{(dif)}(r)$ obtained with the same range resolutions, 300 and 1000 m, provides a significantly less scattered profile than that in Fig. 2.11.

It is known that differentiation is similar to applying a high-pass filter to the signal (Zuev et al., 1983; Beyerle and McDermid, 1999); it amplifies noise, and accordingly, exacerbates the problem of obtaining an accurate inversion result from any profiles corrupted with noise. To compensate for the increase of high-frequency noise constituents in the derived atmospheric parameter, such as the extinction coefficient, it is common to apply low-pass filtering. Selecting an extended range resolution, especially over distant ranges, reduces the high-frequency variations in the retrieved extinction-coefficient profile. However, one should keep in mind that the low-pass filtering does not distinguish between the high-frequency noise and high-frequency backscattering fluctuations, reducing or eliminating both.

The most difficult situation takes place when the low-frequency noise components in the inverted optical depth, $\langle\tau(r_j, r)\rangle$, have a spatial period that is comparable with

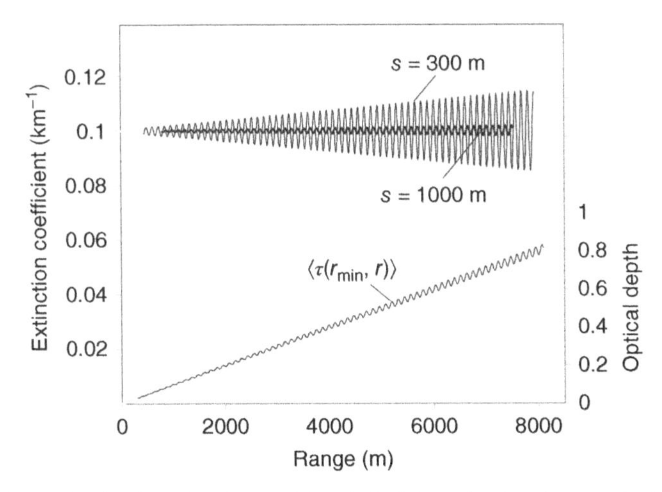

Fig. 2.13 As in Fig. 2.11, but the spatial period of the noise function $\upsilon_{rand}(r)$ is 120 m.

the operative range of the lidar. The electromagnetic low-frequency noise component superimposed on the signal may have different causes, for example, laser discharge (see Section 1.2). Generally, such noise is a slowly decaying ringing voltage induced in the electrical circuits after triggering the light pulse (Ahmad and Bulliet, 1994). The consequence of the presence of such a discharge is illustrated by Fig. 2.14. The synthetic lidar signal is obtained in the same homogeneous atmosphere with $\kappa(r) = 0.1\ \text{km}^{-1}$. However, this time, the low-frequency, sinusoidal noise component is superimposed on the recorded lidar signal. In turn, this component distorts the retrieved optical depth $\langle \tau(r_j, r) \rangle$ shown in the bottom panel of the figure as the thick solid curve. As the backscatter signal is square range corrected, the noise component magnitude increases with range. These noise fluctuations then appear in the extinction-coefficient profile $\kappa^{(\text{dif})}(r)$ derived from the corrupted optical depth $\langle \tau(r_j, r) \rangle$. Unlike previous results shown in the upper panels in Figs. 2.11 and 2.12, the value of the range resolution s being much smaller than the period of the function $\upsilon_{\text{rand}}(r)$ has insignificant influence on the shape of the derived $\kappa^{(\text{dif})}(r)$. As one can see, the shape of these profiles is almost the same as in the noise function $\upsilon_{\text{rand}}(r)$, independent of whether the range resolution 300 or 1000 m is used.

The extinction-coefficient retrieval issues related to the presence of low-frequency noise in the lidar data become especially intricate when the searched atmosphere contains aerosol layers with well-defined boundaries. In such a case, false noise bulges and concavities in the retrieved extinction coefficient may not be discernible from backscattering spatial variations, and there is no reliable way to discriminate the real and the false layering. Fig. 2.15 illustrates the results of numerical simulation under

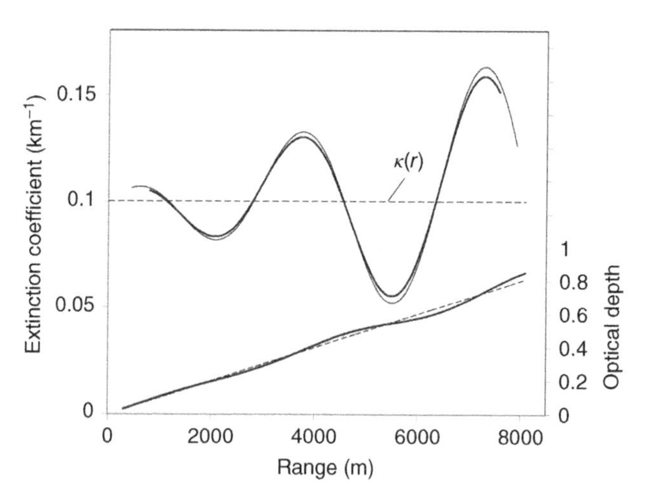

Fig. 2.14 Bottom plot: the synthetic optical depth $\tau(r_j, r)$ (dashed line), and the corresponding optical depth $\langle \tau(r_j, r) \rangle$, corrupted by the low-frequency noise (solid curve). Upper plot: the test range-independent extinction coefficient $\kappa(r)$ (dashed line) and $\kappa^{(\text{dif})}(r)$ extracted from the distorted optical depth, using the range resolution 300 m (the thin solid curve) and 1000 m (the thick solid curve).

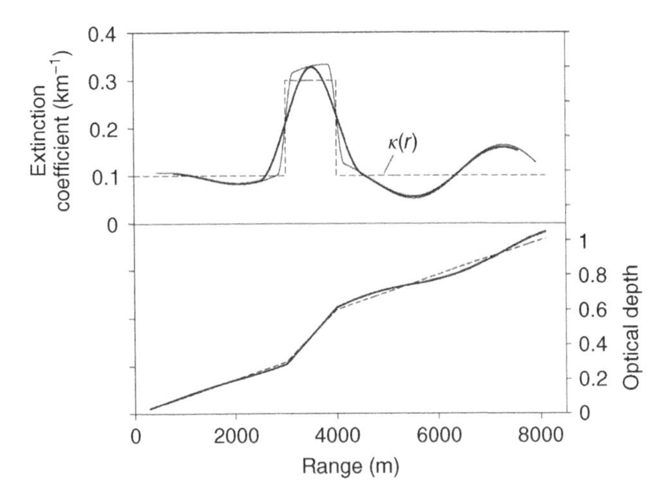

Fig. 2.15 The same profiles as in Fig. 2.14 but determined for the synthetic atmosphere where the polluted layer is present within the height interval from 3000 to 4000 m.

such conditions. Here the test extinction coefficient $\kappa(r)$ and the corresponding optical depth $\tau(r_j, r)$ are shown as dashed curves on the upper and bottom plots, respectively. The profile of the extinction coefficient contains a polluted layer within the height interval from 3000 to 4000 m. As in the previous case, the inverted optical depth is distorted by the low-frequency noise component $\upsilon_{\text{rand}}(r)$ and the extinction coefficient is derived with the range resolution 300 (the thin solid curve) and 1000 m (the thick solid curve). Now both extracted profiles $\kappa^{(\text{dif})}(r)$ show an additional erroneous polluted layer located in the area close to the maximum range.

Thus, the most significant issues in lidar searching occur when the determination of the extinction-coefficient profile is made in a multilayered atmosphere. Let us consider some specifics of the particulate extinction-coefficient profile extraction in such atmospheres using synthetic data of a virtual zenith-directed splitting lidar, corrupted by quasi-random noise. In Fig. 2.16, the black curve is the profile of the noise-corrupted optical depth $\langle\tau(h_j, h)\rangle$ and the gray curve is the product $\langle C\beta_\pi(h)\rangle$. Both the backscatter and optical depth profiles at the wavelength 355 nm have no systematic distortions. The synthetic profile of the extinction coefficient $\kappa(h)$ used in the simulation is shown as the dotted curve in Fig. 2.17; it imitates the vertical profile of the extinction coefficient in a smoke-polluted atmosphere where multiple smoke layering takes place (Kovalev et al., 2009a, 2009b, 2011). In the simulated case, three separate layers exist at the heights 1600–1800 m, 3000–3200 m, and 3300–3400 m. The vertical extinction-coefficient profiles retrieved by differentiating the optical depth $\langle\tau(h_j, h)\rangle$ are extracted using the range resolutions $s = 200$ and $s = 500$ m; these profiles are shown as the thick and thin solid curves, respectively. One can see that neither of these range resolutions allows reliable discrimination of the two adjacent layers with increased backscattering within the height altitude range 3000–3400 m. The most straightforward way to overcome this drawback would be to decrease the range resolution. However, as has been stated in many studies, in practice

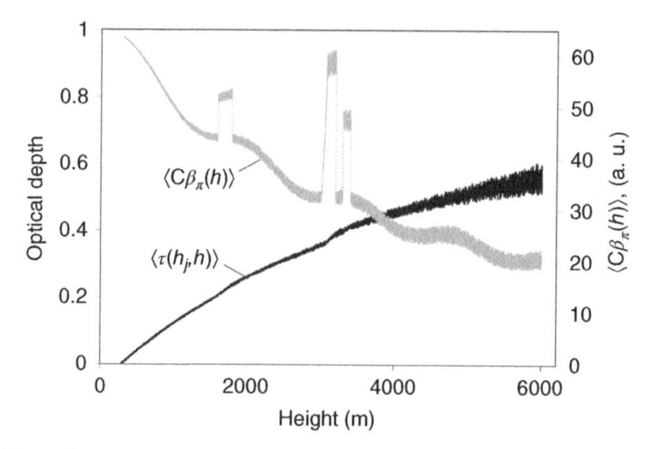

Fig. 2.16 The noise-corrupted optical depth $\langle\tau(h_j,\ h)\rangle$ and the product $\langle C\beta_\pi(h)\rangle$ used in the numerical simulation, results of which are shown in Figs. 2.17 and 2.18.

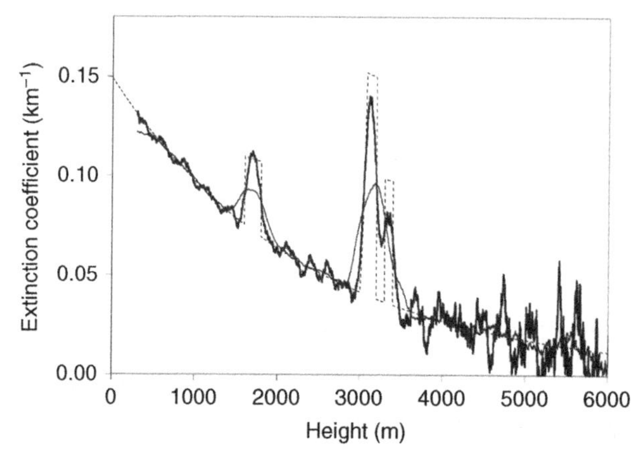

Fig. 2.17 The thick and thin solid curves are the extinction-coefficient profiles $\kappa^{(\mathrm{dif})}(r)$ obtained by differentiating the optical depth profile $\langle\tau(h_j,\ h)\rangle$ in Fig. 2.16 with the range resolutions 200 and 500 m, respectively. The dotted curve is the test profile of the extinction coefficient used for the simulation.

such a decrease does not necessarily improve the inversion result, especially over distant ranges. When an inappropriate small range resolution is used, the increased noise can completely mask the details of the atmospheric profile under investigation. Such a case is illustrated by Fig. 2.18, where the inversion result obtained using the small range resolution $s = 100$ m is shown. One cannot determine whether the sharp spikes within the height interval from 3000 to 3500 m are aerosol layers or noise spikes, and accordingly, the upper boundary of the far-end layering.

Let us summarize. When determining the extinction coefficient through numerical differentiation of a noisy optical depth, the question that always arises is how many

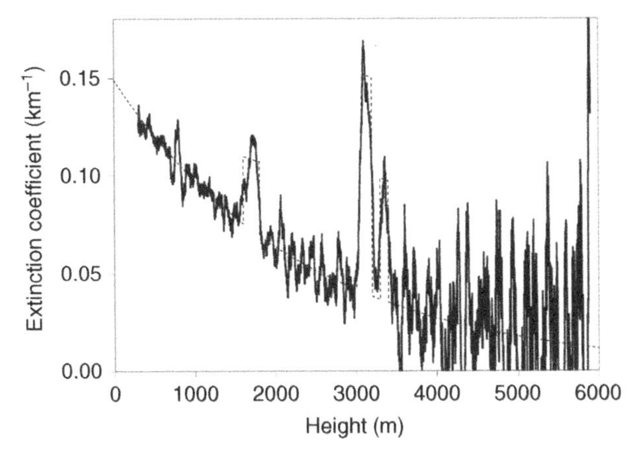

Fig. 2.18 The thick solid curve shows the extinction-coefficient profile $\kappa^{(\mathrm{dif})}(r)$ obtained by differentiating the optical depth profile $\langle\tau(h_j, h)\rangle$ in Fig. 2.16 with the range resolution s = 100 m. The dotted curve is the test extinction coefficient used for the simulation.

details in the retrieved profile of the extinction coefficient may be reliably extracted, and accordingly, what level of filtering and length of the range resolution is appropriate for the processed lidar data. The increase of the range resolution results in greater smoothing of the details in the retrieved profile while decreasing the level of noise. The researcher should establish an optimal range resolution, which would allow revealing the significant details in the extracted profile of the extinction coefficient while avoiding overfitting like that shown in Figs. 2.18. Unfortunately, there is no rigorous theoretical basis, nor are there any rigid methods and rules explaining how such an optimal range resolution should be chosen. The selection of the range resolution is always based on the researcher's intuition and preferences. The harsh reality is that no properly grounded recommendations exist that can guarantee the best retrieval results from the lidar signals. In any signal inversion made using numerical differentiation, some ambiguity remains over whether all the important details of the extinction-coefficient profile are obtained, and simultaneously, whether the local variations in the extracted profile are actual or originated by noise.

2.5 CORRECTION AND EXTRAPOLATION TECHNIQUES FOR THE OPTICAL DEPTH PROFILE DERIVED FROM THE SPLITTING LIDAR DATA

2.5.1 Removal of Erroneous Bulges and Concavities in the Optical Depth Profile: Merits and Shortcomings

As shown in the previous section, in the optical depth profiles derived from splitting lidar data, local distortions such as bulges and concavities generally take place.

These distortions appear as false fluctuations in inverted extinction-coefficient profiles $\kappa^{(dif)}(r)$. Even negative values in the derived profile extinction coefficient can be obtained if inappropriate smoothing of the optical depth profile is used, generally when a small resolution range s is chosen.

When processing the lidar data using numerical differentiation, the common rule is that a higher degree of filtering has to be made over the distant ranges, where the poorest signal-to-noise ratio is commonly obtained. Such filtering is achieved by using a variable range resolution, which increases with range (Godin et al., 1999). However, the improvement achieved by greater signal filtering over distant ranges is generally rather modest. The random noise component $\upsilon_{rand}(r)$ includes not only high-frequency but also low-frequency components, such as those considered in the previous section. These low-frequency noise components cannot be effectively filtered by increasing the range resolution s. In addition to the random noise component $\upsilon_{rand}(r)$, a systematic distortion component $\upsilon_{sys}(r)$ can also be present in the optical depth. The latter may be caused by a systematic distortion in the inverted signal, for example, originating from the nonzero offset ΔB in the backscatter signals. Such distortion components in the optical depth profile will cause corresponding distortions in the derived extinction coefficient $\kappa^{(dif)}(r)$, which dramatically increase with range. As many distortions are hidden, no reliable conclusions about the accuracy of the derived optical depth and extinction-coefficient profiles can be made.

Meanwhile, the commonly used smoothing technique is based on purely statistical principles and does not take into consideration the distortions originating from such distortion components as $\upsilon_{sys}(r)$ and the low-frequency $\upsilon_{rand}(r)$. In many cases, it may be helpful to perform more sophisticated smoothing, not based on statistics, which would reduce obviously erroneous variations in the inverted optical depth. Simple numerical simulations show that when processing lidar data, better inversion results can be achieved if smoothing made before extracting the extinction coefficient from the optical depth combines two procedures: smoothing and "shaping" of the optical depth profile. The last operation, which we refer to as *shaping*, is not based on statistics, or even on rigid mathematics; it is based on ordinary common sense, which permits rejection of the obviously nonphysical results.

This statement needs some clarification. No smoothing reliably provides full elimination of distortions in extended areas where optical depth erroneously decreases with height. To achieve an appropriate correction of such obviously erroneous data points, the above shaping procedure may be performed. The shaping procedure corrects the data points in the concavities where the inverted optical depth has a nonphysical decrease with range. Generally, such concavities in the optical depth profile, which produce erroneous negative values of the extracted extinction coefficient, follow after bumps, which in turn, produce an overestimation of the local extinction coefficient. The shaping procedure eliminates both the concavities and the bumps in the profile of the optical depth, before it is differentiated.

The processing technique, which combines the smoothing and shaping procedures, is actually one of many possible procedures based on common sense, which can be implemented in the methodology of lidar profiling of the atmosphere. As pointed out in Section 1.1, neither the optical depth, nor the extinction-coefficient profile is measured with lidar. Both profiles are simulated on the basis of previous lidar observations, in particular, on specifics of the shape of the recorded lidar signal. Accordingly, the shaping procedure in profiling of the atmosphere can be considered an additional element of the *a posteriori* simulation.

To clarify the essence of the above procedures, let us consider the combined smoothing/shaping procedure for the vertical optical depth shown in the right panel of Fig. 2.19. This noise-corrupted optical depth profile is obtained from signals of a virtual splitting lidar measured in a synthetic multilayer atmosphere. The total (molecular and particulate) extinction coefficient $\kappa(h)$ used in the numerical simulation is shown in the left panel of the figure.

To reduce excessive noise in the original profile $\langle\tau(0, h)\rangle$, a running average of the noise-corrupted optical depth is initially calculated. This procedure yields a preliminary smoothed profile $\tau_{sm}(0, h)$, shown in Fig. 2.20 as the thick solid curve; the dotted curve shows the initial nonsmoothed optical depth profile, the same as that in the right panel of Fig. 2.19. In spite of a relatively large smoothing interval (150 m), smoothing does not eliminate the obvious optical depth distortions in the areas where the optical depth erroneously decreases with height. To appropriately correct the erroneous data points, a shaping procedure has to be performed, which would correct all data points where the smoothed optical depth $\tau_{sm}(0, h)$ has a nonphysical decrease with height.

To determine the shaped profile of the optical depth, $\tau_{sh}(0, h)$, the range-dependent upper and lower limits of the function $\tau_{sm}(0, h)$ should be calculated. The upper limit function $\tau_{up}(0, h)$ is defined as

$$\tau_{up}(0, h) = \max[\tau_{sm}(0, h_{min}); \tau_{sm}(0, h_{min} + \Delta h_d); \tau_{sm}(0, h_{min} + 2\Delta h_d); \ldots ; \tau_{sm}(0, h)], \tag{2.80}$$

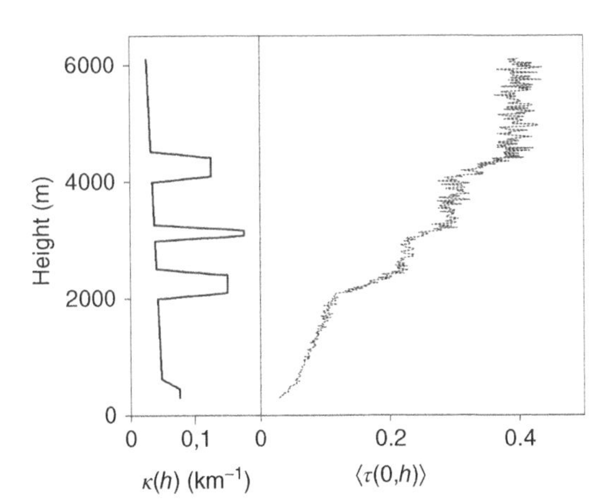

Fig. 2.19 The left panel: The synthetic profile of the total extinction coefficient versus height used for the numerical simulation. The right panel: The noise-corrupted total optical depth $\langle\tau(0, h)\rangle$ versus height, retrieved from the data of a virtual splitting lidar (Kovalev et al., 2006).

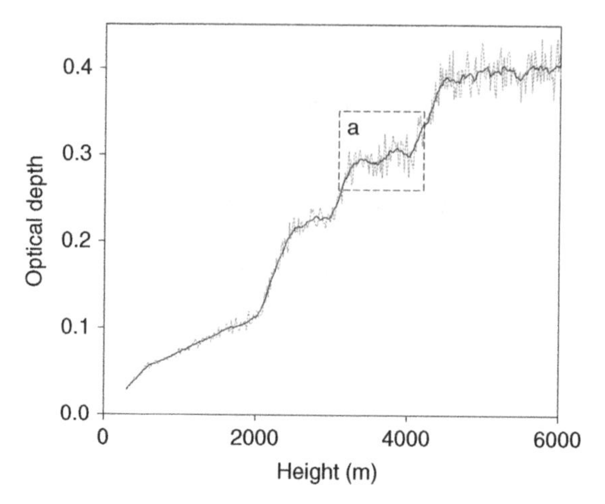

Fig. 2.20 The dotted curve is the total optical depth profile $\langle \tau(0, h) \rangle$, the same as in the right panel of Fig. 2.19. The thick solid curve is the smoothed profile $\tau_{sm}(0, h)$ (adapted from Kovalev et al., 2006).

and the lower limit function $\tau_{low}(0, h)$ is

$$\tau_{low}(0, h) = \min[\tau_{sm}(0, h); \tau_{sm}(0, h + \Delta h_d); \tau_{sm}(0, h + 2\Delta h_d); \ldots ; \tau_{sm}(0, h_{max})], \tag{2.81}$$

where Δh_d is the sampling resolution for the inverted profile of the optical depth, h_{min} and h_{max} are the minimal and maximal heights, established before the shaping procedure is done (see Section 2.5.2). The shaped optical depth is determined as a mean of these two marginal profiles, that is,

$$\tau_{sh}(0, r) = 0.5[\tau_{up}(0, h) + \tau_{low}(0, h)]. \tag{2.82}$$

The initial optical depth $\langle \tau(0, h) \rangle$, the smoothed profile $\tau_{sm}(0, h)$, the marginal profiles $\tau_{sm}(0, h)$ and $\tau_{sm}(0, h)$, and the shaped profile $\tau_{sh}(0, h)$ for the above case are shown in Fig. 2.21. The initial noisy optical depth is shown as the dotted curve, the averaged function, $\tau_{sm}(0, h)$, each point of which is the average of 11 data points, is shown as the curve with the solid data points. For better visualization, only the restricted height interval from 3100 to 4300 m is shown in the figure. Note that the smoothed optical depth $\tau_{sm}(0, h)$ still has local zones where it erroneously decreases with height (the height intervals ~3400–3600 m and ~3900–4000 m). The profiles of the upper and lower limit functions, $\tau_{up}(0, h)$ and $\tau_{low}(0, h)$, found with Eqs. (2.80) and (2.81), are shown as the dashed and dashed-dotted curves, respectively. Note that outside the bad data zones, that is, for the heights lower than 3300 m and higher than 4000 m, the shaping procedure does not change the optical depth profile, here $\tau_{up}(0, h) = \tau_{low}(0, h) = \tau_{sm}(0, h) = \tau_{sh}(0, r)$. In other words, the shaping procedure does not change the data points in the area where the optical depth monotonically increases with height; the only data points corrected are those located within the areas

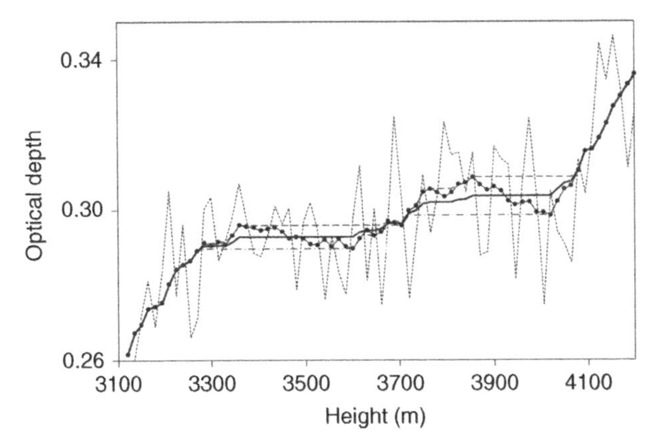

Fig. 2.21 The optical depth profiles within the restricted area *a* in Fig. 2.20. The initial and the smoothed optical depths are shown as the dotted curve and the curve with the solid data points; the limit functions $\tau_{up}(0, h)$ and $\tau_{low}(0, h)$ are shown as dashed and dashed-dotted curves, respectively, and the shaped optical depth $\tau_{sh}(0, h)$ is shown as the thick solid curve. (Adapted from Kovalev et al., 2006).

where the optical depth $\tau_{sm}(0, h)$ decreases with height and in short zones adjacent to these "bad data" areas.

The shaping procedure substantially improves the next inversion results. First, it significantly decreases the false fluctuations in the profile of the extinction coefficient derived with numerical differentiation of the shaped optical depth. Second, the numerical differentiation of the shaped optical depth does not yield erroneous negative values in the retrieved extinction-coefficient profile. This is illustrated in Figs. 2.22 (a) and (b), where the profiles of the total extinction coefficient $\kappa^{(dif)}(h)$, obtained with the numerical differentiation of the differently smoothed optical depths are shown. The profile of $\kappa^{(dif)}(h)$ obtained from the initial noisy optical depth $\langle\tau(0, h)\rangle$ is shown as the dotted curve; the profile obtained from the smoothed optical depth $\tau_{sm}(0, h)$ is shown as the dashed curve; the extinction-coefficient profile obtained from the shaped optical depth $\tau_{sh}(0, h)$ is shown as the bold curve. All the extinction-coefficient profiles were determined using the range resolution $s = 300$ m. Comparing these profiles with the initial test profile of the extinction coefficient, shown in the figure as thin solid curve, one can conclude that the use of the shaping procedure yields minimal discrepancy between the original and retrieved profiles; the procedure significantly reduces the level of random noise and prevents obtaining negative values in the retrieved extinction coefficient.

When the optical depth is corrupted with excessive noise, a two or even three-step combined procedure of smoothing and shaping may be needed. In some cases, additional averaging of the shaped function $\tau_{sh}(0, h)$ may also be helpful, as it may reduce possible inversion errors in the nearest vicinities of "bad" zones.

Let us summarize the merits and shortcomings of replacing the smoothed profile of the optical depth by the shaped profile $\tau_{sh}(0, r)$. The main merit of shaped optical

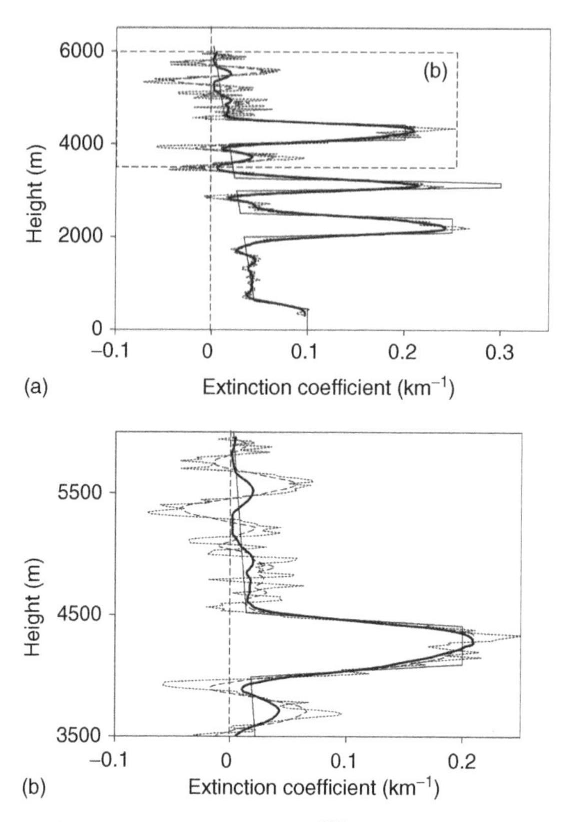

Fig. 2.22 (a) The extinction-coefficient profiles $\kappa^{(\mathrm{dif})}(h)$, retrieved from the initial noisy optical depth profile $\langle\tau(0, h)\rangle$, smoothed $\tau_{\mathrm{sm}}(0, h)$, and the shaped $\tau_{\mathrm{sh}}(0, h)$ (the dotted, dashed, and bold curves, respectively). The thin solid curve is the test profile $\kappa(h)$. (b) The same as in (a) but for better visualization shown in the restricted height interval from 3500 to 6000 m. (Adapted from Kovalev et al., 2006).

depth is the decrease in noise fluctuations in the extracted extinction-coefficient profile in areas of erroneous bumps and concavities. Unlike standard smoothing of the variable function, which is a purely mathematical procedure, shaping should be considered as a procedure of building a simulated profile, in which obviously erroneous fluctuations are removed. The additional advantage of the shaping procedure is that unlike smoothing, it does not change the optical depth profile within the ranges where it has a positive or zero slope. In other words, the shaping procedure removes erroneous negative data points in the extinction-coefficient profile and does not change the extinction coefficient outside "bad zones." The shortcomings of the above methodology are as follows: (i) the shaping procedure does not allow extracting the extinction coefficient within the extended areas of the erroneous bulges and concavities. In these areas, the shaped optical depth does not increase with range, it remains constant. In Fig. 2.21, such an invariable shaped optical depth is obtained over the height intervals 3360–3600 m and 3860–4040 m. Direct numerical differentiation of the optical depth profile in these areas yields extinction coefficients $\kappa^{(\mathrm{dif})}(h) = 0$. Therefore, direct numerical differentiation of the optical depth cannot be used in such areas. However, the alternative methods of indirect differentiation considered in Sections 2.6–2.8 can yield sensible inversion results even within such "bad zones." (ii) The shaping methodology requires mandatory determination of maximum range h_{max}, up to which replacement of the smoothed optical depth profile by the shaped profile is reasonable. In some cases, determining such an optimal maximum height may be a significant challenge. This issue is discussed in the next section.

2.5.2 Implementation of Constraints for the Maximum Range of the Shaped Optical Depth Profile

Far-end lidar signals, being square range corrected, are generally corrupted with extensive noise, which may completely mask important local specifics in the backscatter signal shape. Therefore, before signal inversion is made, a sensible maximum range needs to be established. Generally, this is achieved by selecting some minimally acceptable signal-to-noise ratio in the lidar data. Unfortunately, no standard method for selecting such a fixed minimal signal-to-noise ratio exists except that which is based on pure statistical principles. However, these principles are not valid when determining the maximum range and height for the shaped optical depth.

Let us consider the general principle for selecting h_{max} for the shaped optical depth profile using some typical profile of the optical depth obtained from actual lidar data. In particular, we will determine h_{max} for the smoothed profile of the optical depth, obtained while searching the smoky layers that originated from the I-90 Fire in Montana in August 2005. In Fig. 2.23, this profile, smoothed with sliding height

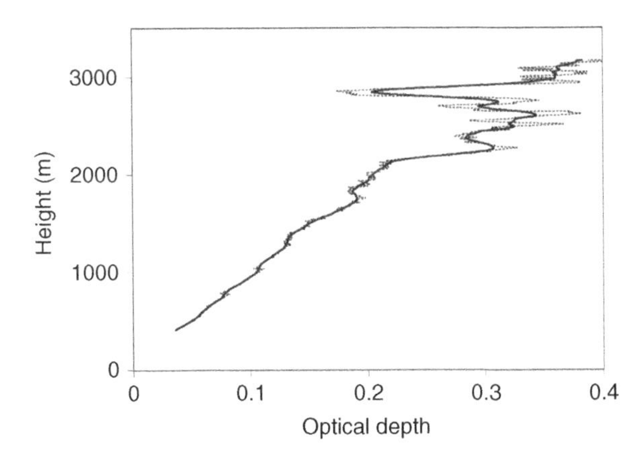

Fig. 2.23 The dotted curve is the initial nonsmoothed optical depth profile, obtained with the FSL lidar at the wavelength 355 nm. The corresponding smoothed profile $\tau_{\mathrm{sm}}(0, h)$ is shown as the bold curve. (Adapted from Kovalev et al., 2007a).

interval 100 m is shown as the bold solid curve. One can see large and obviously erroneous fluctuations on the smoothed profile at heights greater than ~2200 m. When the extinction coefficient needs to be retrieved from such an optical depth profile, the question always emerges on how to select the optimum maximal height h_{max} to exclude, or at least reduce, the large errors in the derived profile of the extinction coefficient due to the presence of the bulges and concavities. The simplest way of making such a selection is to vary the maximum height, reducing it until large errors in the retrieved extinction coefficient disappear. In our case, such an optimal height should be found by the analysis of how close the shaped profile, $\tau_{sh}(0, h)$ is to the smoothed profile $\tau_{sm}(0, h)$ when different h_{max} are chosen.

To clarify this process, let us initially select the maximum height $h_{max} = 3000$ m and determine the corresponding shaped profile $\tau_{sh}(0, h)$ from the smoothed profile $\tau_{sm}(0, h)$ shown in Fig. 2.23. To achieve this purpose, the marginal profiles $\tau_{low}(0, h)$ and $\tau_{up}(0, h)$ should be found [Eqs. (2.80) and (2.81)]. In Fig. 2.24, these profiles and the corresponding shaped profile $\tau_{sh}(0, h)$ are shown. The profiles $\tau_{low}(0, h)$ and $\tau_{up}(0, h)$ are shown as the vertical dashed and dotted-dashed lines; they can be visualized only over the height intervals where they differ from $\tau_{sm}(0, h)$. The shaped profile is shown as the bold solid curve. One can see that the smoothed and shaped profiles significantly diverge within the altitude range from 2000 to 3000 m. This effect occurs because of the large and obviously erroneous spike in the profile $\tau_{sm}(0, h)$ over the heights 2800–2900 m, in the area marked as A in the figure. Obviously, to avoid such a large difference between $\tau_{sh}(0, h)$ and $\tau_{sm}(0, h)$, the maximum height should be selected somewhere below the area *A*. Indeed, the difference between these two profiles becomes less if the maximum height h_{max} is decreased down to 2800 m (Fig 2.25). As before, no noticeable changes between $\tau_{sh}(0, h)$ and $\tau_{sm}(0, h)$ occur over the heights up to, approximately, 2200 m, but then

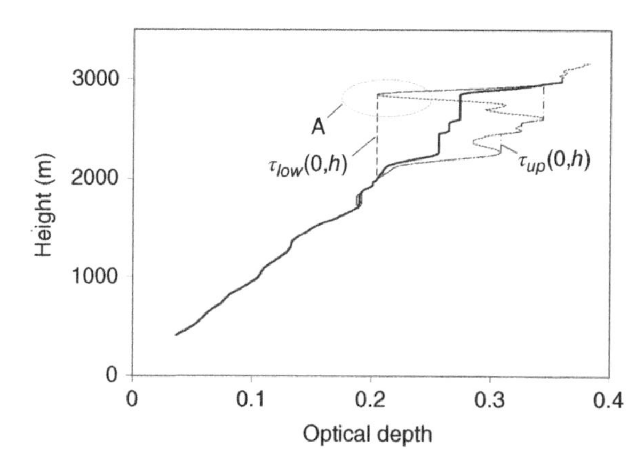

Fig. 2.24 The thin dotted curve is the smoothed optical depth profile $\tau_{sm}(0, h)$, the same as that in Fig. 2.23, and the bold solid curve is the corresponding shaped profile $\tau_{sh}(0, h)$, when the selected $h_{max} = 3000$ m. The vertical dashed and dotted-dashed lines show the marginal profiles $\tau_{low}(0, h)$ and $\tau_{up}(0, h)$ in the local zones where they differ from $\tau_{sm}(0, h)$.

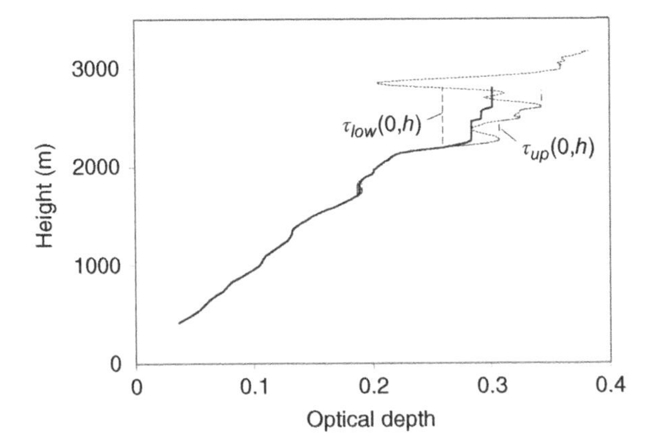

Fig. 2.25 Same as in Fig. 2.24 but the shaped profile $\tau_{sh}(0, h)$ (the bold solid curve) is obtained when the selected h_{max} = 2800 m. (Adapted from Kovalev et al., 2007a).

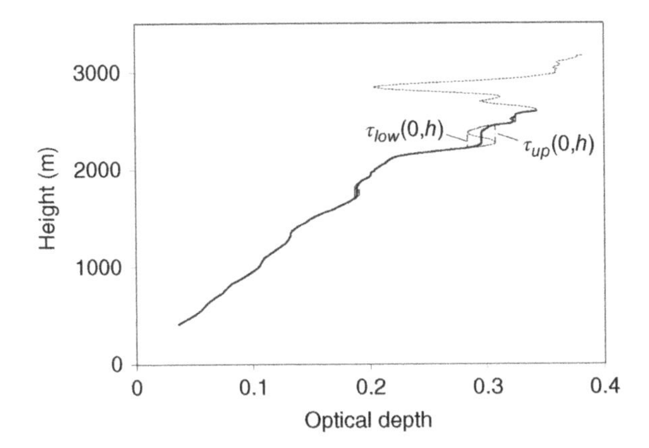

Fig. 2.26 Same as in Fig. 2.24 but calculated with h_{max} = 2600 m.

the shaped profile still significantly diverges from the smoothed one. In Fig. 2.26, the same profiles are shown but obtained when the selected maximum height is decreased to 2600 m. As one can see, now the shaped profile $\tau_{sh}(0, h)$ is close to the smoothed profile $\tau_{sm}(0, h)$ and no significant difference between these profiles take place within the whole altitude range from 400 to 2600 m.

In Fig. 2.27, the shaped optical depths $\tau_{sh}(0, h)$ obtained for three different heights h_{max} are shown. One can see that the profile obtained with h_{max} = 2600 m is the closest to the smoothed curve $\tau_{sm}(0, h)$ and accordingly, it can be considered as an optimal profile. The question that arises is what numerical criteria can be used for selecting the maximal height to obtain the optimal shape of $\tau_{sh}(0, h)$ which is closest to $\tau_{sm}(0, h)$. The simplest method for determining the optimal h_{max} is based on the observation that increased difference between the marginal profiles $\tau_{up}(0, h)$ and

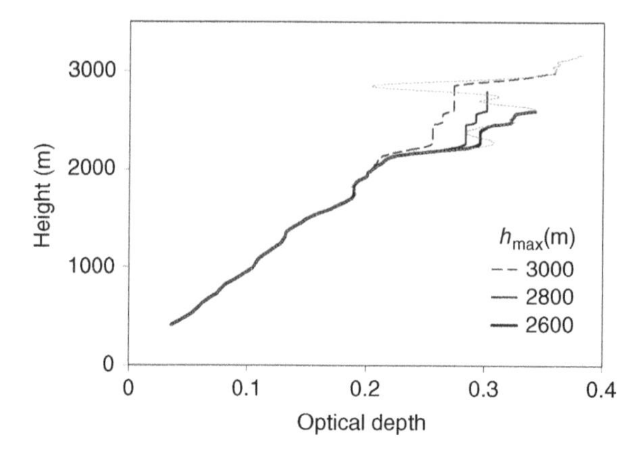

Fig. 2.27 The shaped optical depth profiles obtained with the maximal heights 2600, 2800, and 3000 m (the thick solid, the thin solid, and the dashed curves, respectively). The initial optical depth profile used for determining these shaped profiles is shown as the thin dotted curve.

$\tau_{low}(0, h)$ takes place over heights with large jumps in the profile $\tau_{sm}(0, h)$. This feature allows implementing a simple criterion for estimating the optimal maximum height. The criterion is found using the variable h_{max} in the formula

$$\epsilon(h_{max}) = \frac{\displaystyle\int_{h_{min}}^{h_{max}} [\tau_{up}(0, h) - \tau_{low}(0, h)]dh}{\displaystyle 2\int_{h_{min}}^{h_{max}} \tau_{sh}(0, h)dh}. \tag{2.83}$$

With this criteria, the determination of the optimum maximal height h_{max} is reduced to selecting the maximum height that produces a small value of $\epsilon(h_{max})$. Experimental data show that in most cases, the function $\epsilon(h_{max})$ has a tendency to increase with increase in the selected h_{max}, so smaller values of $\epsilon(h_{max})$ are obtained when h_{max} is not too large. However, there is also another common requirement that has to be met: the maximum height searched by lidar should be as large as possible. In practice, there might be some issue in satisfying the requirements of a small value of $\epsilon(h_{max})$ and a maximal possible lidar-searching height.

In Fig. 2.28, the dependence of $\epsilon(h_{max})$ on h_{max} for the above optical depth profile is shown. When $h_{max} = 2600$ m, the corresponding $\epsilon(h_{max}) = 0.027$ is optimum. When h_{max} becomes larger, a sharp increase of $\epsilon(h_{max})$ occurs. At the height $h_{max} = 2900$ m, the criterion value increases up to $\epsilon(h_{max}) = 0.132$. Analyzing our experimental data, we concluded that, generally, no accurate extinction-coefficient profile can be extracted from the optical depth if $\epsilon(h_{max})$ exceeds $\sim$0.05. In the above case, the height that provides $\epsilon(h_{max}) = 0.05$ is equal to 2700 m. Taking into account a sharp increase of $\epsilon(h_{max})$ close to this height, it is probably better to select $h_{max} = 2600$ m.

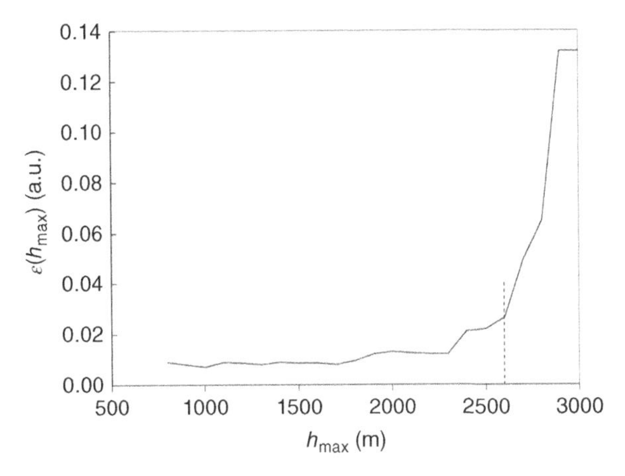

Fig. 2.28 Dependence of $\epsilon(h_{max})$ on selected h_{max} for the shaped profile $\tau_{sh}(0, h)$ when different h_{max} are selected. The optimal h_{max} is shown by the vertical dotted line. (Adapted from Kovalev et al., 2007a).

2.5.3 Modeling the Optical Parameters of the Atmosphere in the Near Zone of Lidar Searching

The operative zone of lidar is always restricted from both sides. In addition to the restricted maximum range, the minimum range r_{min} and the corresponding minimal sounding height h_{min} are also restricted. Below this point, which generally concurs with the minimal range of the lidar complete overlap zone r_0, lidar profiling of the atmosphere meets with some difficulties. In principle, one can determine the profile of the overlap function $q(r)$, in the incomplete overlap zone and then use this calibrated $q(r)$ for backscatter signal inversion. However, such a method has a number of issues (see Section 1.8). Therefore, the extrapolation of the atmospheric profile of interest down to the zone $(0, r_0)$ using lidar data obtained in an adjacent neighboring area is the simplest way. Unlike the procedures related to selection of maximum ranges, considered in the previous section, profiling down requires researchers to focus on the lidar data close to the near end. Such an extrapolation procedure requires the inversion of the lidar signal, at least, in the area closest to the range r_0; that is, the profile of the vertical optical depth over the heights adjacent to h_0, should be initially determined. For better results, the optical depth profile used for the extrapolation should be properly smoothed or, if necessary, shaped before its extrapolation down is made. If possible, available auxiliary data from ground-based instrumentation, such as nephelometer, can also be used.

Let initial profiling of the atmosphere with a ground-based zenith-directed lidar results in the optical depth profile from the minimal height of the complete overlap zone h_0 to the maximal height h_{max}. To perform the extrapolation of the profile down to ground level, some restricted height interval Δh starting at the point h_0 should be selected (Fig. 2.29). The optical depth within the interval from h_0 to $h_0 + \Delta h$ is used as the reference profile for extrapolating this profile down to ground level. Thus, before

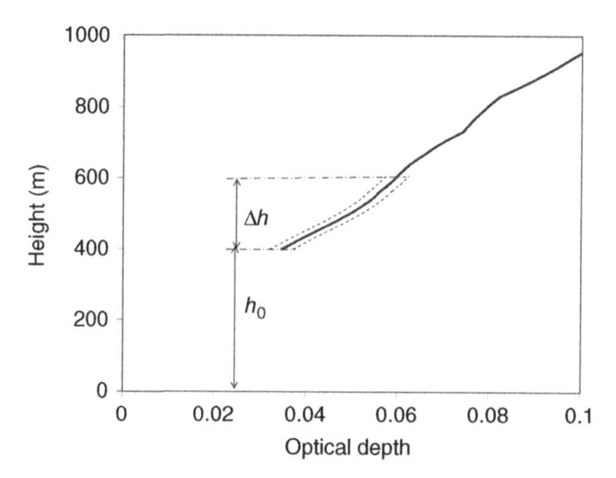

Fig. 2.29 The thick solid curve is the optical depth profile derived from the lidar data in the vicinity of the minimum operative height h_0; Δh is the altitude interval within which this optical depth is used as a reference profile for extrapolating it downward. This part of the profile is marked by two dotted curves.

performing such an extrapolation, the researcher should select the height interval Δh, within which the profile of the optical depth will be used as the reference profile. Then the model shape of optical depth, which presumably is appropriate for interpolating of the optical depth down to ground level is chosen.

Two simple models, which can be used for extrapolating optical depth down to ground, are discussed in the following. The simplest model, which may be sensible in conditions of a relatively polluted atmosphere, is based on the assumption that the total extinction coefficient in the lower atmospheric layer near ground is constant (Kinjo et al., 1999). The alternative models assume a monotonic decrease of the extinction coefficient with height.

Let us consider the first model, that is, assuming that the total extinction coefficient from ground level, ($h = 0$), to the height $h = h_0 + \Delta h$ is constant. Under such a condition, the linear fit of the optical depth profile within the interval from h_0 to $h_0 + \Delta h$ should be found and extrapolated down to $h = 0$. The corresponding extinction coefficient κ_0 within the height interval from ground level to $h = h_0 + \Delta h$ is then found as the slope of the linear fit. This variant is illustrated in Fig. 2.30, where the thick curve shows the lower part of the vertical optical depth extracted from the signals of the FSL scanning lidar at the wavelength 355 nm; these signals were measured in the vicinity of a wildfire in Idaho. The minimum height at which the data points of the optical depth derived from the lidar signals were available was $h_0 = 450$ m.

An obvious question arises - how is the interval Δh for the extrapolation selected? Numerous analyses of our experimental data showed that for the FSL lidar, the optimal interval Δh should be selected from approximately, 200 to 500 m, depending on the height h_0, the optical depth profile derived in the vicinity of this height, and the

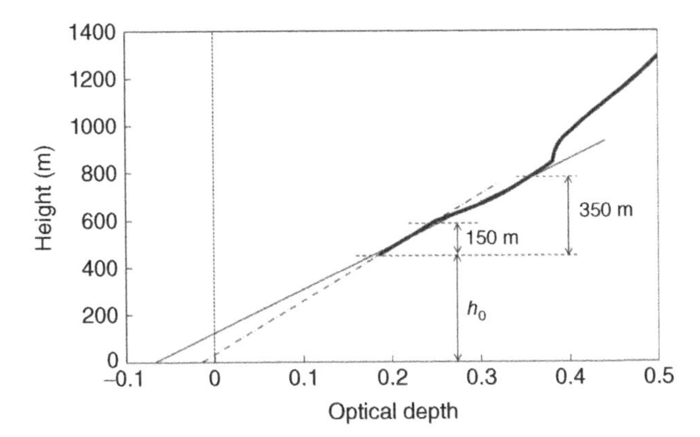

Fig. 2.30 The thick curve shows the lower part of the optical depth profile determined from the lidar signals measured in the vicinity of a wildfire in Idaho. The dot-dashed and solid slope lines are the linear fits of this profile over the altitude intervals Δh equal to 150 and 350 m, respectively.

model profile of the extinction coefficient. In the case under consideration, two altitude intervals Δh equal to 150 and 350 m, are selected, and two linear fits, shown in Fig. 2.30 as the dot-dashed and solid slope lines, are determined. The extinction coefficient found as the slope of the linear fit within the first interval yields $\kappa_0 = 0.54\ \text{km}^{-1}$, whereas the linear fit for the second interval yields $\kappa_0 = 0.44\ \text{km}^{-1}$. To establish which value of κ_0 should be considered as the best estimate, one can use the simple criteria: the extrapolated linear fit should cross the origin (the point of intersection of coordinate axes); or at least, the intersection point of the linear fit with the horizontal axis should be close to the zero point. In the case under consideration, the interception point for the first selected interval, $\Delta h = 150$ m, is much closer to the zero point, than that for the second one, selected with the interval $\Delta h = 350$ m. Therefore, for the selected model where $\kappa_0 = \text{const.}$, the value of $\kappa_0 = 0.44\ \text{km}^{-1}$ looks more sensible than $\kappa_0 = 0.54\ \text{km}^{-1}$. However, one should keep in mind, that this operation should be considered as pure modeling; it provides only a rough estimate of the extinction-coefficient profile in the lower layer of the atmosphere where lidar data either are not available or cannot be properly inverted.

Under the conditions of a clear atmosphere, the assumption of a constant extinction coefficient within the extended atmospheric layer near the ground may not be appropriate. In such cases, neither selected interval of Δh may provide the intersection point of the linear fit with the horizontal axis close to zero. The alternative model, which may be sensible in such a case, is based on the assumption of a monotonic decrease of the total (molecular and particulate) extinction coefficient $\kappa(h)$ with height. For the clear atmosphere, this assumption is more realistic than the assumption of a height-independent extinction coefficient. In such an atmosphere, a simple model can be used in which the extinction coefficient within the height interval from ground level to $h = h_0 + \Delta h$ obeys the formula

$$\kappa_{mod}(h) = \frac{\kappa_{gr}}{1 + \kappa_{gr} h}, \tag{2.84}$$

where κ_{gr} is the total extinction coefficient at ground level. The corresponding model optical depth is determined as

$$\tau_{mod}(0,\ h) = \int_0^h \kappa_{mod}(h')dh' = \ln[1 + \kappa_{gr} h]. \tag{2.85}$$

In Fig. 2.31, the profiles of the model extinction coefficient $\kappa_{mod}(h)$ are shown calculated with Eq. (2.84) for the different extinction coefficients κ_{gr} at ground level. The corresponding profiles of the vertical optical depth $\tau_{mod}(0,\ h)$ are shown in Fig. 2.32.

Let us apply this variant to the same optical depth profile shown in Fig. 2.30. The analysis shows that there might be at least two different zones where the best matches between the derived and the model profiles of the optical depth can be obtained; the first is located between the heights 450 and 600 m, and the second between 900 and 1400 m (Fig. 2.33). The model optical depth determined with Eq. (2.85) within the lower interval matches best with the profile calculated with $\kappa_{gr} = 0.46$ km^{-1} (Curve 1); the second model profile, shown as Curve 2, is obtained with $\kappa_{gr} = 0.5$ km^{-1}. Note that in this variant, the extinction coefficient at ground level κ_{gr} is not taken *a priori*; it is determined by matching the model optical depth $\tau_{mod}(0,\ h)$ with the optical depth $\tau_{sm}(0,\ h)$ within the selected height interval Δh.

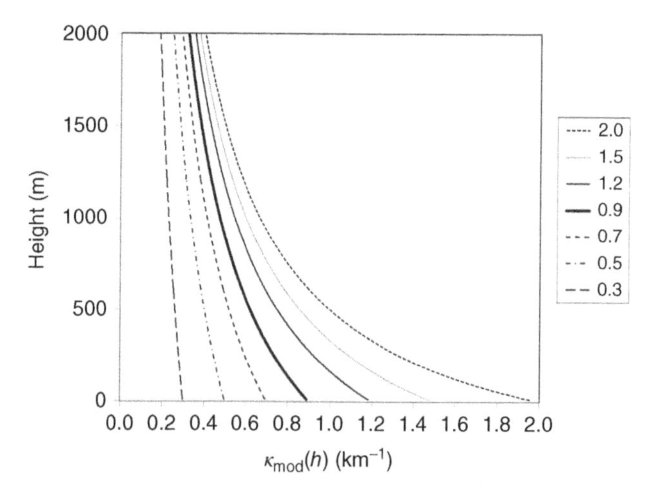

Fig. 2.31 Model profiles of the extinction coefficient $\kappa_{mod}(h)$ calculated for different extinction coefficients κ_{gr} at ground level. The values of the extinction coefficient at ground level are shown in the legend.

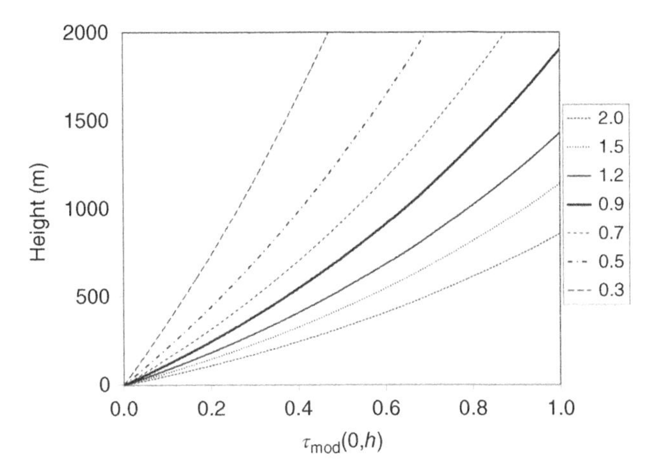

Fig. 2.32 Profiles of the vertical optical depth $\tau_{mod}(0,\ h)$, corresponding to the extinction-coefficient profiles in Fig. 2.31. The values of the extinction coefficient at ground level are shown in the legend.

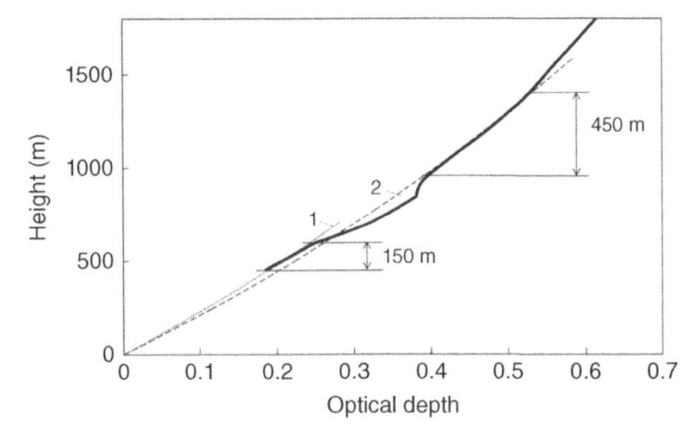

Fig. 2.33 The thick curve is the smoothed vertical optical depth $\tau_{sm}(0,\ h)$, the same as that in Fig. 2.30. Curves 1 and 2 show the profiles $\tau_{mod}(0,\ h)$ obtained when using different height intervals Δh.

The comparison of the two models used for experimental data obtained in the vicinity of the wildfire in Idaho shows that, both models provide comparable inversion results. Recall that when using the assumption of a range-independent extinction coefficient near ground level, we obtained $\kappa_0 = 0.44\ \text{km}^{-1}$ and $\kappa_0 = 0.54\ \text{km}^{-1}$. In other words, for the case under consideration, when a smoke-polluted atmosphere was searched, both models yield the extinction coefficient within the range from $0.44\ \text{km}^{-1}$ to $0.54\ \text{km}^{-1}$.

When profiling the extinction coefficient in the areas outside the lidar operative range, interpolation is a better method than extrapolation as it yields profiles that are more trustworthy. For ground-based lidar, the interpolation can be performed using supplementary *in situ* data obtained at ground level. Let us consider the method of modeling the atmospheric profiles in the lower atmosphere using the same method as above but taking into account the available extinction coefficient at ground level. The principle of such an interpolation is based on simple formulas. Assuming that the absorption at the lidar wavelength can be ignored, the reference value of the extinction coefficient at ground level can be determined as a sum of the particulate extinction coefficient $\kappa_{p,\,\mathrm{neph}}$, determined with the nephelometer and the corresponding molecular scattering coefficient at ground level; that is, the total extinction coefficient at ground level κ_{gr} is found as

$$\kappa_{\mathrm{gr}} = \kappa_{p,\,\mathrm{neph}} + \kappa_{m,\,\mathrm{gr}}, \tag{2.86}$$

where $\kappa_{m,\,\mathrm{gr}}$ is the molecular extinction coefficient at ground level.

Now one can apply the model in which the extinction coefficient within the height interval from ground level to $h = h_0 + \Delta h$ obeys the formula (Kovalev et al., 2007a)

$$k'_{\mathrm{mod}}(h) = k_{\mathrm{gr}} \exp(-\eta h), \tag{2.87}$$

where η is a nonzero constant with dimensions of inverse length; it can be either positive or negative. The corresponding profile of the model optical depth obeys the formula

$$\tau'_{\mathrm{mod}}(0,\, h) = \frac{k_{\mathrm{gr}}}{\eta}[1 - \exp(-\eta h)]. \tag{2.88}$$

The dependencies in Eqs. (2.87) and (2.88) provide a wide variety of possible model profiles. Using positive or negative values of η, one can simulate the atmospheres where the extinction coefficient either decreases or increases with height. The examples of the possible shapes of the model profiles $\tau'_{\mathrm{mod}}(0,\, h)$ and $\kappa'_{\mathrm{mod}}(h)$ are shown in Figs. 2.34 and 2.35, respectively; the values of η are shown in the legend. To use the above combined extrapolation/interpolation simulation, one should, as in the previous case, initially select the height interval from h_0 to $(h_0 + \Delta h)$ within which the profile $\tau'_{\mathrm{mod}}(0,\, h)$ should match the profile $\tau_{\mathrm{sm}}(0,\, h)$ derived from the data of the zenith-directed lidar. The matching operation is completed when the constant η is found that provides the best match between these profiles. The best match can be achieved through finding the minimum of the function

$$\xi = \sum_{h_0}^{h_0+\Delta h} [\tau'_{mod}(0,\, h) - \tau'_{\mathrm{sm}}(0,\, h)]^2. \tag{2.89}$$

over the selected height interval from h_0 to $h_0 + \Delta h$. After matching is achieved, the corresponding model profile of $\kappa'_{\mathrm{mod}}(h)$ is found with Eq. (2.87).

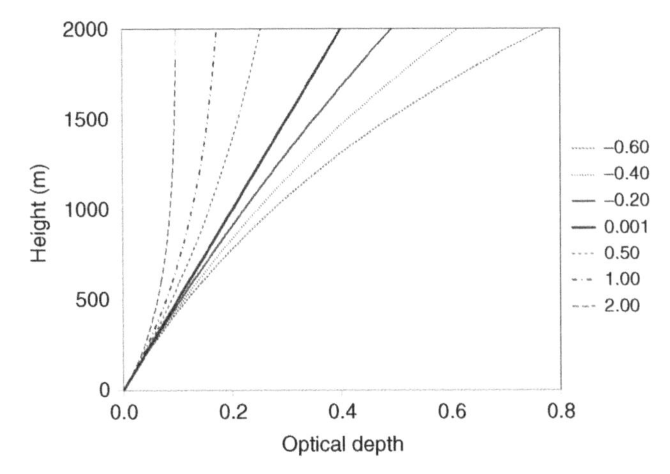

Fig. 2.34 Model profiles of the optical depth $\tau'_{\mathrm{mod}}(0, h)$, calculated for $\kappa_{\mathrm{gr}} = 0.2\ \mathrm{km}^{-1}$ using different η, the values of which are shown in the legend.

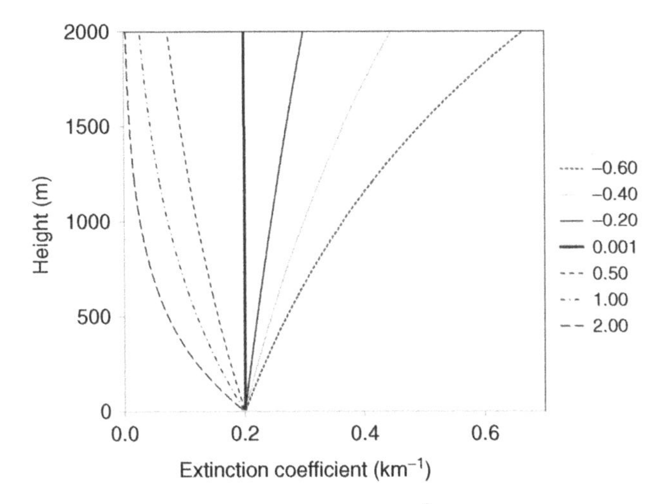

Fig. 2.35 Model profiles of the extinction coefficient $\kappa'_{\mathrm{mod}}(h)$ for the same parameters η and κ_{gr} as in Fig. 2.34.

To illustrate how to match the model optical depth profile $\tau'_{\mathrm{mod}}(0, h)$ over the height interval Δh to the profile retrieved from the lidar data, let us use the scanning lidar data obtained in the vicinity of the wildfire in Montana in August 2005. During monitoring the smoke layers with the lidar, auxiliary data at the lidar site were also recorded. These data included the particulate extinction coefficient, $\kappa_{p,\ \mathrm{neph}}$, measured at ground level with a TSI 3563 integrating nephelometer (Anderson and Ogren, 1998).

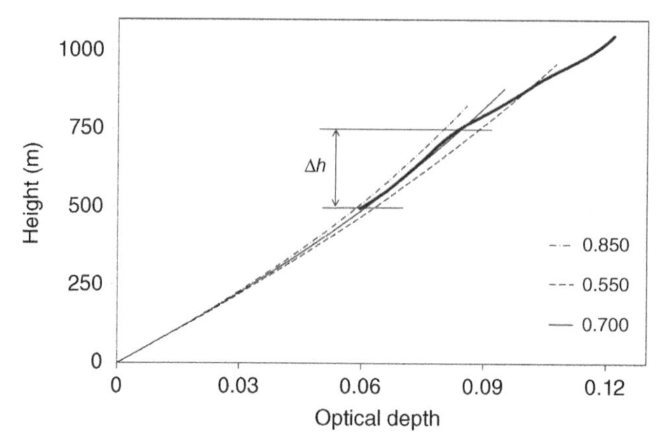

Fig. 2.36 Optical depth profile extracted from the scanning lidar data (thick solid curve) and the model profiles $\tau'_{\mathrm{mod}}(0, h)$ obtained with different values of the constant η, shown in the legend.

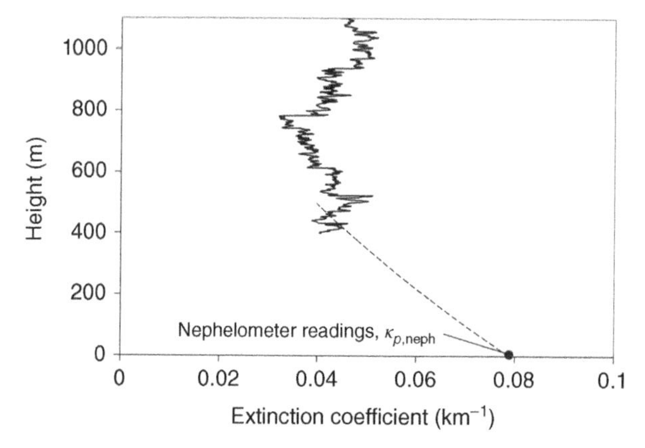

Fig. 2.37 Profile of the extinction coefficient $\kappa^{(\mathrm{dif})}(h)$ retrieved from the nonsmoothed optical depth and the extrapolated extinction-coefficient profile $\kappa'_{\mathrm{mod}}(h)$ (the solid and the dashed curves, respectively).

The results of the *a posteriori* simulations based both on the lidar near-end data and the auxiliary data of the ground-based nephelometer are shown in Figs. 2.36 and 2.37. The optical depth profile $\tau_{\mathrm{sm}}(0, h)$ extracted from the lidar data is shown in Fig. 2.36 as the thick solid curve. The matching model profiles $\tau'_{\mathrm{mod}}(0, h)$ were calculated within the height interval Δh located between the heights 500 and 750 m (Fig. 2.36). The particulate extinction coefficient measured by the nephelometer and used for determining the total κ_{gr} was $\kappa_{p,\,\mathrm{neph}} = 0.079\ \mathrm{km}^{-1}$. As can be seen in the figure, the best match between the derived and the modeled optical depth profiles over the selected interval Δh takes place with $\eta = 0.7$, whereas larger and smaller values of η yield larger systematic differences with the data points of the optical depth $\tau_{\mathrm{sm}}(0, h)$

within this interval. The total extinction-coefficient profile $\kappa'_{mod}(h)$ calculated using $\eta = 0.7$, and the nonsmoothed noisy profile of the total extinction coefficient derived from the lidar data are put together in Fig. 2.37. The extinction coefficient decreases from ground level up to the heights, approximately, 750–800 m, and shows some tendency to slightly increase at higher altitudes. Such a profile is typical in the vicinity of wildfires.

2.6 PROFILING OF THE EXTINCTION COEFFICIENT USING THE OPTICAL DEPTH AND BACKSCATTER-COEFFICIENT PROFILES

2.6.1 Theoretical Basics and Methodology

Straightforward extraction of the extinction coefficient from the optical depth profile requires the use of direct numerical differentiation, which often yields poor inversion accuracy. Another principal drawback of the extinction coefficient retrieval by direct numerical differentiation is that only the optical depth profile extracted from the two-way transmittance term is used to obtain the extinction coefficient. Valuable information about the particulate loading contained in the backscatter term is generally discarded.

Meanwhile, the use of the backscatter-coefficient profile when extracting the extinction coefficient from the lidar signal puts additional constraints on the erroneous fluctuations in the extracted profile. In many cases, these constraints narrow the scope of possible distortions in the derived extinction coefficient. Such an alternative technique, where both the optical depth and the backscatter-coefficient profiles are used for extracting the extinction coefficient was considered in the study by Kovalev (2006). However, the solution still utilizes numerical differentiation. Solutions where profiling of the extinction coefficient was made without numerical differentiation were proposed by Pahlow et al. (2004) and later by Shcherbakov (2007); the latter author included in his solution an additional smoothness constraint taken *a priori*. Alternative variants of the determination of the profile of the extinction coefficient using both the backscatter and optical depth profiles were considered in studies by Kovalev et al. (2011a, 2011b, 2012). Using these techniques, the retrieval of the extinction coefficient can be made without direct numerical differentiation.

In this section, a key principle of such a technique is considered. In this technique, the particulate optical depth profile extracted from the splitting lidar signal is not differentiated. It is used only to determine the optimum values of the stepwise lidar ratio over the fragmented zones of the searched atmosphere. The sequence of the retrieval steps in such a technique is as follows. Initially the total operative range of the lidar is fragmented into a number of restricted zones. For each zone, the constant lidar ratio is found that provides best agreement between the profile of the initial optical depth derived from the signals of the splitting lidar, and the auxiliary optical depth profile obtained by using the backscatter coefficient. The particulate extinction coefficient over each zone is found after determining the constant lidar ratio that provides the best match between two optical depth profiles in such a restricted zone. The alternative technique significantly suppresses the noise in the profile of the derived extinction coefficient, reducing the noise level down to that in the backscatter coefficient.

Let us consider the principle of the retrieval of the particulate extinction coefficient $\kappa_p(h)$ from the profiles of the particulate optical depth $\tau_p(h_i, h)$ and the particulate backscatter coefficient $\beta_{\pi, p}(h)$ extracted from a ground-based zenith-directed splitting lidar. As follows from the definition of the lidar ratio $S_p(h)$, the relationship between $\kappa_p(h)$ and $\beta_{\pi, p}(h)$ obeys the formula

$$\kappa_p(h) = S_p(h)\beta_{\pi, p}(h). \tag{2.90}$$

Accordingly, the dependence of the particulate optical depth $\tau_p(h_i, h)$ on the extinction coefficient $\kappa_p(h)$ can be written in the form

$$\tau_p(h_i, h) = \int_{h_i}^{h} \kappa_p(x)\, dx = \int_{h_i}^{h} S_p(x)\beta_{\pi, p}(x)dx, \tag{2.91}$$

where h_i is some starting point. Eq. (2.91) can be rewritten as

$$\tau_p(h_i, h) = \overline{S_p(h_i, h)} \int_{h_i}^{h} \beta_{\pi, p}(x)dx; \tag{2.92}$$

where $\overline{S_p(h_i, h)}$ is the column-integrated lidar ratio over the altitude range from h_i to h, defined as

$$\overline{S_p(h_i, h)} = \frac{\int_{h_i}^{h} S_p(x)\beta_{\pi, p}(x)dx}{\int_{h_i}^{h} \beta_{\pi, p}(x)dx}. \tag{2.93}$$

If the function $S_p(h)$ is constant within a fragmented height interval from h_i to $h_{i+1} = h_i + \Delta h_i$, then the function $\overline{S_p(h_i, h)}$ is also constant within this interval. If the backscatter coefficient $\beta_{\pi, p}(h) = \text{const.}$ within the interval Δh_i, the column-integrated lidar ratio $\overline{S_p(h_i, h)}$ is equal to the mean value of $S_p(h)$ for this interval. Thus, only simultaneous sharp changes in both functions $S_p(h)$ and $\beta_{\pi, p}(h)$ will cause a significant difference between the mean and column-integrated lidar ratios.

As with any mean value, the column-integrated lidar ratio is generally a relatively smooth function, the variation in which is significantly less than that in the nonintegrated function $S_p(h)$. If the column-integrated lidar ratio has no sharp change over the height interval Δh_i, one can apply the assumption $\overline{S_p(h_i, h)} \approx \text{const.}$ for this interval. Accordingly, the optical depth profile $\tau_p(h_i, h)$ in Eq. (2.92) can be approximated by the function $\tau_p^{(i)}(h_i, h)$, defined as (Mattis et al., 2004; Cattrall et al., 2005; Cadet et al., 2005; Ansmann, 2006)

$$\tau_p^{(i)}(h_i,\ h) = S_p^{(i)} \int_{h_i}^{h} \beta_{\pi,\ p}(x)dx, \tag{2.94}$$

where the symbol $S_p^{(i)}$ is used as an abbreviation of the quantity $\overline{S_p(h_i,\ h)}$, which is constant within the height interval from h_i to $h_i + \Delta h_i$. After selection of $S_p^{(i)}$ that equalizes $\tau_p^{(i)}(h_i,\ h)$ and initial $\tau_p(h_i,\ h)$ within the interval Δh_i, the piecewise extinction coefficient within the interval can be derived as

$$\kappa_p^{(i)}(h) \approx S_p^{(i)} \beta_{\pi,\ p}(h). \tag{2.95}$$

As mentioned, the basic idea of the above retrieval technique is to extract the extinction coefficient from the backscatter-coefficient profile rather than from the optical depth. The optical depth $\tau_p(h_i,\ h)$ obtained from the splitting lidar signal is not used for direct extraction of the extinction coefficient with numerical differentiation, as in the conventional method described in Section 2.1; it is used only as a reference function to find its best match from the optical depth $\tau_p^{(i)}(h_i,\ h)$.

To find the quantity $S_p^{(i)}$ that equalizes $\tau_p^{(i)}(h_i,\ h)$ and $\tau_p(h_i,\ h)$ within the interval Δh_i, different $S_p^{(i)}$ have to be tested. Because the initial lidar ratio $S_p^{(i)}$ is selected arbitrarily, the optical depth profiles, $\tau_p^{(i)}(h_i, h)$ and $\tau_p(h_i,\ h)$, commonly diverge within the altitude interval Δh_i and need to be equalized. Such equalizing yields much more stable results if excessive fluctuations in the reference optical depth $\tau_p(h_i,\ h)$ are initially suppressed as discussed in Section 2.5.1. In other words, to obtain a more stable inversion result, it is preferable to compare the auxiliary optical depth $\tau_p^{(i)}(h_i, h)$, calculated with Eq., (2.94), with the shaped function $\tau_{p,\ \mathrm{sh}}(h_i,\ h)$, rather than with the original $\tau_p(h_i,\ h)$.

Finding the stepwise lidar ratio $S_p^{(i)}$ which minimizes the difference

$$\Delta\tau_p(h_i, h) = \tau_p^{(i)}(h_i, h) - \tau_{p,\mathrm{sh}}(h_i, h), \tag{2.96}$$

within the selected interval Δh_i is the key operation in this method. In practice, the acceptable difference $\Delta\tau_{p,\ \max}$ between these optical depths may be initially established, so that the ratio $S_p^{(i)}$ is found that validates the inequality

$$|\Delta\tau_p(h_i,\ h)| \leq \Delta\tau_{p,\ \max}, \tag{2.97}$$

for any point h within the interval Δh_i.

The requirements for the selection of the optimal height interval Δh_i are contradictory. When making vertical soundings, the difference between the local lidar ratio $S_p(h)$ and the column-integrated lidar ratio $\overline{S_p(h_i, h)}$ commonly increases with the increase of Δh_i, so that the smaller interval generally provides a smaller difference.

On the other hand, the column-integrated lidar ratio $\overline{S_p(h_i, h)}$ becomes less variable when the larger interval Δh_i is used. Taking Eqs. (2.91) and (2.94) into consideration, Eq. (2.97) can be rewritten as

$$S_p^{(i)} \int_{h_1}^{h} \beta_{\pi,p}(x)dx - \int_{h_1}^{h} S_p(x)\beta_{\pi,p}(x)dx \leq \Delta\tau_{p,\max}. \tag{2.98}$$

Accordingly, the absolute value of the difference between the selected $S_p^{(i)}$ and the actual column-integrated $\overline{S_p(h_i, h)}$ obeys the formula

$$|\Delta S_p^{(i)}(h)| = |S_p^{(i)} - \overline{S_p(h_i, h)}| \leq \frac{\Delta\tau_{p,\max}(h)}{\int_{h_i}^{h} \beta_p(x)dx}. \tag{2.99}$$

The general flow diagram for equalizing optical depths $\tau_p^{(i)}(h_i, h)$ and $\tau_{p,\mathrm{sh}}(h_i, h)$ and next determining the piecewise particulate extinction-coefficient profile within the interval (h_i, h_{i+1}) is shown in Fig. 2.38. In the procedure, the profiles of the vertical optical depth $\tau(0, h)$ and the product $[C\beta_\pi(h)]$ obtained from the lidar signals are used as inputs. To calculate the particulate component $\beta_{\pi,p}(h)$ from the product $[C\beta_\pi(h)]$, the molecular backscatter coefficient $\beta_{\pi,m}(h)$ should be known

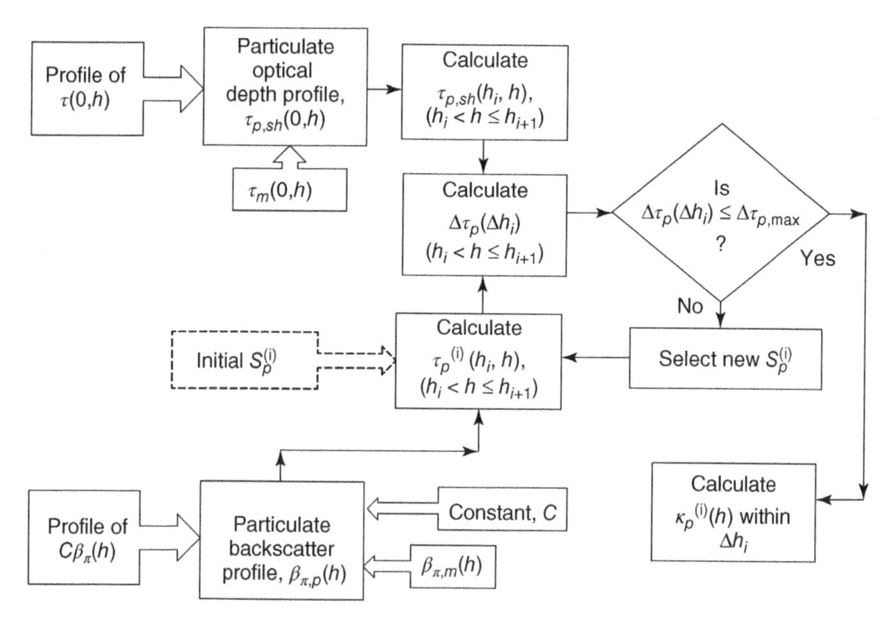

Fig. 2.38 Simplified schematic for determining the extinction coefficient $\kappa_p^{(i)}(h)$ within the interval Δh_i. (Adapted from Kovalev et al., 2007c).

and the lidar equation constant C needs to be determined. The vertical molecular extinction-coefficient profile is used to determine the molecular optical depth $\tau_m(0, h)$ and separate the shaped particulate component $\tau_{p,\,sh}(0, h)$ from the total optical depth $\tau(0, h)$. After the difference between the optical depths $\tau_{p,\,sh}(h_i, h)$ and $\tau_p^{(i)}(h_i, h)$ is minimized, so that $|\Delta\tau_p(h_i, h)| \leq \Delta\tau_{p,\,max}$, the particulate extinction-coefficient profile is calculated as the product of the selected lidar ratio and backscatter coefficient $\beta_{\pi,\,p}(h)$.

After these steps are completed and the extinction coefficient $\kappa_p^{(i)}(h)$ for the interval from h_i to h_{i+1} is found, the next fragmented interval from h_{i+1} to h_{i+2} is examined, and the value of the next constant lidar ratio $S_p^{(i+1)}$ that makes the inequality in Eq. (2.97) valid is found; accordingly, this permits calculation of the piecewise extinction coefficient within this interval. Then the next fragmented interval is selected and so on. When all the values of the stepwise lidar ratios $S_p^{(i)}$, $S_p^{(i+1)}$, … , $S_p^{(N)}$ have been found, the piecewise continuous extinction-coefficient profile is determined within the total altitude range from h_{min} to h_{max}.

Let us summarize the features of the above technique. For the inversion of the splitting lidar signal, the backscatter coefficient $\beta_\pi(h)$ and optical depth $\tau(0, h)$ over the total operative altitude range of the lidar needs to be initially determined. Then the total altitude range from h_{min} to h_{max} is fragmented into separate height intervals. The first interval starts at $h_1 = h_{min}$ and ends at $h_2 = h_1 + \Delta h_1$, and the last one starts at h_{N-1} and ends at $h_N = h_{max}$. For each interval, the constant lidar ratio is found that provides the best agreement between the shaped profile of the particulate optical depth $\tau_{p,\,sh}(h_i, h)$ and that obtained with Eq. (2.94). The extinction coefficient $\kappa_p^{(i)}(h)$ in each height interval is calculated as the product $S_p^{(i)}\beta_{\pi,\,p}(h)$.

2.6.2 Distortions in the Derived Particulate Extinction Coefficient Due to Inaccuracies in the Involved Parameters

The above technique for determining the extinction coefficient does not require direct numerical differentiation of the noisy optical depth profile. However, unlike the latter, it requires knowledge of the profile of the particulate backscatter coefficient $\beta_{\pi,\,p}(h)$ within the heights from h_{min} to h_{max}. To determine this profile from the product $[C\beta_\pi(h)]$, the constant C needs to be estimated. In some cases, this can be done beforehand in laboratory conditions. Otherwise, the constant has to be estimated from the lidar sounding data. In clear atmospheres, the constant can be found by using the assumption of pure molecular scattering at high altitudes, generally somewhere close to the maximal height, h_{max}. If at such a reference height h_{ref}, the particulate component $\beta_{\pi,\,p}(h_{ref}) \approx 0$, the estimate of the constant $\langle C\rangle$ can be found as

$$\langle C\rangle \approx \frac{[C\beta_\pi(h_{ref})]}{\beta_{\pi,\,m}(h_{ref})}. \tag{2.100}$$

Unlike conventional methods, here the constant C is determined from the product $[C\beta_\pi(h)]$; the lidar is not used for the estimate. This feature can provide a more accurate estimate of the constant as compared to that found from the backscatter signal at a distant range. To get an accurate and reliable estimate of the constant C it is useful to check whether minimal aerosol loading takes place at the ranges close to the maximum height, where h_{ref} is commonly selected. To check this, an auxiliary function $\gamma(h)$ within the total height interval $h_{min} \leq h \leq h_{max}$ is calculated, that is,

$$\gamma(h) = \frac{[C\beta_{\pi}(h)]}{\beta_{\pi,\,m}(h)}. \tag{2.101}$$

The maximal constant $\langle C \rangle$ that provides positive $\beta_{\pi,\,p}(h)$ over the lidar total altitude range, should obey the formula

$$\langle C \rangle \leq \min \gamma(h). \tag{2.102}$$

After $\langle C \rangle$ is found, the profile of the particulate backscatter-coefficient profile is estimated using the formula

$$\langle \beta_{\pi,\,p}(h) \rangle = \frac{[C\beta_{\pi}(h)]}{\langle C \rangle} - \beta_{\pi,\,m}(h). \tag{2.103}$$

Selecting the constant $\langle C \rangle$ to be larger than the minimum value of $\gamma(h)$ will result in the appearance of local zones in which nonphysical negative values of $\langle \beta_{\pi,\,p}(h) \rangle$ are obtained.

Let us consider the difference in the errors in the retrieved extinction coefficient when the exact formula [Eq. (2.90)] and the approximate solution [Eq. (2.95)] are used. When the exact formula is used, the fractional error in the retrieved extinction coefficient $\delta\kappa_p(h)$ may be quantified using the simple formula

$$\delta\kappa_p(h) = \sqrt{[\delta\beta_{\pi,\,p}(h)]^2 + [\delta S_p(h)]^2}, \tag{2.104}$$

which assumes that both error components in Eq. (2.90) are not correlated and obey statistical laws. However, this objective is not valid when determining $\kappa_p^{(i)}(h)$ with Eq. (2.95), where two correlated distortion components should be taken into consideration. The first uncertainty component, the same as in Eq. (2.104), may be present in the distorted backscatter-coefficient profile $\langle \beta_{\pi,\,p}(h) \rangle$ used in Eq. (2.95) instead of the true $\beta_{\pi,\,p}(h)$. The distortion depends on the accuracy of determining the lidar solution constant C. When $\langle C \rangle \neq C$, a shifted profile of the backscatter coefficient $\langle \beta_{\pi,\,p}(h) \rangle = \beta_{\pi,\,p}(h)[1 + \delta\beta_{\pi,\,p}(h)]$ with the relative error $\delta\beta_{\pi,\,p}(h) \neq 0$ is obtained. The other uncertainty is related to the possible difference between the actual lidar ratio $S_p(h)$ and the constant quantity $S_p^{(i)} = S_p(h)\lfloor 1 + \delta S_p(h) \rfloor$ used for the retrieval of $\kappa_p^{(i)}(h)$.

A simple analysis shows that the accuracy in the derived extinction coefficient determined with Eq. (2.95) does not depend dramatically on the error in the estimated constant C, at least, in the areas with a moderate gradient in the extinction-coefficient profile. This observation follows from the specifics of the inversion procedure, in which the difference between two optical depths, the shaped $\tau_{p,\,\mathrm{sh}}(h_i,\ h)$ and the auxiliary $\tau_p^{(i)}(h_i,\ h)$ is minimized before $\kappa_p^{(i)}(h)$ is calculated. The initial optical depth $\tau_p(h_i,\ h)$ from which the shaped profile is obtained, obeys the formula

$$\tau_p(h_i,\ h) = \int_{h_i}^{h} \kappa_p(x)dx = \int_{h_i}^{h} S_p(x)\beta_{\pi,\,p}(x)dx, \tag{2.105}$$

whereas the optical depth $\tau_p^{(i)}(h_i,\ h)$ can be rewritten as

$$\tau_p^{(i)}(h_i,\ h) = \int_{h_i}^{h} \kappa_p^{(i)}(x)dx = \int_{h_i}^{h} S_p(x)\beta_{\pi,\,p}(x)\ [1 + \delta S_p(x)]\ [1 + \delta\beta_{\pi,\,p}(x)]\ dx. \tag{2.106}$$

After the difference between these optical depth profiles is minimized, $\tau_p^{(i)}(h_i,\ h) \approx \tau_{p,\,\mathrm{sh}}(h_i,\ h)$. It follows from this equality that the relative errors $\delta S_p(h)$ and $\delta\beta_{\pi,\,p}(h)$ are related. Indeed, equalizing the right sides of Eqs. (2.105) and (2.106), one obtains

$$\lfloor 1 + \delta S_p(h) \rfloor\ \lfloor 1 + \delta\beta_{\pi,\,p}(h) \rfloor \approx 1. \tag{2.107}$$

As follows from Eq. (2.107), the uncertainties $\delta\beta_{\pi,\,p}(h)$ and $\delta S_p(h)$ are correlated and the dependences between these obey simple formulas

$$\delta S_p(h) = \frac{-\delta\beta_{\pi,\,p}(h)}{1 + \delta\beta_{\pi,\,p}(h)}, \tag{2.108}$$

and

$$\delta\beta_{\pi,\,p}(h) = \frac{-\delta S_p(h)}{1 + \delta S_p(h)}. \tag{2.109}$$

To illustrate the influence of an inaccurately estimated constant C on the inversion result, let us perform a numerical simulation using the synthetic profiles of $[C\beta_\pi(h)]$ and two-way total transmittance, $T_\Sigma^2(h_{\min}, h)$. Both profiles, presumably retrieved from the same set of data of a virtual splitting lidar within the complete overlap zone are shown in Fig. 2.39. The shape of the noise-corrupted profile of the product $[C\beta_\pi(h)]$ versus height is shown in the left panel and the profile of the two-way transmittance profile $T_\Sigma^2(h_{\min}, h)$, in the right panel. Both synthetic profiles, recorded at the wavelength 355 nm, are corrupted by quasi-random noise.

Before starting the inversion procedure, one should select the number and location of the intervals Δh_i within which the column-integrated lidar ratio is taken as constant. (The principle and specifics of the selection of these intervals is considered in Section 2.7.2). For this numerical simulation, eight overlapping intervals Δh_i within the total altitude range from $h_{\min}$ to $h_{\max}$ were selected. In Figs. 2.40 and 2.41, the extracted piecewise continuous profiles of the extinction coefficient $\kappa_p^{(i)}(h)$ are shown as the dotted curves, whereas the test profile of the extinction coefficient is shown as the solid curve. In Fig. 2.40, the extinction coefficient $\kappa_p^{(i)}(h)$ is retrieved using the

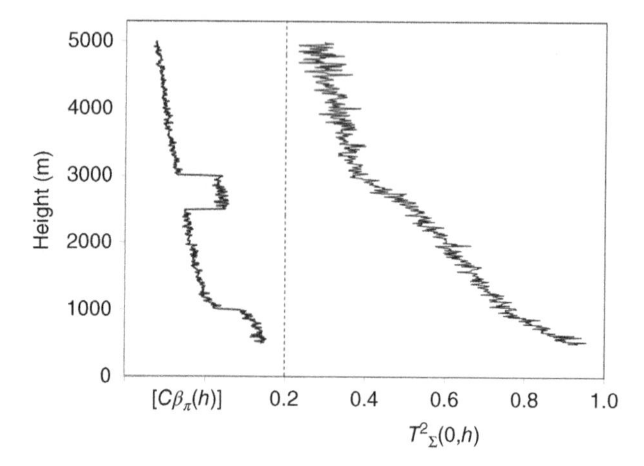

Fig. 2.39 The synthetic profiles of $[C\beta_\pi(h)]$ and two-way total transmittance $T^2_\Sigma(0, h)$ used in the numerical simulation, inversion results of which are shown in Figs. 2.40 and 2.41. (Adapted from Kovalev et al., 2011b).

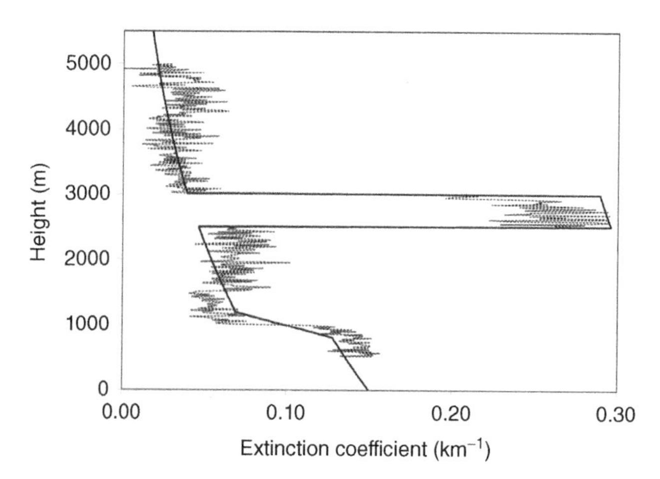

Fig. 2.40 The solid curve is the test profile of the synthetic extinction coefficient $\kappa_p(h)$ used in the numerical simulation. The dotted curve is the piecewise continuous profile of the particulate extinction coefficient $\kappa_p^{(i)}(h)$, extracted from the profiles $[C\beta_\pi(h)]$ and $[T^2_\Sigma(0, h)$ with the estimated constant $\langle C \rangle = 1.08\ C$. (Adapted from Kovalev et al., 2011b).

maximal constant $\langle C \rangle$, estimated using Eq. (2.102). This constant proved to be as much as 8% higher than the actual constant C. The increased value of the estimated constant $\langle C \rangle$ was obtained because the condition of purely molecular scattering at the reference altitude (5000 m) was not properly met. Nevertheless, there is no significant systematic difference between the test and derived extinction-coefficient profiles.

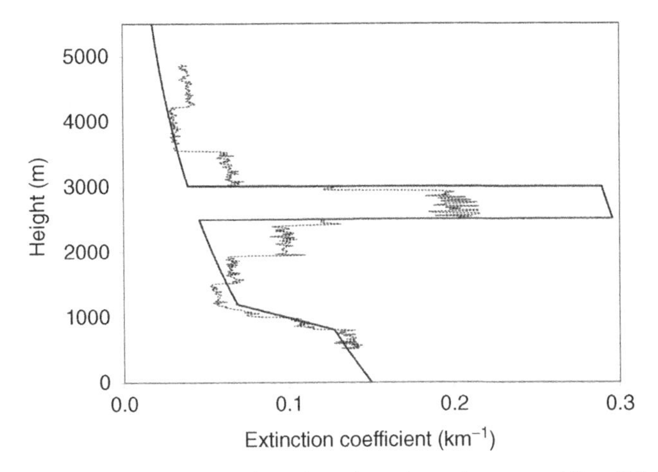

Fig. 2.41 The same as in Figure 2.40 but where the estimated constant $\langle C \rangle = 0.5\ C$. (Adapted from Kovalev et al., 2011b).

Fig. 2.41 shows the inversion result obtained from the same test profiles $[C\beta_\pi(h)]$ and $[T_\Sigma^2(0, h)$ but when the constant is significantly underestimated. In particular, the estimated constant $\langle C \rangle$ is two times less than the actual C. One can see that even such an underestimated constant does not significantly corrupt the retrieved profile of the extinction coefficient in the near zone up to the height of ~1900 m. The most significant distortions of the extracted $\kappa_p^{(i)}(h)$ occur in the vicinity of sharp changes of the extinction coefficient, which take place at heights greater than ~1900 m.

Let us summarize. An inexact estimate of the constant in the product $[C\beta_\pi(h)]$ results in shifted backscatter coefficient $\langle \beta_{\pi,\ p}(h) \rangle$. This shift in turn, produces an error in the lidar ratio $S_p^{(i)}$ and then in the retrieved piecewise extinction coefficient $\kappa_p^{(i)}(h)$. The maximal distortions take place in the vicinity of sharp changes of the extinction coefficient. The impact of the shifted estimate of the constant C is somewhat equivalent to the inaccuracy in the selected lidar ratio in the elastic lidar solution. Similar to that, the distortion in the extinction coefficient derived from lidar data is less when strong aerosol heterogeneity is absent; the distortions are significantly increased in heterogeneous areas with sharp boundaries.

2.6.3 Extraction of the Particulate Extinction Coefficient by Minimizing the Discrepancy between the Alternative Piecewise Transmittances

A simpler and more practical variant of the inversion of the splitting lidar signal was proposed in the study by Kovalev et al. (2011b). Using the same assumption of a range-independent lidar ratio within the fragmented height intervals, one can extract the particulate extinction coefficient equalizing the alternative transmittances within this interval. In principle, such a variant is similar to equalizing the optical depth profiles $\tau_p^{(i)}(h_i, h)$ and $\tau_{p,\ \mathrm{sh}}(h_i,\ h)$ by minimizing the difference between these down to some established $\Delta\tau_{p,\ \max}$ [Eq. (2.97)]. However, such a variant has two issues.

First, the values of the possible optical depths can be significantly different for the different intervals Δh_i and this obstacle significantly complicates the procedure of equalizing $\tau_p^{(i)}(h_i, h)$ and $\tau_{p,\,\mathrm{sh}}(h_i,\ h)$. The second issue is related to the selection of $\Delta\tau_{p,\,\max}$, the optimum value of which depends on many factors. It depends on the length of the fragmented interval Δh_i; on the value and the shape of the actual optical depth within the intervals, which can significantly vary; on the level of noise in the intermediate data; on the presence of the bulges and concavities in inverted signal; etc. If the selected $\Delta\tau_{p,\,\max}$ is overestimated, the inversion results may be quite inaccurate; if underestimated, the condition in Eq. (2.97) is difficult, if not impossible, to meet.

These issues are reduced if the discrepancy between the alternative piecewise transmittances rather than between the optical depths is minimized. Unlike the large scope of the possible values of the optical depth within the fragmented intervals Δh_i, the transmittances over such intervals are close to unity and vary slightly. Accordingly, equalizing these can be done much faster and more easily than when operating with optical depths.

Let us consider this variant in detail. In the complete overlap zone, the vertical square-range-corrected backscatter signal versus height can be written as

$$P(h)h^2 = C\beta_\pi(h)T_\Sigma^2(0,\ h), \tag{2.110}$$

where $T_\Sigma^2(0,\ h)$ is the total two-way transmittance in the vertical direction. Using the splitting lidar data, one can separate the product $[C\beta_\pi(h)]$ and extract the corresponding profile of $T_\Sigma^2(0, h)$ within the lidar operative range as

$$T_\Sigma^2(0,\ h) = \frac{P(h)h^2}{[C\beta_\pi(h)]}. \tag{2.111}$$

As in Section 2.6.1, the altitude range from $h_{\min}$ to $h_{\max}$, where $T_\Sigma^2(0,\ h)$ is determined, should be divided into a number of fragmented intervals Δh_i. Within each such interval, the piecewise profile of the particulate transmittance $T_p^2(h_i,\ h)$ is determined using the corresponding two-way transmittance profile $T_\Sigma^2(0,\ h)$. The particulate transmittance, determined within the fragmented interval, Δh_i, can be found as

$$T_p^2(h_i,\ h) = \frac{T_\Sigma^2(h_i,\ h)}{T_m^2(h_i,\ h)}, \tag{2.112}$$

where h_i is the starting point of the interval Δh_i, $T_p^2(h_i,\ h)$ and $T_m^2(h_i,\ h)$ are the two-way particulate and molecular transmissions within this interval, respectively. Taking into consideration Eq. (2.111), Eq. (2.112) can be rewritten as

$$T_p^2(h_i,\ h) = \frac{[C\beta_\pi(h_i)]}{[C\beta_\pi(h)]}\frac{P(h)h^2}{P(h_i)h_i^2}\frac{T_m^2(0,\ h_i)}{T_m^2(0,\ h)}. \tag{2.113}$$

Eq. (2.113) yields the piecewise two-way transmission profile obtained from the square-range-corrected signal, $P(h)h^2$ and the product $[C\beta_\pi(h)]$. The alternative transmittance $\langle T_p^2(h_i, h)\rangle$ within the same interval Δh_i, which matches with $T_p^2(h_i, h)$, is found by determining the lidar ratio, $S_p^{(i)}$, that equalizes these piecewise transmittance profiles. To trigger this procedure, the initial $S_p^{(i)}$ is arbitrarily selected, as in Section 2.6.2. The corresponding profile of the particulate extinction coefficient within the fragmented interval is calculated as

$$\kappa_p^{(i)}(h) = S_p^{(i)}\beta_{\pi,\,p}(h), \tag{2.114}$$

where the backscatter extinction coefficient, $\beta_\pi(h)$ is found from the product $[C\beta_\pi(h)]$. The corresponding two-way transmittance profile within this interval is then found as

$$\langle T_p^2(h_i, h)\rangle = \exp\left[-2\int_{h_i}^{h} \kappa_p^{(i)}(x)dx\right]. \tag{2.115}$$

The profile obtained with Eq. (2.115) is compared with that obtained with Eq. (2.113). As the initial lidar ratio $S_p^{(i)}$ in Eq. (2.114) is selected arbitrarily, the profiles $T_p^2(h_i, h)$ and $\langle T_p^2(h_i, h)\rangle$ may initially diverge and need to be equalized. The equalization can be achieved by determining the proper constant value of the lidar ratio $S_p^{(i)}$. To determine how close the piecewise profiles $T_p^2(h_i, h)$ and $\langle T_p^2(h_i, h)\rangle$ are to each other, a simple criterion may be used, which compares the slopes of the linear fits of these transmittance profiles. As the maximal transmittance does not exceed unity, and both transmittance profiles decrease with range, the linear fits of these within the same interval Δh_i can be written as

$$T_p^2(h_i, h) = 1 - B_1(h - h_i), \tag{2.116}$$

and

$$\langle T_p^2(h_i, h)\rangle = 1 - B_2(h - h_i), \tag{2.117}$$

where B is a positive nonzero value. Note that owing to possible distortions, the constant in Eq. (1.117) can not be equal to unity, but only close to it however, this quantity is not taken into consideration. The only task is to find the lidar ratio $S_p^{(i)}$ that minimizes the criterion

$$\Lambda(\Delta h_i) = (B_1 - B_2)^2 = \min. \tag{2.118}$$

After the criterion $\Lambda(\Delta h_i)$ is minimized, the corresponding $\kappa_p^{(i)}(h)$, calculated with Eq. (2.114), is taken as the piecewise extinction coefficient within the interval, Δh_i. These operations in Eqs. (2.113–2.118) are repeated for each fragmented zone Δh_i within the total height interval $h_{\min} - h_{\max}$.

The use of the linear fit for the transmission profile $T_p^2(h_i, h)$ makes it possible to check, and if necessary, to correct the maximal operative height $h_{\max}$ of the lidar signal used for the inversion. Such a possibility follows from the obvious requirement that $B_1 > 0$, which may not be satisfied in areas where the lidar signal is significantly corrupted. If the inequality is not met, the linear fit of the two-way transmission within this fragmented interval will have a nonphysical increase with height; in most cases, this occurs in areas close to h_{max}. The only possibility of satisfying the requirement of a positive B_1 in this area is to decrease the maximal height h_{max} and accordingly, the length of the corresponding interval until $B_1 > 0$. Note also that when $B_1 = 0$, this does not necessarily mean that no particulate loading exists within this interval. The zero slope of the linear fit of the function $T_p^2(h_i, h)$ can be caused by the presence of "bad zones" within the lidar operative range (see Section 2.5.1).

Let us summarize the general points of the alternative inversion technique considered in this section. As in the previous section, the basic idea is to extract the extinction coefficient from the product $[C\beta_\pi(h)]$ rather than to use direct numerical differentiation of the optical depth derived from the lidar data. Instead of using differentiation, the alternative piecewise profiles of two-way transmittance are equalized. The first profile is extracted using the two-way transmittance term obtained from the signal of a splitting lidar; the second profile is extracted using the assumption of a constant lidar ratio over the fragmented intervals within the lidar operative range.

2.7 PROFILING OF THE EXTINCTION COEFFICIENT WITHIN INTERVALS SELECTED *A PRIORI*

2.7.1 Determination of Piecewise Continuous Profiles of the Extinction Coefficient and the Column Lidar Ratio Using Equal Length Intervals

To utilize the inversion technique considered in the previous section, the total operative range of the zenith-directed splitting lidar should be divided into a number of intervals, within each of which the lidar ratio is assumed constant, or at least, only slightly variable relative to its mean value. The selection of such intervals is the key point of the technique under consideration. Different principles and variants can be proposed, however, it is impossible to give unique criteria for selecting the number and the length of such intervals. Unfortunately, similar issues are quite typical for lidar profiling of the atmosphere. As shown in Section 2.4, the same problem arises when selecting the optimum range resolution for numerical differentiation.

When there are no sharp changes in the aerosol loading within the lidar operative range, the simplest option is to divide the total altitude range from $h_{min} - h_{max}$ into *a priori* selected equal intervals. To clarify the inversion procedure, let us perform a numerical simulation, where the piecewise continuous extinction-coefficient profile is extracted using arbitrarily selected equal intervals. Suppose that a virtual zenith-directed lidar, which allows splitting the backscatter and transmission terms in the lidar equation, is working in a synthetic atmosphere. The simulated noise-corrupted profiles of the particulate backscatter coefficient and particulate optical depth obtained from the data of the virtual lidar are shown in Fig. 2.42. The profile of the backscatter coefficient $\beta_{\pi,p}(h)$, which contains the turbid aerosol layer at the heights 2000–3000 m, is shown on the left side of the figure; the

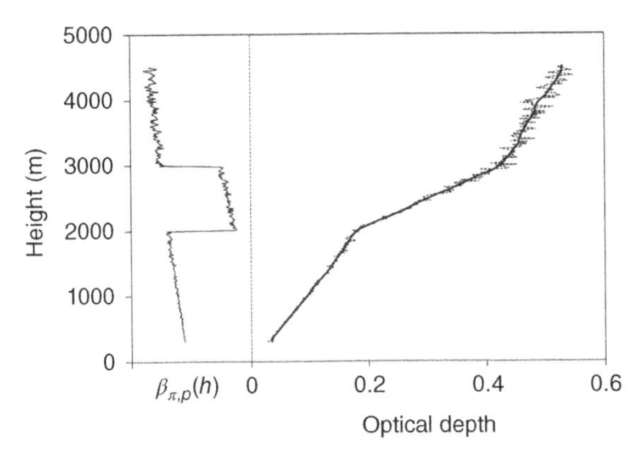

Fig. 2.42 The noise-corrupted backscatter coefficient $\beta_{\pi,\,p}(h)$ (left side) and the profiles of $\tau_p(0,\ h)$ and $\tau_{p,\,\mathrm{sh}}(0,\ h)$ (thin dotted and thick solid curves on the right side) obtained from virtual lidar data at the wavelength 532 nm and used for the numerical simulation (Kovalev et al., 2007c).

corresponding profiles of the optical depths, the initial $\tau_p(0,\ h)$, and the shaped $\tau_{p,\,\mathrm{sh}}(0,\ h)$, are shown on the right side of the figure as the thin dotted and thick solid curves, respectively. The inversion is done using the technique discussed in Section 2.6.1, but minimizing the term $\Delta\tau_p(h_i, h)$ without establishing the acceptable maximal difference $\Delta\tau_{p,\,\max}$. The total altitude range over the heights from $h_{\min} = 300$ to $h_{\max} = 4500$ m is divided into four equal intervals, $\Delta h_i = 1050$ m, within each of which the column-integrated lidar ratio is assumed to be invariable, that is, $S_p^{(i)} = \mathrm{const}$. For each height interval, the profile of the optical depth $\tau_p^{(i)}(h_i, h)$ is determined using initially arbitrarily selected $S_p^{(i)}(h)$ in Eq. (2.94); then the optical depth is compared to the shaped profile $\tau_{p,\,\mathrm{sh}}(h_i,\ h)$. To perform such a comparison, the simple criterion

$$\epsilon(\Delta h_i) = \sum_{h_i}^{h_{i+1}} [\tau_p^{(i)}(0,\ h) - \tau_{p,\,\mathrm{sh}}(0,\ h)]^2, \tag{2.119}$$

is calculated and minimized by determining the lidar ratio $S_p^{(i)}$ that yields minimum $\epsilon(\Delta h_i)$ for the selected interval.

The results of the numerical simulation are shown in Figs. 2.43 and 2.44. The particulate extinction coefficient $\kappa_p^{(i)}(h)$, extracted with Eq. (2.95), is shown in Fig. 2.43 as the thick solid curve; the test profile of $\kappa_p(h)$ is shown as the thin solid curve. The test profiles of the lidar ratio $S_p(h)$ and the retrieved piecewise continues lidar ratio $S_p^{(i)}(h)$ are shown as thin and thick solid curves in Fig. 2.44. Note that unlike the extinction-coefficient profile, no sharp changes in the test lidar-ratio profile take place even in the turbid area; within the whole searched atmospheric layer, the ratio increases monotonically.

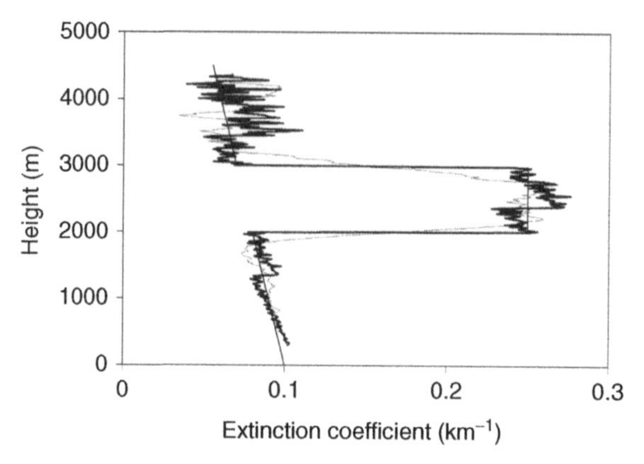

Fig. 2.43 The test profile of $\kappa_p(h)$ (thin solid curve) and the piecewise continuous extinction coefficient $\kappa_p^{(i)}(h)$ (thick solid curve) retrieved with the method under consideration. The profile of $\kappa_p^{(\mathrm{dif})}(h)$ derived with numerical differentiation is shown as the dotted curve (Kovalev et al., 2007c).

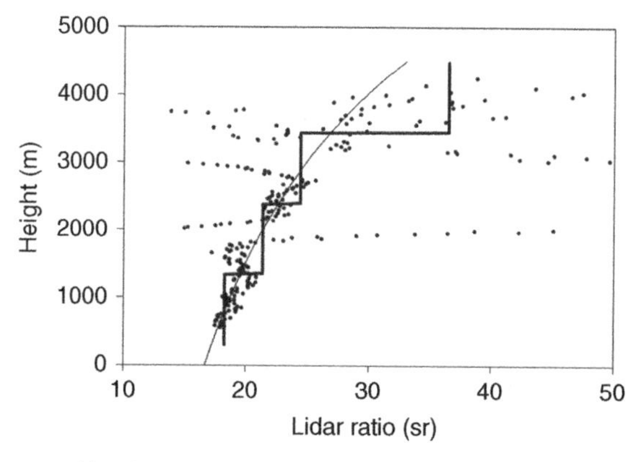

Fig. 2.44 The test profile of the lidar ratio $S_p(h)$ (thin solid curve), and the retrieved piecewise continuous lidar ratio $S_p^{(i)}(h)$ (thick solid curve). The lidar ratio calculated as the ratio of the extinction coefficient $\kappa_p^{(\mathrm{dif})}(h)$ to the backscatter coefficient $\beta_{\pi,\,p}(h)$ is shown by the scattered filled circles (Kovalev et al., 2007c).

The local values of the lidar ratio $S_p(h)$ can also be calculated as a ratio of the extinction coefficient $\kappa_p(h)$ to the backscatter coefficient $\beta_{\pi,\,p}(h)$. However, in practice there is an obvious inconsistency, which should be kept in mind. When calculating the lidar ratio in this way, the profile of some mean $\kappa_p^{(\mathrm{dif})}(h)$, extracted with the numerical differentiation of the optical depth $\tau_p(0,\ h)$, is commonly used, rather than the required local $\kappa_p(h)$. Such an inconsistency may result in an extremely noisy profile of the calculated lidar ratio. The profile of the extinction coefficient $\kappa_p^{(\mathrm{dif})}(h)$ derived from the optical depth $\tau_p(0,\ h)$ using conventional numerical differentiation with the sliding range resolution $s = 500$ m is shown in Fig. 2.43 as the dotted curve. The corresponding profile of the lidar ratio $S_p(h)$, extracted as the ratio of $\kappa_p^{(\mathrm{dif})}(h)$ to $\beta_{\pi,p}(h)$ is shown in Fig. 2.44 as solid dots. One can see that in spite of a relatively good agreement between the extracted profile of the extinction coefficient $\kappa_p^{(\mathrm{dif})}(h)$ and the test profile $\kappa_p(h)$, the data points of the corresponding $S_p(h)$ are extremely scattered and their accuracy is not comparable with the piecewise lidar ratio, obtained through the minimization of the function $\epsilon(\Delta h_i)$.

To show how the technique with arbitrarily selected equal intervals works in real atmospheres, let us consider the lidar data obtained when profiling smoky layers originating from Montana's I-90 Fire in August 2005. These data were obtained with the FSL scanning lidar at the wavelength 355 nm. The conventional multiangle inversion method was used to split the profiles of the optical depth $\tau(0, h)$ and the product $[C\beta_\pi(h)]$. The solution constant C was found from the profile $[C\beta_\pi(h)]$ using the assumption of an aerosol-free atmosphere at high altitudes. The shape of the extracted vertical profile of the backscatter coefficient $\beta_{\pi,p}(h)$ is shown in the left panel of Fig. 2.45, and the corresponding profiles of the aerosol optical depths, the original $\tau_p(0, h)$ and the shaped, $\tau_{p,\mathrm{sh}}(0, h)$, are shown in the right panel as the thin dashed and thick solid curves, respectively. The total altitude range over the heights from 400 to 3200 m, where these profiles were obtained, was divided into four equal intervals each of length 700 m. For each interval, the constant column-integrated lidar ratio was determined that minimized corresponding criteria $\epsilon(\Delta h_i)$ in Eq. (2.119). The vertical profile of the corresponding particulate extinction coefficient is shown in Fig. 2.46 as the thick solid curve. The dotted curve is the profile of $\kappa_p^{(\mathrm{dif})}(h)$ obtained through numerical differentiation of the optical depth $\tau_p(0, h)$, shown in the right panel of Fig. 2.45, with a sliding range resolution of 500 m. As one can see, these profiles at the near end agree well with the extinction coefficient recorded with the ground-based

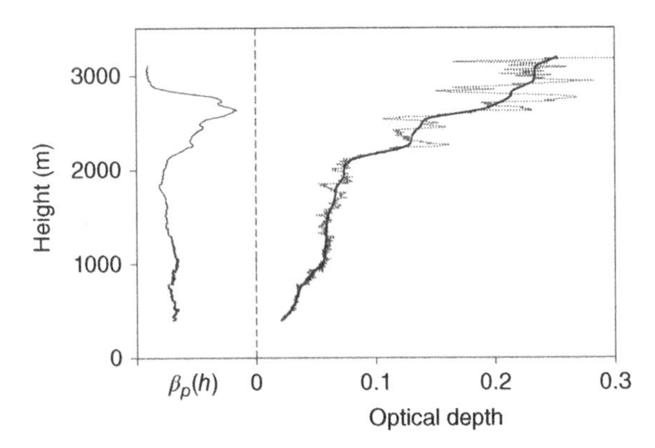

Fig. 2.45 In the left panel, the vertical profile of the backscatter coefficient $\beta_{\pi,p}(h)$ is shown derived from the lidar data obtained during Montana's I-90 Fire. In the right panel, the profiles of the corresponding particulate optical depth $\tau_p(0, h)$ and shaped $\tau_{p,\mathrm{sh}}(0, h)$ are shown as the thin dashed and thick solid curves, respectively (Kovalev et al., 2007c).

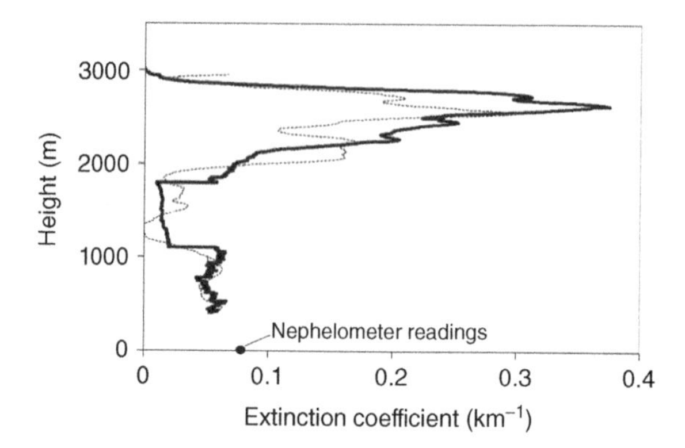

Fig. 2.46 The thick solid curve is the piecewise continuous profile of $\kappa_p^{(i)}(h)$ retrieved from the profiles shown in Fig. 2.45 when using equal intervals. The profile of the extinction coefficient $\kappa_p^{(\mathrm{dif})}(h)$ obtained through conventional numerical differentiation is shown as the dotted curve (Kovalev et al., 2007c).

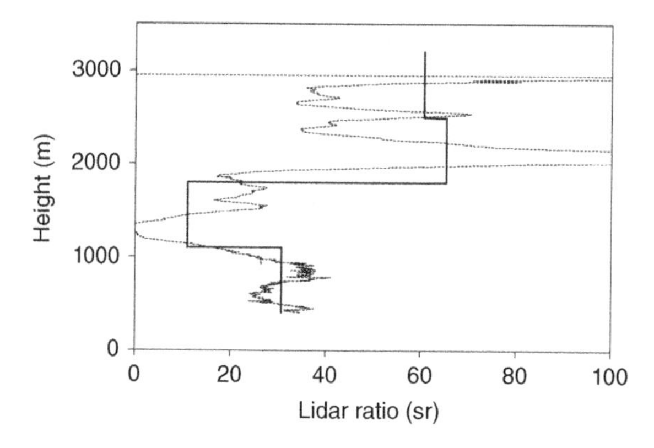

Fig. 2.47 The solid lines show the piecewise continuous lidar ratio $S_p^{(i)}(h)$ retrieved using the same intervals as in Fig. 2.46. The lidar ratio $S_p(h)$ obtained from the extinction coefficient $\kappa_p^{(\mathrm{dif})}(h)$ is shown as the dotted curve (Kovalev et al., 2007c).

nephelometer. The corresponding piecewise column-integrated lidar ratio $S_p^{(i)}(h)$ is shown in Fig. 2.47 as the thick solid curve; the lidar ratio $S_p(h)$, extracted as the ratio of the extinction coefficient $\kappa_p^{(\mathrm{dif})}(h)$ to the backscatter coefficient $\beta_{\pi,\,p}(h)$ is shown as the dotted curve.

When using arbitrarily selected intervals, the derived profile of the extinction coefficient is generally acceptable when the lidar ratio has a moderate monotonic change, such as the test lidar ratio shown in Fig. 2.44. If the changes in $S_p(h)$ are large and its boundaries are sharp, the inversion result can be significantly worse. Let us perform a numerical simulation, using the synthetic profile of the optical depth which is the same as that in the right side of Fig. 2.42, but now assuming that the lidar ratio in the turbid area significantly differs from that in the clear air area. In particular, let us

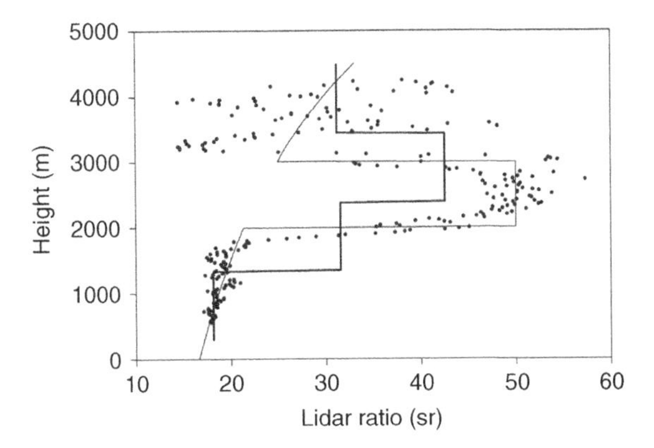

Fig. 2.48 Test profile of the lidar ratio $S_p(h)$ (thin solid curve), the retrieved piecewise continuous lidar ratio $S_p^{(i)}(h)$ (the thick solid curve), and the lidar ratio calculated as the ratio of the extinction coefficient $\kappa_p^{(\mathrm{dif})}(h)$ to the particulate backscatter coefficient (the filled circles).

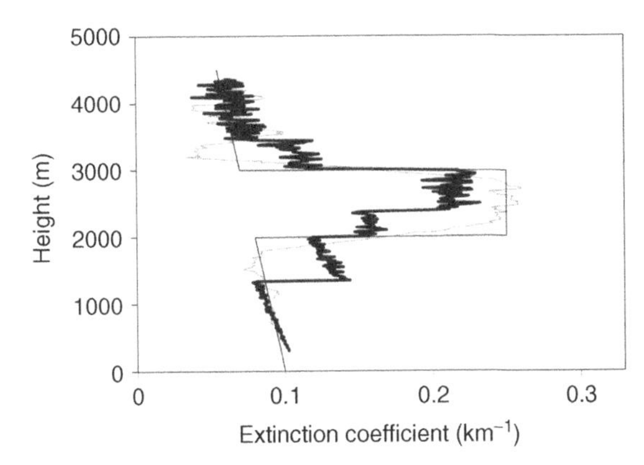

Fig. 2.49 The test profile of $\kappa_p(h)$ (thin solid curve), the retrieved piecewise continuous extinction coefficient $\kappa_p^{(i)}(h)$ (thick solid curve), and the extinction coefficient $\kappa_p^{(\mathrm{dif})}(h)$ derived through numerical differentiation (dotted curve).

assume the same monotonic profile of $S_p(h)$ as in Figs. 2.44 outside the turbid area and with increased lidar ratio $S_p(h) = 50$ sr, within the turbid area. For the inversion, the total altitude range is divided by four equal intervals of length 1035 m, within each of which the lidar ratio $S_p^{(i)}(h)$ is determined by the minimization of $\epsilon(\Delta h_i)$. The test profiles and the inversion results for such a situation are shown in Figs. 2.48 and 2.49. The sharp spatial changes in the test $S_p(h)$, whose boundaries differ from the boundaries of the selected segment zones Δh_i significantly worsen the inversion result as compared with the case with no sharp changes. As can be seen, not only does the difference between the test and retrieved lidar ratios and the corresponding extinction-coefficient profiles increase but also the boundaries of the turbid layers in the retrieved profiles are significantly shifted. Such an inconsistency between shapes $S_p(h)$ and $S_p^{(i)}(h)$, and between $\kappa_p(h)$ and $\kappa_p^{(i)}(h)$ takes place in spite of all the quantities $\epsilon(\Delta h_i)$ having been properly minimized.

As follows from the above numerical simulations, the inversion technique with the use of arbitrarily selected intervals Δh_i works well when the actual lidar ratio $S_p(h)$ in the atmosphere monotonically varies within the operative range without sharp changes. If such sharp changes occur, the assumption $S_p^{(i)}$ = const. within the *a priori* selected interval Δh_i may not be valid, and large errors may be incurred both in $\kappa_p^{(i)}(h)$ and $S_p^{(i)}(h)$. To avoid such increased discrepancies between the actual and derived profiles, other methods of selection of the intervals Δh_i considered below and in Section 2.8 can be used.

2.7.2 Determination of the Piecewise Continuous Profiles of the Extinction Coefficient and the Column Lidar Ratio Using Range-Dependent Overlapping Intervals

In the previous section, the inversion variant was considered in which the total profiling range is segmented into equal intervals Δh_i. It was shown that such a simple method may not be the best option and may cause significant distortions in the derived profile of the extinction coefficient.

Some improvement can be achieved by using overlapping segments. The use of overlapping segments reveals potential differences in the retrieved data within the same areas. This in turn provides a more reliable estimation of the actual uncertainty in the inversion result, which cannot be disclosed when using common statistical methods.

One should also keep in mind the presence of random noise in the inverted signal, which dramatically increases with range. Therefore, the corresponding increase of the intervals Δh_i over distant ranges may be helpful, the same as the increase of the range resolution when performing numerical differentiation.

The procedure for establishing the variable length of the overlapping intervals is straightforward. After the total altitude range $h_{min} - h_{max}$ is established, the length of the first interval Δh_1, which starts at h_{min} and the extension factor ϵ for the next intervals are selected. When the factor $\epsilon = 1$, is selected, $\Delta h_1 = \Delta h_2 = \ldots = \Delta h_N$. The selection $\epsilon > 1$ increases the length of the following intervals. If the length of the first interval is Δh_1, the next interval $\Delta h_2 = \epsilon \Delta h_1$, then $\Delta h_3 = \epsilon^2 \Delta h_1$, and the last interval, $\Delta h_N = \epsilon^{N-1} \Delta h_1$. In Fig. 2.50(a), the sequence of eight intervals Δh_i which create seven overlapping piecewise intervals is shown. The selected ϵ is equal to 1.1, and the overlap factor $m = 0.5$. The first range interval starts at $h_1 = h_{min}$ and ends at $h_1' = h_{min} + \Delta h_1$; the second starts at $h_2 = h_{min} + m\Delta h_1$, and ends at $h_2' = h_2 + \Delta h_2$. The third interval starts at $h_3 = h_1'$ and ends at $h_3' = h_3 + \Delta h_3$, and so on. The last interval starts at $h_8 = h_6'$ and ends at $h_8' = h_{max}$. In order not to complicate the figure, only the ranges h_5 and h_5' are marked. Fig. 2.50 (b), which gives detailed positions of adjacent overlapping intervals, can help in understanding the principle.

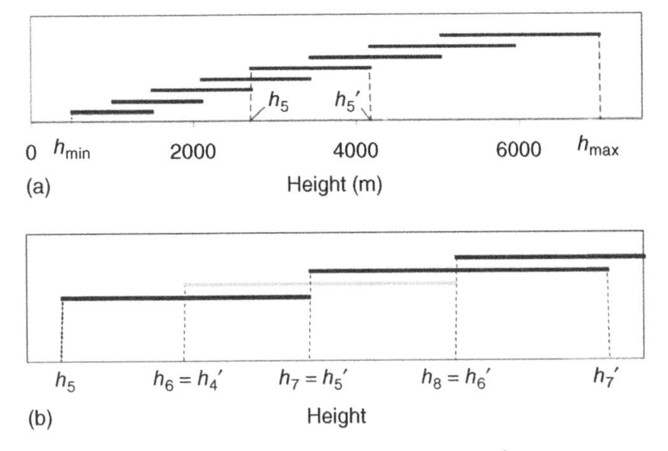

Fig. 2.50 (a) Schematic of the eight overlapping intervals $h_i - h_i'$ whose lengths increase with height. The beginning and the end of the fifth interval are marked as h_5 and h_5' (Kovalev et al., 2011b). (b) Detailed schematic of the fifth, sixth, and seventh overlapping intervals, which clarifies the symbols used in the text.

The extinction coefficient and the piecewise lidar ratio in the overlapping areas can be derived using either simple or weighted averages. The weight function within each overlapping zone (h'_i, h'_{i+1}) used below is calculated with the formula

$$w(h'_i, h'_{i+1}) = \left\{ \sqrt{\frac{\sum_{h'_i}^{h'_{i+1}} \left[\tau_p^{(i)}(0, h) - \tau_{p,\,\mathrm{sh}}(0, h)\right]^2}{n(h'_i, h'_{i+1})}} \right\}, \qquad (2.120)$$

where $n(h'_i, h'_{i+1})$ is the number of data points within the selected overlapped interval (h'_i, h'_{i+1}) [see Fig. 2.50 (b)]. Thus, for the interval (h'_4, h'_5), the weight functions are calculated for the second part of the fifth interval and for the first part of sixth interval; for the interval (h'_5, h'_6), the weight functions are calculated for the second part of the sixth interval, and for the first part of seventh interval, etc.

Let us retrieve the piecewise continuous profiles of the extinction coefficient $\kappa_p^{(i)}(h)$ and the lidar ratio $S_p^{(i)}(h)$ using the same test profiles, $S_p(h)$ and $\kappa_p(h)$, as in Figs. 2.48 and 2.49 but now performing the inversion using the overlapping intervals whose lengths increases with height. For the inversion, the eight overlapping intervals within the altitude range from $h_{\mathrm{min}} = 300$ to $h_{\mathrm{max}} = 4500$ m are selected. The length of the first interval $\Delta h_1 = 700$ m and the extension factor $\epsilon = 1.13$. The inversion results

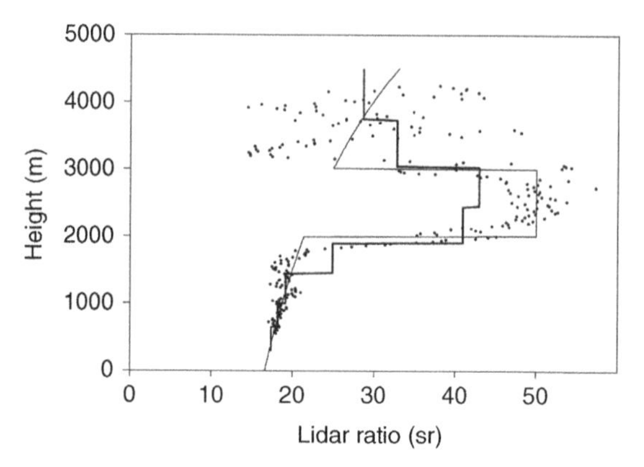

Fig. 2.51 The thin solid curve is the same test profile of $S_p(h)$ as in Fig. 2.48. The piecewise average profile, $S_p^{(i)}(h)$, obtained using eight overlapping intervals is shown as thick curve. The filled circles show the lidar ratio obtained as the ratio of the extinction coefficient $\kappa_p^{(\mathrm{dif})}(h)$ to the backscatter coefficient.

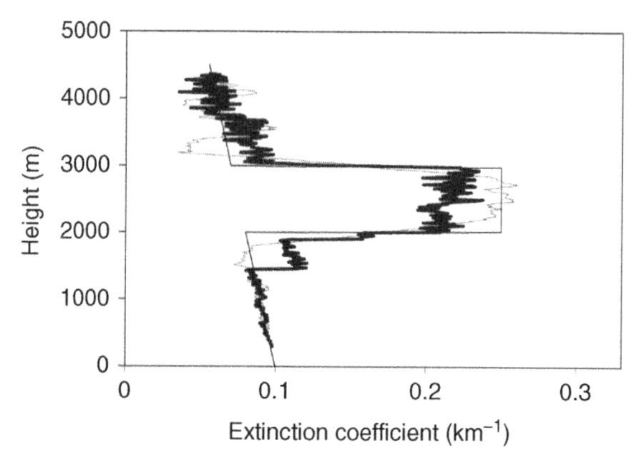

Fig. 2.52 Same as in Fig. 2.49 but where for the retrieval, eight overlapping intervals are used. The profile of $\kappa_p^{(i)}(h)$ averaged over each overlapping interval is shown as the thick solid curve and the extinction coefficient $\kappa_p^{(\mathrm{dif})}(r_i)$ retrieved with numerical differentiation is shown as the dotted curve.

retrieved using simple and weighted averages are shown in Figs. 2.51–2.54. The piecewise lidar ratio $S_p^{(i)}(h)$ and the extinction coefficient $\kappa_p^{(i)}(h)$, averaged within each overlapping area without using weight functions, are shown in Figs. 2.51 and 2.52. The profile of the lidar ratio $S_p^{(i)}(h)$ is still distorted in the areas adjacent to the sharp boundaries of the layer with increased backscattering; over heights 1455–3750 m, the difference between retrieved $S_p^{(i)}(h)$ and test $S_p(h)$ varies within the range ± 20 % and more. However, these distortions are less than the distortions in the lidar ratio

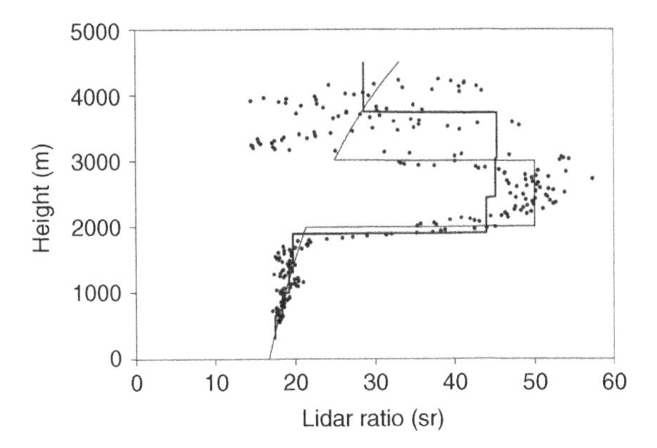

Fig. 2.53 Same as Fig. 2.51 but where for the retrieval of $S_p^{(i)}(h)$, the weighted average within each overlapping area is used. Dotted points show the lidar ratio obtained as the ratio of the extinction coefficient $\kappa_p^{(\mathrm{dif})}(h)$ to the backscatter coefficient.

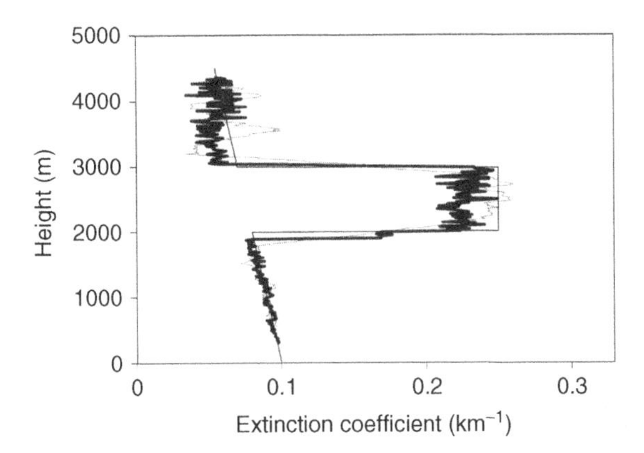

Fig. 2.54 Same as Fig. 2.52 but where the weighted average within each overlapping area is used for the retrieval of the extinction coefficient.

determined as the ratio of the extinction coefficient $\kappa_p^{(\mathrm{dif})}(h)$, obtained with numerical differentiation, to the backscatter coefficient $\beta_{\pi,\,p}(h)$.

The lidar ratio and the extinction coefficient retrieved from the same simulated data but using the weighted averages are shown in Figs. 2.53 and 2.54. Comparing Figs. 2.53 and 2.54 with Figs. 2.51 and 2.52, one can conclude that within and close to the layer of increased backscatter with sharp boundaries, the profile of the extinction coefficient retrieved with the weighted average is more accurate than that retrieved with the simple average. The significant difference between the test lidar ratio $S_p(h)$ and retrieved $S_p^{(i)}(h)$ for the heights ~3100–3800 m is caused by the influence of increased noise over these altitudes.

Let us summarize. The numerical simulations and the experimental data show that the extraction of the piecewise extinction coefficient from the backscatter coefficient assuming invariable column-integrated lidar ratio within restricted height intervals and using the optical depth as a constraint can yield less noisy extinction-coefficient profiles as compared to that obtained with numerical differentiation. In each fragmented interval, the shape of the derived extinction-coefficient profile replicates the relative shape of the backscatter coefficient. Accordingly, the level of the random noise in the derived extinction coefficient is proportional to that in the backscatter-coefficient profile used for the inversion and can be reduced by standard smoothing of the inverted backscatter-coefficient profile.

As with conventional numerical differentiation, the increase in the length of the intervals Δh_i with range reduces the influence of random noise at distant ranges. The derived profile of the extinction coefficient $\kappa_p^{(i)}(h)$ obtained using overlapping intervals, whose lengths increase with height, is generally more accurate than the profile found using equal range intervals. However, in the areas of heterogeneous aerosol loading located at distant ranges, increasing the lengths of the intervals Δh_i may result in the loss of significant details in the profile of the derived extinction coefficient. As to the derived piecewise lidar ratio, the difference between it and the actual profile over distant ranges may also be significant. However, even then, the profile of the lidar ratio is not as scattered as the profile of $S_p(h)$ calculated using the particulate extinction coefficient derived through numerical differentiation.

There are different ways that allow improving the quality of the profiling of the extinction coefficient using the above inversion technique. First, a proper analysis of the shape of the inverted square-range-corrected signal may allow for better selection of the segmented zone for signal inversion. Such an analysis may help to establish the location of the layers with increased backscattering and help to select the boundaries of the segmented zone that correspond with the boundaries of these layers. Second, improvement of this inversion technique may be achieved by using more sophisticated techniques also based on better selection of the segmented zones Δh_i. Such a variant of extracting the piecewise continuous profile of the extinction coefficient, based on the estimate of the uncertainty in the inverted optical depth, is considered in the next section.

2.8 DETERMINATION OF THE EXTINCTION-COEFFICIENT PROFILE USING UNCERTAINLY BOUNDARIES OF THE INVERTED OPTICAL DEPTH

When using the data processing technique discussed in Section 2.7, the most challenging issue is the selection of the length of adjacent (or overlapping) intervals Δh_i. If the arbitrarily selected segmented zone Δh_i includes both the area of a relatively clear atmosphere and a heterogeneous layer with increased backscattering, the piecewise extinction coefficient $\kappa_p^{(i)}(h)$ extracted in this zone may be significantly distorted (see Fig. 2.49). This distortion originates from the assumption that the lidar ratio within the segmented zone is constant, which is not valid in the above case. Therefore, sometimes it is useful to swap the arbitrary selection of the length of the segmented intervals within the operative lidar range by determination of the heights where the slope of the optical depth $\tau_{p,\,sh}(0,\ h)$ abruptly changes. Such an inversion technique is based on the assumption that the heights, where the sharp changes in the slope of the particulate optical depth take place, concur with the heights of the changes in the lidar ratio. Accordingly, to determine the location and the length of the consequent intervals, $(h_i,\ h_{i+1})$, $(h_{i+1},\ h_{i+2})$, etc., one should determine the locations of the sharp change in the slope of the optical depth profile. In other words, one should determine the locations where the increments of the optical depth become noticeably different. To avoid numerical differentiation for determining these slope changes, a more robust method can be used. In this section, a variant of the retrieval technique is presented which allows determining the segmented zones Δh_i using the uncertainty boundaries of the inverted optical depth as constraints. As in the previous section, the piecewise lidar ratio is then found that provides the best match between the initial profile of the optical depth extracted from the splitting lidar signal and the auxiliary profile, determined by integrating the product $S_p^{(i)}(h)\beta_{\pi,\,p}(h)$ within the segmented interval [Eq. 2.94)].

The need for determining the uncertainty boundaries of the inverted particulate optical depth is a significant issue of this technique. As mentioned, when using the conventional numerical differentiation technique, one assumes that common standard statistics provide a proper estimation of the uncertainty boundaries in the extracted atmospheric profile (Whiteman, 1999; Rocadenbosch et al., 2000; Volkov et al., 2002; Shcherbakov, 2007; Adam et al., 2007). However, as shown in the previous sections, systematic errors of an unknown sign and magnitude can have prevailing influence on the accuracy of the inversion result, especially over distant ranges. Unfortunately, there is no commonly accepted technique for addressing the problem of estimation of uncertainty boundaries in the corrupted optical depth profile. Under such circumstances, the only way to solve this issue is to use some empirical estimates of such an uncertainty, even if these have no proper mathematical basis. However, if we consider profiling of the optical parameters of the atmosphere as an *a posteriori* simulation, the selection of modeling details is the researcher's prerogative.

2.8.1 Computational Model for Estimating the Uncertainty Boundaries in the Particulate Optical Depth Profile Extracted from Lidar Data

As in all previous variants, the particulate backscatter coefficient $\beta_{\pi,\,p}(h)$ and the particulate optical depth $\tau_p(0,\ h)$ within the heights from h_{min} to h_{max} should be initially determined. As discussed in Section 2.6, this is done by estimating the constant C, and using the molecular extinction vertical profile. After the profiles $\beta_{\pi,\,p}(h)$ and $\tau_p(0,\ h)$ are separated, two auxiliary profiles are extracted from the initial, not smoothed, optical depth $\tau_p(0,\ h)$. First, the shaped particulate optical depth profile $\tau_{p,\,sh}(0,\ h)$ is determined as discussed in Section 2.5. Second, two auxiliary profiles are extracted for determining (or more exactly - for simulating) the optical depth uncertainty boundaries. These are determined from the unsmoothed optical depth $\tau_p(0,\ h)$. The only exception is the case where the original optical depth is extremely noisy. In such a case, the option "to smooth or not to smooth" and the selection of the level of smoothing is the prerogative of the researcher. If the noise is moderate, the nonsmoothed $\tau_p(0, h)$ is used for simulating the lower and upper optical depth boundaries. Two supplementary profiles are calculated using the formulas

$$\tau_{p,1}(0, h) = \min[\tau_p(0, h); \tau_p(0, h + \Delta h_d); \tau_p(0, h + 2\Delta h_d); \ldots ; \tau_p(0, h_{\max})], \tag{2.121}$$

and

$$\tau_{p,2}(0, h) = \max[\tau_p(0, h_{\min}); \tau_p(0, h_{\min} + \Delta h_d); \tau_p(0, h_{\min} + 2\Delta h_d); \ldots ; \tau_p(0, h)], \tag{2.122}$$

where Δh_d is the sampling resolution for the profile $\tau_p(0, h)$. The absolute difference between these profiles and their mean is

$$\Delta\tau_p(h) = 0.5\lfloor\tau_{p,2}(0, h) - \tau_{p,1}(0, h)\rfloor. \tag{2.123}$$

To clarify the principles of estimating the uncertainty boundaries in the optical depth profile, let us consider the inversion methodology in a simple numerical simulation using the input data shown in Fig. 2.55. The test profiles of the particulate extinction coefficient and the particulate lidar ratio used for the simulation are shown in the left and central panels of the figure. Note the sharp changes in these profiles at the heights 2000 and 3000 m. These profiles are used to generate the synthetic noise-corrupted profile of the optical depth $\tau_p(0, h)$ obtained from the signal of the virtual splitting

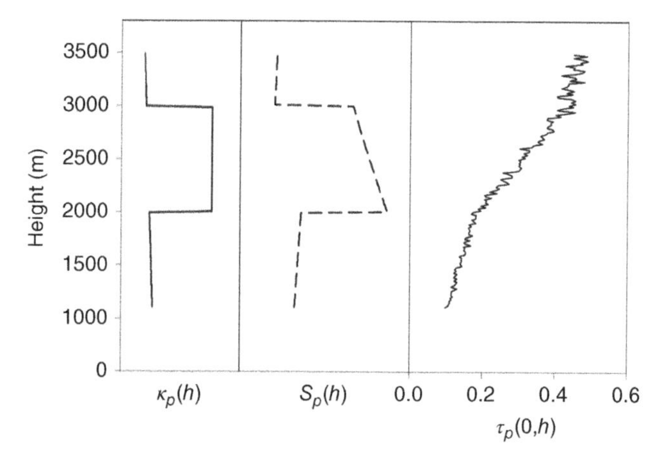

Fig. 2.55 Synthetic optical profiles used for the numerical simulation. Left panel: Test profile of the particulate extinction coefficient $\kappa_p(h)$. Center panel: Test profile of the lidar ratio $S_p(h)$. Right panel: Noise-corrupted profile of optical depth $\tau_p(0, h)$ extracted from the signal of the artificial splitting lidar.

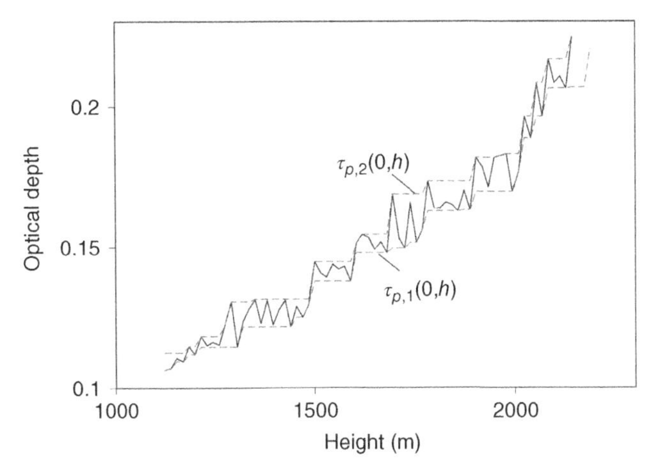

Fig. 2.56 Simulated noise-corrupted profile of optical depth $\tau_p(0,\ h)$ (solid curve) and the corresponding profiles $\tau_{p,\ 1}(0,\ h)$ and $\tau_{p,\ 2}(0,\ h)$.

lidar. This profile is shown as the noisy solid curve in the right panel. Two bounding optical depth profiles, $\tau_{p,\ 1}(0,\ h)$ and $\tau_{p,\ 2}(0,\ h)$, calculated with Eqs. (2.121) and (2.122), are shown as thin dashed curves in Fig. 2.56; the solid curve in the figure is the noise-corrupted profile of the optical depth $\tau_p(0,\ h)$, the same as that in the right panel of Fig. 2.55. The corresponding profile $\Delta\tau_p(h)$ calculated with Eq. (2.123) is shown as the solid curve in Fig. 2.57. As can be expected, the profile extracted from the nonsmoothed optical depth $\tau_p(0,\ h)$ is extremely noisy. Nevertheless, it shows the tendency of increased magnitude with height. To make this tendency more distinct,

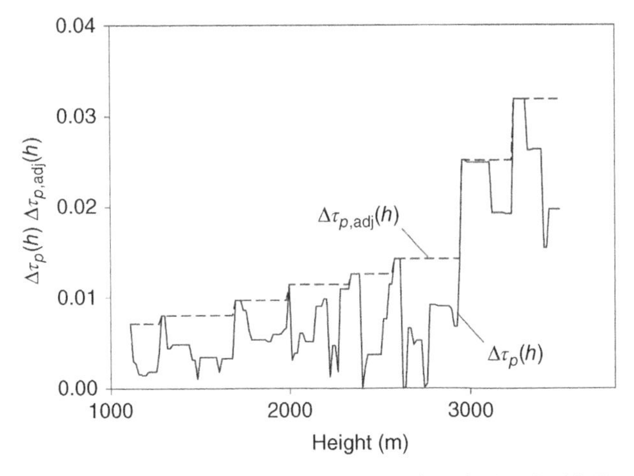

Fig. 2.57 The profiles $\Delta\tau_p(h)$ and the adjusted $\Delta\tau_{p,\ \mathrm{adj}}(h)$ calculated with Eqs. (2.123) and (2.124), respectively.

special smoothing of the function $\Delta\tau_p(h)$ can be done. Such smoothing transforms the noisy profile $\Delta\tau_p(h)$ into the adjusted profile $\Delta\tau_{p,\text{ adj}}(h)$, which increases with height without concavities, that is, which has no negative increments. This can be achieved if the profile $\Delta\tau_p(h)$ is adjusted with the formula

$$\Delta\tau_{p,\text{ adj}}(h) = \max[\Delta\tau_p(h_{\min}); \Delta\tau_p(h_{\min} + \Delta h_d); \Delta\tau_p(h_{\min} + 2\Delta h_d); \ \dots \ ; \Delta\tau_p(h)]. \tag{2.124}$$

The adjusted profile $\Delta\tau_{p,\text{ adj}}(h)$, which is shown in Fig. 2.57, is taken as a simulated uncertainty level for the optical depth from which the extinction coefficient is to be extracted. Obviously, such an estimate of the uncertainty boundaries has no rigid mathematical basis. The formulas in Eqs. (2.121)–(2.124) present some mathematical model for the estimation of uncertainty for the optical depth profile, which in turn, is the result of a simulation, but made on the basis of lidar observations. (Note that the determination of uncertainty boundaries with statistical estimates of random noise in the optical depth is nothing other than the application of another model solution. Such a model solution assumes that the optical depth is extracted from ideal lidar signals with no systematic distortions).

Linking the profile $\Delta\tau_{p,\text{ adj}}(h)$ with the shaped profile $\tau_{p,\text{ sh}}(0, h)$, one can obtain the necessary constraints, that is, the upper and lower uncertainty boundaries for $\tau_{p,\text{ sh}}(0, h)$. These boundary profiles are determined as (Kovalev et al., 2007b)

$$\tau_{p,\text{ up}}(0, h) = \tau_{p,\text{ sh}}(0, h) + \Delta\tau_{p,\text{ adj}}(h), \tag{2.125}$$

and

$$\tau_{p,\text{ low}}(0, h) = \tau_{p,\text{ sh}}(0, h) - \Delta\tau_{p,\text{ adj}}(h). \tag{2.126}$$

Practice shows that the profiles $\tau_{p,\text{ up}}(0, h)$ and $\tau_{p,\text{ low}}(0, h)$ determined with Eqs. (2.125) and (2.126), should generally be corrected in the nearest zone starting at $h_{\min}$. The numerical simulations and atmospheric experiments made by the author have revealed that close to $h_{\min}$, the calculated function $\Delta\tau_{p,\text{ adj}}(0, h)$ might be extremely small. Correction can be recommended when the calculated $\Delta\tau_{p,\text{ adj}}(0, h)$ in the interval nearest to $h_{\min}$ is less than 0.05. When determining the piecewise extinction coefficient in this area, the selection of the range-independent $\Delta\tau_{p,\text{ adj}}(0, h) \approx 0.05$ provides the more stable inversion result.

In Fig. 2.58, the initial noise-corrupted optical depth $\tau_p(0, h)$, the shaped optical depth $\tau_{p,\text{ sh}}(0, h)$, and the upper and lower boundaries of its uncertainty, $\tau_{p,\text{ up}}(0, h)$ and $\tau_{p,\text{ low}}(0, h)$, calculated with Eqs. (2.125) and (2.126), are shown. The profiles are shown in the restricted height intervals from ~1000 to 2200 m for better visualization of the details. The initial noisy optical depth $\tau_p(0, h)$ is shown as the thin solid curve, and the profile of the shaped profile $\tau_{p,\text{ sh}}(0, h)$ is shown by the empty diamonds. Note that the extreme points of the initial noisy optical depth $\tau_p(0, h)$ may be beyond the boundary functions $\tau_{p,\text{ up}}(0, h)$ and $\tau_{p,\text{ low}}(0, h)$.

Unlike the retrieval technique considered in Section 2.7, in this inversion variant, the value of $S_p^{(i)}$ is found that maximizes the interval Δh_i within which the shaped

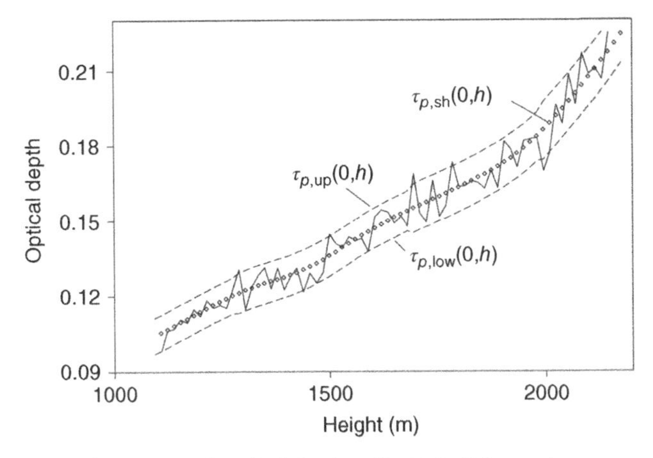

Fig. 2.58 Initial noise-corrupted optical depth $\tau_p(0,\ h)$ (solid curve), same as in the right panel of Fig. 2.55, and the related functions: the shaped profile $\tau_{p,\ \mathrm{sh}}(0,\ h)$ and the upper and lower uncertainty boundaries, $\tau_{p,\ \mathrm{up}}(0,\ h)$ and $\tau_{p,\ \mathrm{low}}(0,\ h)$ (Kovalev et al., 2007b).

profile $\tau_{p,\ \mathrm{sh}}(h_i,\ h)$ and the alternative $\tau_p^{(i)}(h_i,\ h) = S_p^{(i)} \int_{h_i}^{h} \beta_{\pi,\ p}(x)\, dx$ remain close to each other. The latter condition is taken as being satisfied if the profile of the optical depth $\tau_p^{(i)}(h_i,\ h)$ does not extend beyond the area restricted by the boundary profiles $\tau_{p,\ \mathrm{up}}(0,\ h)$ and $\tau_{p,\ \mathrm{low}}(0,\ h)$, shown in Fig. 2.58 as dashed curves.

2.8.2 Essentials of the Data Processing Technique

To determine the piecewise continuous extinction-coefficient profile, the starting point h_i of the interval Δh_i and an arbitrary lidar ratio $S_p^{(i)}$ within it are chosen. Using the ratio and the profile of the particulate backscatter coefficient $\beta_p(h)$, the profile of the optical depth $\tau_p^{(i)}(0,\ h)$ is found with the formula

$$\tau_p^{(i)}(0,\ h) = \tau_{p,\ \mathrm{sh}}(0,\ h_i) + \left[S_p^{(i)} \int_{h_i}^{h} \beta_p(h')dh' \right], \qquad (2.127)$$

where $\tau_{p,\ \mathrm{sh}}(0,\ h_i)$ is the shaped optical depth at the starting point, h_i. Unlike the previous variant in Section 2.7, the length of the interval Δh_i is not selected, so that the upper boundary of the interval is initially limited only by the maximal height $h_{\max}$.

The simplified flowchart in Fig. 2.59 clarifies the principle of determining the maximum interval within which the replacement of the actual column lidar ratio $\overline{S_p(h_i, h)}$ [Eq. (2.93)] by the selected constant $S_p^{(i)}$ does not shift the derived profile $\tau_p^{(i)}(0, h)$ outside the boundaries restricted by $\tau_{p,\ \mathrm{up}}(0,\ h)$ and $\tau_{p,\ \mathrm{low}}(0,\ h)$. Under such a condition, the shapes of the profiles $\tau_{p,\ \mathrm{sh}}(0,\ h)$ and $\tau_p^{(i)}(0, h)$ are close to each other, so that the replacement of the profile $\tau_{p,\ \mathrm{sh}}(0,\ h)$ by $\tau_p^{(i)}(0,\ h)$ will not significantly distort the extracted piecewise extinction coefficient $\kappa_p^{(i)}(h)$, within the interval $\Delta h_{i,\ \max}$.

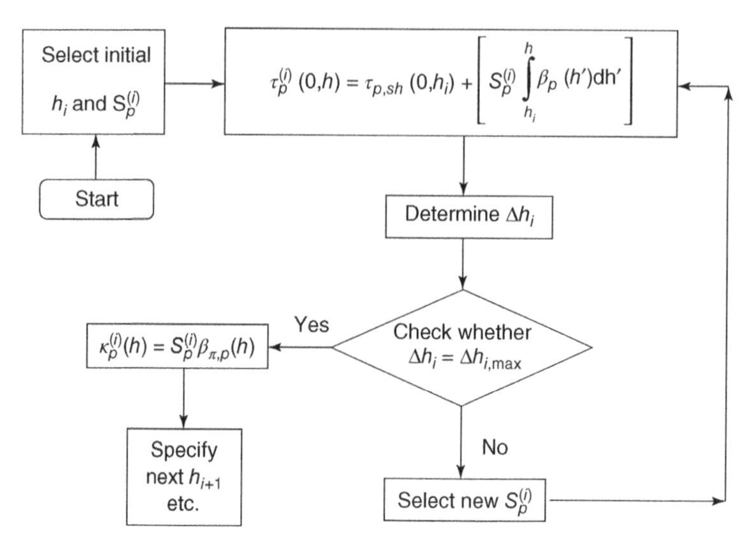

Fig. 2.59 Simplified flowchart for determining the maximum length of the interval $\Delta h_{i,\,\mathrm{max}}$ and extracting the piecewise profile $\kappa_p^{(i)}(h)$ within this interval. (Adapted from Kovalev et al., 2007b).

The procedure of extraction of the piecewise extinction coefficient is as follows. After the initial starting point $h_i = h_{\mathrm{min}}$ and an arbitrary value of $S_p^{(i)}$ are chosen, the corresponding profile $\tau_p^{(i)}(0, h)$ is calculated and compared to the profile $\tau_{p,\,\mathrm{sh}}(0,\ h)$. The goal is to determine how well the profiles $\tau_p^{(i)}(0, h)$ and $\tau_{p,\,\mathrm{sh}}(0,\ h)$ match each other and at what distance they start to significantly diverge. In particular, the maximum height is established at which the optical depth $\tau_p^{(i)}(0, h)$ goes outside the area restricted by the boundaries, $\tau_{p,\,\mathrm{up}}(0,\ h)$ or $\tau_{p,\,\mathrm{low}}(0,\ h)$, that is, where it intersects any of these boundaries. The value of $S_p^{(i)}$ is varied until the maximal distance $h_{i,\,\mathrm{max}}$ is found. After the maximum distance, and accordingly, the maximum interval $\Delta h_{i,\,\mathrm{max}}$ is established, the piecewise extinction coefficient for this interval is determined.

The principle underlying the determination of the maximum interval $\Delta h_{i,\,\mathrm{max}}$ within which the profile $\tau_p^{(i)}(0,\ h)$ does not extend beyond $\tau_{p,\,\mathrm{up}}(0,\ h)$ or $\tau_{p,\,\mathrm{low}}(0,\ h)$ is illustrated in Figs. 2.60 and 2.61. After the initial lidar ratio $S_p^{(i)}$ is selected, the corresponding profile of $\tau_p^{(i)}(0,\ h)$ over the heights $h \geq h_{\mathrm{min}}$ is calculated. In practice, it is convenient to start this procedure with some minimum value of the lidar ratio $S_p^{(i)}$, selecting it, for example, as being equal to the molecular lidar ratio, and then consequently increasing it until the best value is found. The profile $\tau_p^{(i)}(0, h)$, which is shown in Fig. 2.60 by the solid triangles, is obtained using initial $S_p^{(i)} = 8\pi/3$ sr. It intersects the lower boundary profile $\tau_{p,\,\mathrm{low}}(0,\ h)$ at the point (a), which is quite close to the starting point $h_{\mathrm{min}} = 1110$ m. In other words, the profiles $\tau_{p,\,\mathrm{sh}}(0,\ h)$ and $\tau_p^{(i)}(0,\ h)$ diverge at an extremely short distance, matching each other only within the short interval $\Delta h_i \approx 105$ m. Obviously, the selected lidar ratio $S_p^{(i)} = 8/3\pi$ sr is too small and needs to be increased. Indeed, increasing the value of $S_p^{(i)}$ shifts the

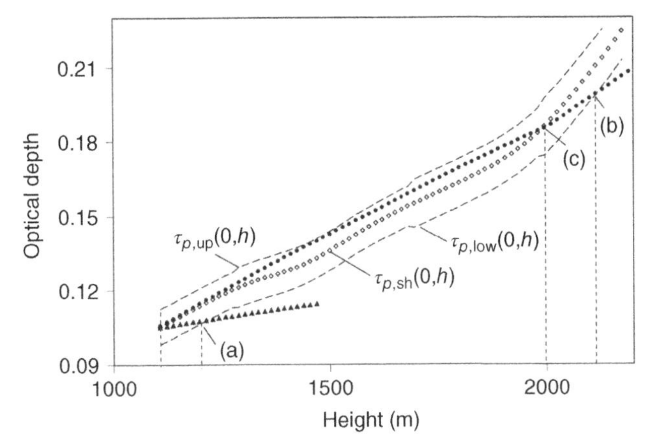

Fig. 2.60 Shaped profile $\tau_{p,\,\mathrm{sh}}(0,\,h)$ and the profiles $\tau_{p,\,\mathrm{up}}(0,\,h)$ and $\tau_{p,\,\mathrm{low}}(0,\,h)$ within the first interval $\Delta h_{i,\,\max}$. The solid triangles show the profile of $\tau_p^{(i)}(0,\,h)$ derived with $S_p^{(i)} = 8/3\pi$ sr, whereas the solid dots show the same profile derived with $S_p^{(i)} = 31$ sr (Kovalev et al., 2007b).

intersection point further from the starting point h_i, so that $\tau_p^{(i)}(0,\,h)$ and $\tau_{p,\,\mathrm{sh}}(0,\,h)$ match each other over larger distances. By incrementally increasing $S_p^{(i)}$, one should find the value that results in the maximal possible interval $\Delta h_i = \Delta h_{i,\,\max}$ over which the calculated optical depth $\tau_p^{(i)}(0,\,h)$ remains within the area restricted by the boundaries $\tau_{p,\,\mathrm{low}}(0,\,h)$ and $\tau_{p,\,\mathrm{up}}(0,\,h)$. For the case shown in the figure, the maximum length takes place with the selection of $S_p^{(i)} \approx 31$ sr; the profile of $\tau_p^{(i)}(0,\,h)$ that corresponds to this $S_p^{(i)}$ is shown in the figure by the solid dots. The intersection point is shifted to the point (b) at the height, $h_{i,\,\max} = 2130$ m, that is, $\Delta h_{i,\,\max} = 1020$ m. Any further increase in $S_p^{(i)}$ will sharply decrease rather than increase Δh_i and degrade the profile match. Even a slight increase of $S_p^{(i)}$ will result in a significant reduction of the intersection height. In particular, if the selected $S_p^{(i)}$ becomes even slightly larger than 31 sr, the profile $\tau_p^{(i)}(0,\,h)$ intersects the upper boundary $\tau_{\mathrm{up}}(0,\,h)$ at a lower height, reducing the interception height from 2130 down to ~1500 m. Thus, the selection of the lidar ratio $S_p^{(i)} = 31$ sr provides the maximal height interval, $\Delta h_{i,\,\max}$.

After the maximum length $\Delta h_{i,\,\max}$ for the first interval is established, the maximum length for the next interval Δh_{i+1} should be determined. Numerous numerical and atmospheric experiments made by the author have shown that the best inversion results are obtained when the segmented zone, where the corresponding piecewise extinction coefficient $\kappa_p^{(i)}(h)$ is determined, is a little less than the maximum length $\Delta h_{i,\,\max}$; accordingly, the starting height h_{i+1} for the next interval should be less than the previous $h_{i,\,\max}$, that is,

$$h_{i+1} = h_{i,\,\max} - \Delta h^*. \tag{2.128}$$

To determine the shift Δh^*, one should find the intersection point of the profile $\tau_p^{(i)}(0,\ h)$, with the shaped function $\tau_{p,\ \mathrm{sh}}(0,\ h)$ closest to the point (b). As one can see in Fig. 2.60, this intersection point (c) is shifted down to $\Delta h^* = 135$ m. Thus, the end of the first interval, which coincides with the starting point for the second interval, should be selected at the point (c), that is, at the height 1995 m.

Now one should find the value of $S_p^{(i+1)}$ which provides the maximum length $\Delta h_{i+1,\ \max}$ for the second segmented zone. This is done using the same method as above. The results of determining the second consecutive profile $\tau_p^{(i+1)}(0,\ h)$ are shown in Fig. 2.61. The location of the intersection point (b) for the second interval is 3210 m and the point (c) is located at $h = 3045$ m. Thus, the second interval extends over the heights from 1995 to 3045 m. These steps for determining the maximum length of each interval are repeated until the entire piecewise continuous profile $\tau_p^{(i)}(0,\ h)$ and the corresponding extinction coefficient is found over the entire operative range from $h_{\min}$ to $h_{\max}$. The piecewise profiles $\kappa_p^{(i)}(h)$ are found over the segmented zones between the adjacent intersection points labeled as (c) in Figs. 2.60 and 2.61.

One special feature of this variant should be mentioned. When determining the lidar ratio $S_p^{(N)}$, for the uppermost interval, the procedure of determining the maximum length of the interval may not provide a unique value for $S_p^{(N)}$. Therefore, for the last interval, the difference between $\tau_p^{(i)}(0, h)$ and $\tau_{p,\ \mathrm{sh}}(0,\ h)$, should be minimized the same way as it was done when arbitrarily selected intervals were used. The lidar ratio $S_p^{(N)}$ for the N-th is found by determining the profile of $\tau_p^{(i)}(0,\ h)$ that is the closest to $\tau_{p,\ \mathrm{sh}}(0,\ h)$; this is achieved by using the criterion $\epsilon(\Delta h_i)$ [Eq. (2.119) in Section 2.7].

Thus, the basic principle for the determination of the extinction-coefficient piecewise continuous profile discussed in this section is similar to that discussed in Section 2.7. However, there are two differences between these two variants. First, in the variant given here, the intervals are not arbitrarily selected, and their selection is based on the estimates of the uncertainty in the optical depth, $\tau_p(0,\ h)$. Second, the

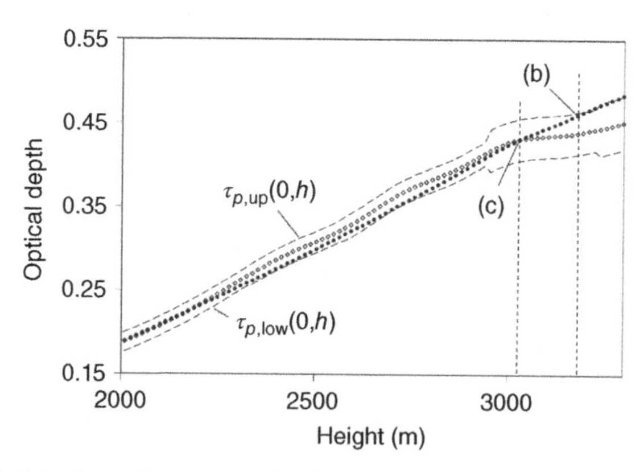

Fig. 2.61 Optical depth profiles, same as in Fig. 2., but determined within the second interval. The solid dots show the profile of $\tau_p^{(i)}(0,\ h)$ calculated with $S_p^{(i)} = 56.5$ sr.

lidar ratio that provides the proper match between $\tau_{p,\,sh}(0,\ h)$ and $\tau_p^{(i)}(0,\ h)$ within the maximal interval, is found under the condition that the optical depth $\tau_p^{(i)}(0,\ h)$ remains within the area restricted by the boundary profiles $\tau_{low}(0,\ h)$ and $\tau_{up}(0,\ h)$.

To go over the main results of the above numerical simulation, the piecewise continuous profiles of the lidar ratio $S_p^{(i)}$ and the particulate extinction coefficient $\kappa_p^{(i)}(h)$ for the total altitude range from $h_{min} = 1100$ to $h_{max} = 3500$ m are presented in Fig. 2.62. The retrieved profiles in both panels are shown as thick solid curves, whereas the initial test profiles are shown as dashed curves. The value of $S_p^{(i)}$ for the last interval, over the heights from 3045 to 3500 m, equal to 20 sr, is determined using the criterion $\epsilon(\Delta h_i)$. In both piecewise continuous profiles, the boundaries of increased aerosol loading between the heights 2000 and 3000 m are well defined. This observation is not valid for the extinction coefficient $\kappa_p^{(dif)}(r)$ obtained with numerical differentiation of the optical depth $\tau_p(0,\ h)$, shown in the right panel as empty circles. One can see that this profile, obtained with the sliding height resolution $s = 450$ m, does not properly define the boundaries of the increased aerosol loading; it erroneously extends the thickness of the vertical layer ~400 m relative to the true value. The decrease of the range resolution below 450 m is not helpful. It increases the noise fluctuation in the derived extinction coefficient $\kappa_p^{(dif)}(r)$, making it impossible to discern these boundaries.

It is interesting to compare two inversion variants, the variant given in this section and that considered in Section 2.7. Let us invert the same test profiles, $S_p(h)$ and $\kappa_p(0,\ h)$, given in Figs. 2.48 and 2.49. This time, for the inversion, we will use the condition that the data points of the retrieved $\tau_p^{(i)}(0,\ h)$ do not go outside the area defined by the corresponding boundary profiles $\tau_{p,\,up}(0,\ h)$ and $\tau_{p,\,low}(0,\ h)$. Under

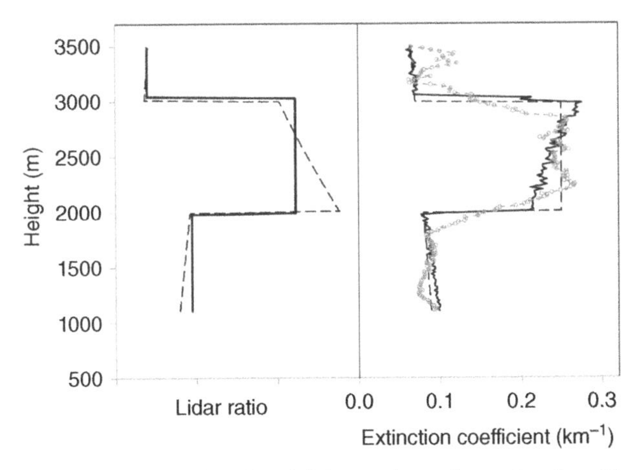

Fig. 2.62 Thick solid curves in the left and right panels are the retrieved profiles of the piecewise continuous lidar ratio $S_p^{(i)}(h)$ and the extinction coefficient $\kappa_p^{(i)}(h)$, respectively. The dashed curves in both panels are the test profiles used for the simulations. The empty circles in the right panel show the profile extinction coefficient $\kappa_p^{(dif)}(r)$.

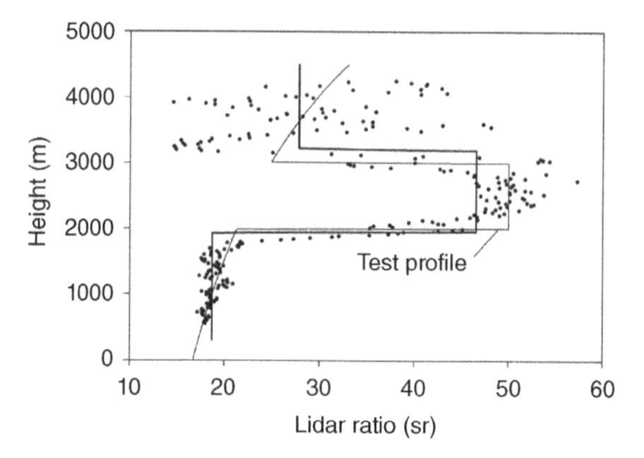

Fig. 2.63 The test profile of $S_p(h)$ (thin solid curve), the retrieved lidar ratio $S_p^{(i)}(h)$, obtained using the maximum attainable segmented intervals (thick curve) and the lidar ratio, obtained as the ratio of $\kappa_p^{(\mathrm{dif})}(h)$ to $\beta_{\pi,p}(h)$ (Kovalev et al., 2007c).

such a condition, the total altitude range is segmented into three unequal intervals, 300–1965, 1965–3210, and 3210–4500 m. The corresponding ratios $S_p^{(i)}$ are 18.7, 48.5, and 27.8 sr, respectively; the value of $S_p^{(i)}$ for the uppermost zone is obtained by minimizing the criterion $\epsilon(\Delta h_i)$. The inversion results are shown in Figs. 2.63 and 2.64. In Fig. 2.63, the retrieved piecewise continuous function $S_p^{(i)}(h)$ is shown as the thick solid curve, whereas the thin solid curve is the initial test profile of the lidar ratio $S_p(h)$. Comparing the inversion results in Fig. 2.63 with those in Fig. 2.48, where equal, arbitrarily selected segmented zones Δh_i were used, one can clearly see the advantage of the variant when determining the maximum attainable intervals. The

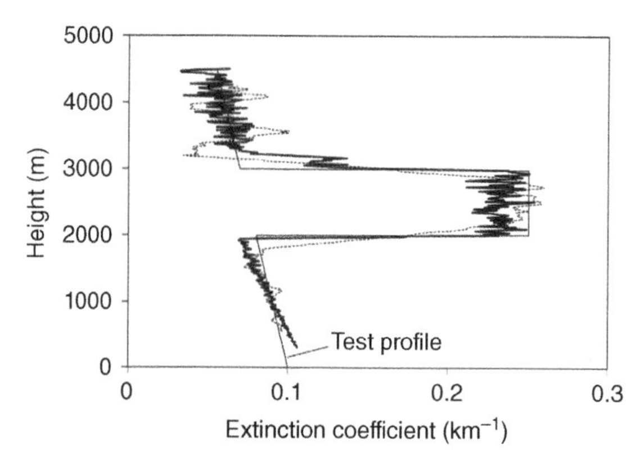

Fig. 2.64 The vertical profiles of the particulate extinction coefficient. The thin solid curve is the test profile used in the numerical simulation, and the thick solid curve shows $\kappa_p^{(i)}(h)$ obtained when using the maximum attainable segmented intervals. The dotted curve is the profile of $\kappa_p^{(\mathrm{dif})}(h)$ (Kovalev et al., 2007c).

same conclusion follows when comparing the profiles of the piecewise continuous extinction coefficient $\kappa_p^{(i)}(h)$ shown in Figs. 2.49 and 2.64. Thus, comparing the two inversion methods, one can state that in the layered atmosphere, the method that determines the maximum attainable intervals within the lidar operative range may provide more accurate inversion results than the variant with the arbitrarily selected intervals.

The dotted curve in Fig. 2.64 is the profile of the extinction coefficient $\kappa_p^{(\mathrm{dif})}(h)$ derived through numerical differentiation of the optical depth $\tau_p(0, h)$ with the height resolution $s = 500$ m. This profile is very close to the extracted piecewise continuous profile $\kappa_p^{(i)}(h)$ shown as the thick solid curve. Meanwhile, the data points of the corresponding lidar ratio, calculated as the ratio of the extinction coefficient $\kappa_p^{(\mathrm{dif})}(h)$ to the backscatter coefficient $\beta_{\pi, p}(h)$ are extremely scattered, especially over heights greater than 3000 m; these data points are shown in Fig. 2.63 as the filled circles.

2.8.3 Examples of Experimental Data obtained in the Clear Atmospheres

Let us consider specifics of the inversion results obtained from experimental data when using maximum attainable segmented intervals within the lidar operative range. To demonstrate how robust this method is when being utilized in a clear cloudless atmosphere, we present the example of lidar experimental data corrupted with an extremely high level of random noise. These data at the wavelength 355 nm were obtained in the multiangle mode, during lidar test and adjustment. To extract the height profiles of the optical depth $\tau_p(0, h)$ and the relative backscatter coefficient $C\beta_\pi(h)$, the multiangle inversion method, discussed in the study by Adam et al., 2007, was used. The constant C was determined using the assumption of a particulate-free atmosphere at the altitude ~4000 m. The extracted profiles of the particulate optical depth $\tau_p(0, h)$ with no smoothing are shown as the dotted curve in Fig. 2.65. Curve 1 is the shaped profile of the optical depth, $\tau_{p,\,\mathrm{sh}}(0, h)$, the thin dashed curves 2 and 3 are the estimated uncertainty boundaries $\tau_{p,\,\mathrm{up}}(0, h)$ and $\tau_{p,\,\mathrm{low}}(0, h)$, respectively. Because of extremely excessive noise in the optical depth $\tau_p(0, h)$, these boundary profiles are significantly spread out the profile, $\tau_{p,\,\mathrm{sh}}(0, h)$. This objective restricts the number of the segmented intervals $\Delta h_{i,\,\max}$ within the heights from 0.23 to 3.5 km to two. The corresponding piecewise continuous extinction-coefficient profile $\kappa_p^{(i)}(h)$, given without any smoothing, is shown in Fig. 2.66 as the thick solid curve. In spite of increased noise, the profile looks sensible for this clear and cloudless atmosphere. The nephelometer readings show good agreement with the profile of $\kappa_p^{(i)}(h)$ at the near end. For comparison, the extinction-coefficient profile $\kappa_p^{(\mathrm{dif})}(h)$ obtained through numerical differentiation is also shown in the figure. This profile, shown as the dotted curve, has extremely doubtful spatial fluctuations, unphysical negative values over the heights ~2200 – 2800 m, and sharp increase (not shown in the figure) over higher altitudes. Meanwhile, the corresponding profile of $\beta_{\pi, p}(h)$ does not show any increase in particulate loading over these altitudes; the extinction coefficient $\kappa_p^{(i)}(h)$ also has no tendency to increase over high altitudes.

This inversion result needs some comments related to the specifics of extracting the piecewise extinction coefficient when using the maximal attainable intervals $\Delta h_{i,\,\max}$.

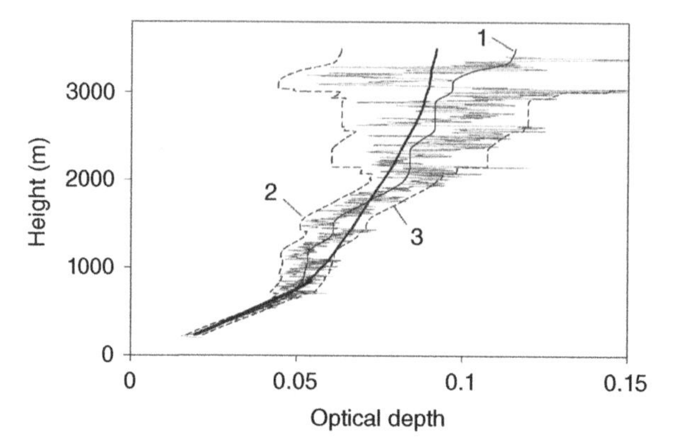

Fig. 2.65 The noisy dotted curve is the nonsmoothed profile of the particulate optical depth $\tau_p(0,\ h)$ obtained in a clear cloudless atmosphere. Curve 1 is the shaped profile $\tau_{p,\ \mathrm{sh}}(0,\ h)$ and the thick solid curve is the piecewise continuous profile $\tau_p^{(i)}(0,\ h)$. Curves 2 and 3 show the profiles of $\tau_{p,\ \mathrm{low}}(0,\ h)$ and $\tau_{p,\ \mathrm{up}}(0,\ h)$ (Kovalev et al., 2007b).

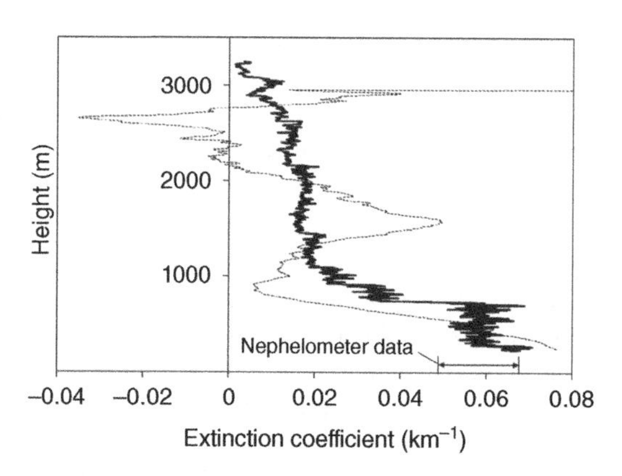

Fig. 2.66 The piecewise continuous profile $\kappa_p^{(i)}(h)$ (thick solid curve) derived using the boundaries $\tau_{p,\ \mathrm{low}}(0,\ h)$ and $\tau_{p,\ \mathrm{up}}(0,\ h)$ shown in Fig. 2.65 and the extinction-coefficient profile obtained through numerical differentiation of $\tau_p(0,\ h)$ (dotted curve). The range of variations of the ground-based nephelometer data during the lidar measurement is shown in the bottom of the figure (Kovalev et al., 2007b).

When the level of noise in the inverted optical depth profiles is small, the shape of the derived extinction coefficient primarily depends on the fluctuations in the optical depth. This observation is not valid if the noise in the original profile $\tau_p(0,\ h)$ is large, as in Fig. 2.65, so that the spread of the functions $\tau_{p,\ \mathrm{low}}(0,\ h)$ and $\tau_{p,\ \mathrm{up}}(0,\ h)$ over distant ranges is also large. In this case, the influence of the fluctuations in the optical depth on the extracted extinction coefficient $\kappa_p^{(i)}(h)$ decreases whereas the influence of these on the profile $\beta_{\pi,\ p}(h)$ increases. One can say that the inversion algorithm "does not trust" the shape of $\tau_{p,\,sh}(0,\ h)$, when the uncertainty boundaries in it are too large; in such a case, it relies more on the shape of $\beta_{\pi,\,p}(h)$. In short, when extracting $\kappa_p^{(i)}(h)$ from the extremely noisy optical depth, the influence of the profile of $\beta_{\pi,\,p}(h)$ dominates.

In Fig. 2.67, the lidar ratio profiles, extracted from these experimental data using different retrieval methods are shown. The thick solid Curve 1 is the piecewise continuous lidar ratio $S_p^{(i)}(h)$, the values of which within the two intervals are 37.6 sr and 27.9 sr, respectively. The profile of the lidar ratio $S_p(h)$ determined as the ratio of the extinction coefficient $\kappa_p^{(\mathrm{dif})}(r_i)$ to the backscatter coefficient $\beta_{\pi,\,p}(h)$ is shown as Curve 2. As usual, the lidar ratio extracted in this way is extremely noisy, here even over lower heights. Meanwhile, it follows from Eqs. (2.92) and (2.93) that there is another way to extract the profile of the lidar ratio with numerical differentiation, which provides a significantly less scattered profile of $S_p(h)$. Such a profile can be extracted from the column-integrated lidar ratio. The differentiation of Eq. (2.93), followed by simple transformations, yields the following dependence between the local and column-integrated lidar ratios (Kovalev et al., 2007b):

$$S_p(h) = \overline{S_p(h_i,\ h)} + \frac{\int_{h_i}^{h} \beta_{\pi,\,p}(h')dh'}{\beta_{\pi,\,p}(h)} \frac{d}{dh}[\overline{S_p(h_i,\ h)}], \tag{2.129}$$

where h_i is the starting point of the integrated ratio $\overline{S_p(h_i,\ h)}$. To reduce noise fluctuations in the derived $S_p(h)$, the functions $\overline{S_p(h_i,\ h)}$ in Eq. (2.129) should be replaced by the smoothed column-integrated lidar ratio $\overline{S_{p,\,\mathrm{sm}}(h_i,\ h)}$. Such a smoothed profile

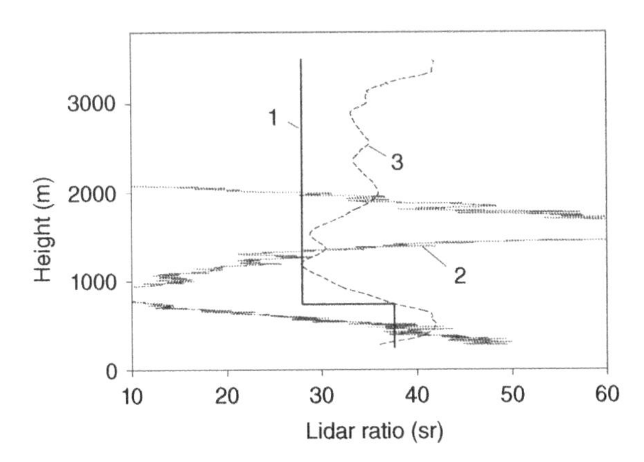

Fig. 2.67 Piecewise continuous lidar ratio $S_p^{(i)}(h)$ (Curve 1) and the profile of $S_p(h)$ found as the ratio of the $\kappa_p^{(\mathrm{dif})}(r_i)$ to the backscatter coefficient $\beta_{\pi,\,p}(h)$ (Curve 2). The profile of the lidar ratio $S_{p,\,\mathrm{sm}}(h)$ obtained by combining Eqs. (2.129) and (2.130) is shown as Curve 3 (Kovalev et al., 2007b).

$\overline{S_{p,\,\mathrm{sm}}(h_i,\ h)}$ can be found as

$$\overline{S_{p,\,\mathrm{sm}}(h_i,\ h)} = \frac{\tau_{p,\,\mathrm{sh}}(h_i,\ h)}{\int_{h_i}^{h} \beta_{\pi,\,p}(h')dh'}. \tag{2.130}$$

Replacing both functions $\overline{S_p(h_i,\ h)}$ in the right side of Eq. (2.129) by the smoothed function $\overline{S_{p,\,\mathrm{sm}}(h_i,\ h)}$, one can obtain the local profile of $S_{p,\,\mathrm{sm}}(h)$, which is much smoother than that obtained as the ratio of $\kappa_p^{(\mathrm{dif})}(h)$ to the backscatter coefficient. The profile of the lidar ratio $S_{p,\,\mathrm{sm}}(h)$ obtained in this way is shown in Fig. 2.67 as Curve 3. Up to the height ~1600 m, the difference between the lidar ratios $S_{p,\,\mathrm{sm}}(h)$ and $S_p^{(i)}(h)$ does not exceed ~10%, but then sharply increases up to ~25%, following the sharp increase in the uncertainty boundaries of the optical depth $\tau_p(0,\ h)$.

Finally, let us consider the case when relatively moderate noise is present in the optical depth, $\tau_p(0,\ h)$. For the inversion, the experimental data obtained when profiling the smoky layers in the vicinity of Montana's I-90 Fire are used. For convenience of comparing the inversion results, the same experimental profiles as in Section 2.7 are considered here. The nonsmoothed optical depth profile $\tau_p(0,\ h)$ obtained in the vicinity of the wildfire is shown in Fig. 2.68 as the dotted curve and the corresponding shaped profile $\tau_{p,\,\mathrm{sh}}(0,\ h)$, as the solid curve; the uncertainty boundaries $\tau_{p,\,\mathrm{up}}(0,\ h)$ and $\tau_{p,\,\mathrm{low}}(0,\ h)$ are shown as the dashed curves. In Fig. 2.69, the profiles, $\tau_p(0,\ h)$ and $\tau_{p,\,\mathrm{sh}}(0,\ h)$, are shown together with the profile of the piecewise continuous optical depth $\tau_p^{(i)}(0,\ h)$, shown as the bold curve. The comparison of the corresponding piecewise continuous profiles $\kappa_p^{(i)}(h)$ and $S_p^{(i)}(h)$ with the profiles in Fig. 2.46 and 2.47,

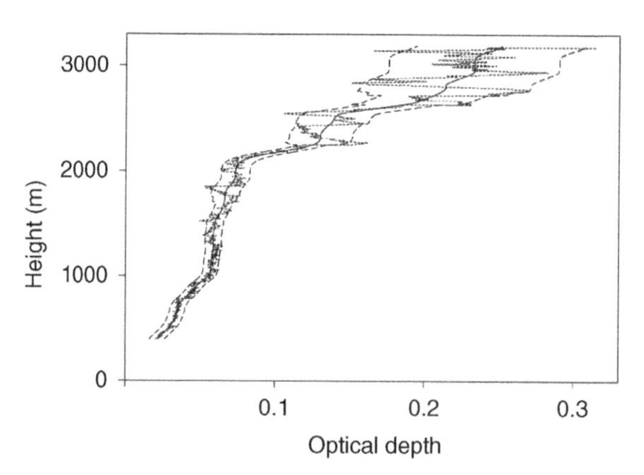

Fig. 2.68 Noise-corrupted optical depth, $\tau_p(0,\ h)$ (dotted curve) and the shaped profile $\tau_{p,\,\mathrm{sh}}(0,\ h)$ (solid curve) extracted from the lidar data in the vicinity of the Montana I-90 Fire. The dashed curves show the estimated upper and lower uncertainty boundaries $\tau_{p,\,\mathrm{up}}(0,\ h)$ and $\tau_{p,\,\mathrm{low}}(0,\ h)$.

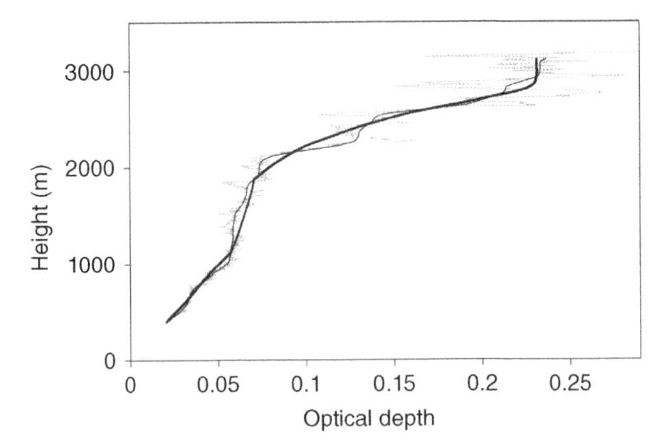

Fig. 2.69 Dotted and thin solid curves are the same as in Fig. 2.68. The thick solid curve is the piecewise continuous optical depth $\tau_p^{(i)}(0, h)$, derived using the uncertainty boundaries, shown in Fig. 2.68.

obtained with equal segmented intervals shows that, unlike the previous case, these alternative profiles are close to each other. Such a result is obtained as the boundaries of the arbitrarily selected intervals Δh_i coincide with the boundaries determined by using the method discussed in Section 2.8.2. The profiles $S_p^{(i)}(h)$ determined by two alternative methods are shown in Fig. 2.70. One can see that the boundaries of the two lower intervals Δh_1 and Δh_2 determined by the two methods almost coincide, whereas the upper segmented intervals Δh_3 and Δh_4, located above the height ~2000 m, have different extent, probably due to increased signal noise in these distant areas. The profiles of $\kappa_p^{(i)}(h)$ determined by the two methods are practically identical over the whole altitude range from $h_{\min}$ to $h_{\max}$.

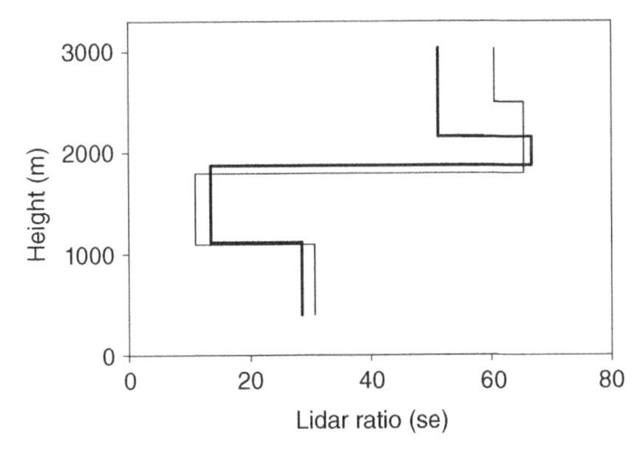

Fig. 2.70 The thin and thick solid curves are the profiles of the piecewise continuous lidar ratio $S_p^{(i)}(h)$, derived from the same experimental data when using the alternative inversion variants.

Let us summarize. Unlike the technique considered in Section 2.7, in the inversion technique considered in this section, the number of the segmented intervals within the total operative altitude range, their location, and length are determined by analysis of the shape of the inverted optical depth and its uncertainty boundaries. The method of profiling of the atmosphere considered in this section is based on determining the maximal length of the segmented intervals within the total altitude range from h_{min} to h_{max}. Such an approach minimizes the erroneous changes in the derived profiles caused by the difference between the boundaries of increased backscattering in the examined atmosphere and the boundaries of the selected intervals Δh_i.

2.9 MONITORING THE BOUNDARIES AND DYNAMICS OF ATMOSPHERIC LAYERS WITH INCREASED BACKSCATTERING

There are different ways to select the length of the intervals when profiling the extinction coefficient using the optical depth and backscatter-coefficient profiles. The simplest way uses arbitrarily selected intervals, Δh_i. However, as shown in Section 2.7, inappropriate selection of the interval length and location can produce significant distortion in the extracted extinction coefficient $\kappa_p^{(i)}(h)$. Such distortions appear if the particulate backscattering has large variations within the selected interval; for example, when the air is clear in a near zone of the interval and increased backscattering takes place over the far end. To avoid significant distortion in the derived profile of the extinction coefficient, it is desirable to select the length and location of the intervals Δh_i that would match the length and location of the atmospheric layers with different levels of backscattering. If the requirement of the absence of large backscatter gradients within the interval is satisfied, one can expect that the initial requirement of an invariable lidar ratio within this interval to also be met. The inversion technique based on estimating the uncertainty boundaries in the inverted optical depth profile (Section 2.8) produces a better match between the boundaries of the selected intervals and the actual boundaries of the atmospheric layers. However, such a technique is only applicable when layers with different levels of backscattering and corresponding changes in the slope of the inverted optical depth profile are well defined.

To satisfy the requirement that the boundaries of layers with different levels of backscattering and selected intervals Δh_i should coincide, initial analysis of the recorded signals can be very helpful. Lidar is extremely appropriate for determining the location of intense backscatter regions in the atmosphere. It can easily detect the boundary between different atmospheric layers and reliably discriminate regions with high levels of backscattering from regions of clear atmosphere. Obviously, the analysis of the lidar signal shape for initial determining the locations of backscatter layers may be considered as the additional way to properly select the lengths and locations of the intervals Δh_i within the lidar operative range.

Different methodologies have been proposed for identifying regions and boundaries of intense backscatter from lidar signals. Over the years, a variety of the lidar data processing methods have been tested and used. The most advanced techniques for determining the boundary between clear air and areas with increased particulate loading were developed in numerous studies of the boundary layer (Kunkel et al., 1977; Melfi et al., 1985; Hooper and Eloranta, 1986; Boers and Melfi, 1987; Piironen and E. W. Eloranta, 1995; Flamant et al., 1997; Menut et al., 1999; Brooks, 2003; Baars et al., 2008; etc). However, there is no possibility to establish a standard criterion for determining the aerosol layer boundary which would be acceptable for different tasks of atmospheric sounding. The selection of concrete criteria for determining the boundaries of atmospheric heterogeneous layers is always highly subjective. Initially, the boundary layer height was found by establishing a threshold level for the backscatter signal intensity (Melfi et al., 1985; Boers and Melfi, 1987). A more popular technique for determining the boundary-layer height was based on determining either the first- or the second-order derivative of the square-range-corrected signal (Menut et al., 1999, Flamant et al., 1997). When using the gradient method, the boundary location was determined as the height where the examined parameter of interest, the derivative of the square-range-corrected lidar signal, has either a maximum value or decreases from the maximum value down to a fixed, user-defined level. Alternative data processing techniques were focused on the behavior of the variance or covariance profiles of the lidar signal (Hooper and Eloranta, 1986; Piironen and Eloranta, 1995). Then the wavelet covariance transform technique became popular, which at present is assumed to be most practical (Brooks, 2003; Baars et al., 2008). The wavelet technique requires the selection of concrete parameters, and this is a significant challenge when a region of intense backscatter with a poorly defined boundary is examined (Brooks, 2003). The bottom line is that there is no commonly accepted technique for remotely monitoring the location and temporal and spatial changes of regions of increased backscatter with poorly defined boundaries, such as is found in the troposphere. Consequently, researchers are continuously looking for new ways suited to large dataset applications (Mao et al., 2011) to solve the issue.

In the following, a technique which allows defining the boundaries of even weak aerosol heterogeneity in the troposphere is considered. The technique is based on principles used for determining the cloud-base height when the ceiling has no well-defined lower boundary.

2.9.1 Methodology

The recorded lidar signal $P_\Sigma(r)$ is the sum of two components, the range-dependent backscatter signal $P(r)$, and the range-independent offset B, which originates from the background component of the lidar signal and the electronic offset:

$$P_\Sigma(r) = P(r) + B. \tag{2.131}$$

Unlike the commonly used gradient techniques, the technique for determining the areas of increased backscattering proposed in the studies by Kovalev et al. (2009b, 2011a) does not require initial separation of these two components in the lidar signal. This specific significantly simplifies data processing. The recorded signal is directly transformed into the auxiliary function $y(\nu)$, defined as

$$y(\nu) = P_\Sigma(\nu)\nu = [P(\nu) + B]\nu, \tag{2.132}$$

where ν is the new independent variable. Two variants of defining this variable can be used for lidar searching, depending on the lidar wavelength. If the lidar operates at the wavelength at which the molecular component as compared to the particulate component is minor, as happens when profiling at the wavelength, $\lambda = 1064$ nm, the variable ν is defined as the squared range, that is, $\nu = r^2$. If the molecular component is large, or at least, comparable with the particulate component, the molecular attenuated backscatter $\beta_{\pi,m}(r)T_m^2(0, r)$ should also be taken into consideration. This situation can occur when lidar sounding in clear atmospheres is made at wavelengths close to the UV range of the spectra, for example, at $\lambda = 355$ nm, and the particulate component is relatively small.

The variable ν is then defined as (Kovalev et al., 2009b)

$$\nu = \frac{r^2}{\beta_{\pi,m}(r)T_m^2(0, r)}. \tag{2.133}$$

The lidar signal is more sensitive to variations in backscattering than to changes in optical depth profile. Therefore, in order to determine the heterogeneity boundaries, it is preferable to use a lidar operating at the wavelength 1064 nm rather than at shorter wavelengths. Such a case is analyzed in the following. Accordingly, here the independent variable ν is defined as the squared range and the molecular terms in Eq. (2.133) are omitted.

The procedure of finding the boundaries of aerosol layering begins by determining the sliding linear fit $Y(\nu)$ of the function $y(\nu)$, which is then extrapolated to $\nu = 0$. The intercept point of the linear fit with the vertical axis can be determined as

$$Y_0(\nu) = Y(\nu) - \frac{\Delta Y}{\Delta \nu}\nu. \tag{2.134}$$

The term $\Delta Y/\Delta \nu$ is calculated using a variable sliding step $\Delta \nu$. Numerical differentiation is made with a constant step $\Delta r = r_{i+1} - r_i = \text{const}$. Accordingly, the corresponding variable step $\Delta \nu$ can be defined as

$$\Delta \nu = r_{i+1}^2 - r_i^2 = \Delta r^2\left(1 + \frac{2r_i}{\Delta r}\right). \tag{2.135}$$

The intercept function $Y_0(\nu)$ is then adjusted and defined as the absolute value of the ratio

$$Y_0^*(\nu) = \left|\frac{Y_0(\nu)}{\nu + \epsilon_\nu \nu_{max}}\right|, \tag{2.136}$$

where ν_{max} is the maximum value of the variable ν over the selected range, that is, $\nu_{max} = [r_{max}]^2$, and ϵ_ν is a positive nonzero constant, whose value may range from ~0.02 to ~0.05. The component $\epsilon_\nu \nu_{max}$ in the denominator of the equation is included to suppress the excessive increase of the function $Y_0^*(\nu)$ in the region of small $\nu \to 0$, which, in our case, is not the region of interest. Therefore, the selection of the value of ϵ_ν is not critical.

After the function $Y_0^*(v)$ is determined, the next step is to transform it into a function of range or height. In this section, the data processing technique for zenith-directed lidar is considered, so that in all of the formulas that follow, the functions will be defined as functions of height. Accordingly, the function $Y_0^*(v)$ is transformed into the function $Y_0^*(h)$, calculated within the total altitude range from h_{min} to h_{max}. The data points of $Y_0^*(h)$ are found for the heights $h_1 = h_{min}$, $h_2 = h_{min} + \Delta h, \ldots, h_j = h_{min} + (j-1)\Delta h, \ldots$, and $h_M = h_{max}$, where Δh is the selected height resolution and M is the total number of discrete height intervals within the interval (h_{min}, h_{max}). To find the location of aerosol layer boundaries and to determine whether they are sharp or dispersed, it is convenient to analyze the ratio function normalized to unity, that is,

$$R_Y(h) = \frac{Y_0^*(h)}{Y_{0,\,max}^*(h)}. \tag{2.137}$$

Here, $Y_{0,\,max}^*(h)$ is the maximal value of $Y_0^*(h)$ within the total altitude range from h_{min} to h_{max}; accordingly, the ratio $R_Y(h)$ can vary between zero and unity.

2.9.2 Determining the Boundaries of Layers Having Increased Backscattering

To clarify the general principles of monitoring the boundaries and dynamics of the atmospheric layers with increased backscattering, the simplest variant for processing the data of zenith-directed lidar is considered. The specifics of the data processing technique are analyzed, using the lidar signals obtained during the CalNex-LA 2010 experiment. The experiment was focused on studying the pollution sources in the Los Angeles urban area (see http://www.esrl.noaa.gov/csd/calnex/whitepaper.pdf). The FSL zenith-pointed lidar operated daily, from May 20 to May 31, 2010, starting sounding at 8:00 and finishing at 18:00–18:30, local time. During operation, the recorded signal was averaged for 1 min (1800 pulses) every 15 min. These signals were used to determine the locations of areas of intense backscatter and temporal changes of their boundaries during the day.

Let us consider details of the data processing technique using typical backscatter signals obtained during this experiment. Such a typical 1-min averaged square-range-corrected signa, $P(h)h^2$ versus height is shown in Fig. 2.71. The only conclusion that can be made by visual analysis of the signal is that the sounded atmosphere contains two separated layers of different thickness and the aerosol pollution tends to decrease with height. Let us apply the processing technique given above to this signal. In the left panel of Fig. 2.72, the corresponding function $R_Y(h)$ obtained with Eq. (2.137) is shown as the thin solid curve; the thin dotted curve in the right panel is the initial square-range-corrected signal. Four spikes of the function $R_Y(h)$ in the left panel coincide with the locations of sharp changes in the square range-corrected signal, that is, show the locations of boundaries of multiple layering. Using these locations, one can define the boundaries of the layers, and

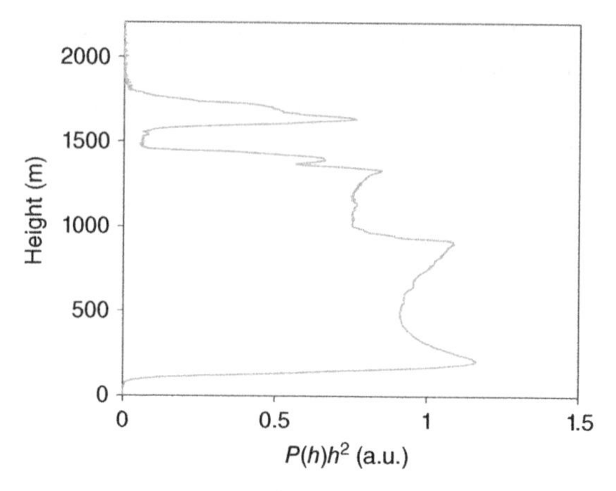

Fig. 2.71 Typical 1-min averaged square-range-corrected signal $P(h)h^2$ versus height obtained during the CalNex-LA 2010 experiment.

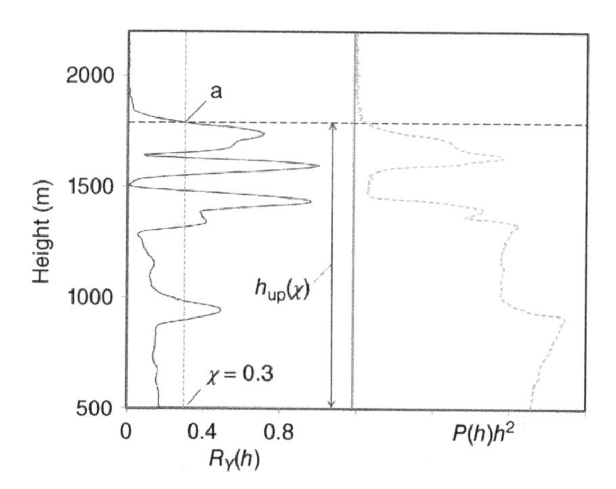

Fig. 2.72 Right panel: The square-range-corrected signal, same as in Fig. 2.71. Left panel: The normalized ratio function $R_Y(h)$, extracted from the signal on the right panel.

this knowledge can provide the basis for a more educated selection of the intervals required for using the methodology discussed in Section 2.7. In our case, such intervals can be selected within the heights from 500 to ~900 m, from 1000 to 1400 m, and from 1400 m to the maximal height, h_{max}. The use of shorter intervals at high altitudes needs to be investigated separately.

Let us examine the function, $R_Y(h)$, focusing on the upper boundary of layering that should be determined by lidar. To determine this boundary, some level of $R_Y(h)$, denoted by the symbol χ, should be selected and the maximum height should be found where $R_Y(h)$ is equal to the selected χ. In Fig. 2.72, the point where $R_Y(h)$ intersects the selected level $\chi = 0.3$ is shown; the corresponding height $h_{\mathrm{up}}(\chi) = 1790$ m is marked *a*.

In the case of a poorly defined boundary, the derived height $h_{\mathrm{up}}(\chi)$ of the polluted layer will strongly depend on the selected level, χ. To minimize this issue, a special methodology should be used. Instead of a fixed χ, a number of discrete levels, χ_1, χ_2, χ_2, etc., should be selected, and the location of the intersection points of these levels with $R_Y(h)$, that is, the corresponding heights, $h_{\mathrm{up}}(\chi_1)$, $h_{\mathrm{up}}(\chi_2)$, etc., should be found. In Fig. 2.73, the heights $h_{\mathrm{up}}(\chi)$ are shown for the set of χ selected with the increment 0.05. As can be expected, selecting different χ yields different heights $h_{\mathrm{up}}(\chi)$. When changing χ from 0.1 to 0.2, the retrieved height $h_{\mathrm{up}}(\chi)$ decreases from 1820 to 1800 m, whereas when χ is changed from 0.7 to 0.75, the height decreases from 1745 to 1615 m. In other words, to apply this technique in practice, simple and sensible principles for selecting χ should be used.

The analysis of possible techniques for the determination of boundaries of areas of increased backscattering shows that an optimal solution can be achieved by taking advantage of the principles used for determining the cloud-base height. When the ceiling has a poorly defined lower boundary, the cloud-base height is defined as the lowest level of the atmosphere where cloud properties are detectable (see Website of the ARM Climate Research Facility, U.S. Department of Energy, http://www.arm.gov/measurements/cloudbase). Such a definition is very practical, so the analogous definition was used in the study by Kovalev et al. (2011a); here the upper height of the region of intense backscatter with a poorly defined boundary between this region and the clear air above was determined as the maximal height where aerosol heterogeneity is detectable, that is, where it can be discriminated from noise. To determine such a boundary with lidar, two alternative methods considered below can be used.

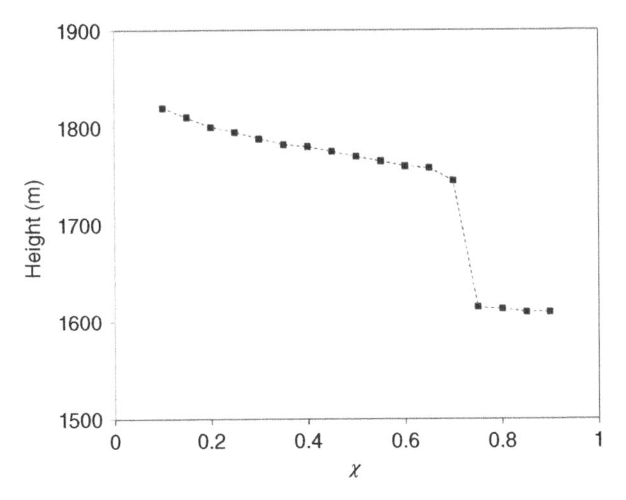

Fig. 2.73 Dependence of the height, $h_{up}(\chi)$, on selected χ calculated for the signal shown in Fig. 2.71.

A Determination of the Maximum Height of the Area of Increased Backscattering by Selecting the Optimal Level of the Normalized Ratio Function To show how this variant can be applied to zenith measurements, let us consider some results obtained during the CalNex-LA experiment. The basic principle for determination of the optimal level χ_{opt} is clarified in Figs. 2.74–2.78. In Fig. 2.74, three typical vertical profiles of the lidar signals measured at 8:15, 11:00, and 14:00 local time on May 25, 2010 are shown; the corresponding functions $Y_0^*(h)$ obtained from these signals are shown in Fig. 2.75. These functions were converted into the ratio function $R_Y(h)$, that is, normalized to unity, and then used to determine optimal quantities χ_{opt} and the corresponding upper boundaries of layering of interest $h_{\text{up}}(\chi_{\text{opt}})$.

The methodology of these procedures is as follows. Initially, the zero level, $\chi_0 = 0$ is selected. Because of the presence of nonzero noise in the signals and in the corresponding functions $R_Y(h)$, the height $h_{\text{up}}(\chi_0)$ will always be equal to the selected h_{max}; in our case, $h_{\text{up}}(\chi_0) = h_{\text{max}} = 5000$ m. Then the next consecutive level χ is selected and analyzed. It is convenient to select some fixed range resolution $\Delta\chi$, so that the next level would be $\chi_1 = \Delta\chi$, then $\chi_2 = 2\Delta\chi$, ... , $\chi_N = N\Delta\chi$ (note the maximal $\chi_N \leq 1$). In the case under consideration, we selected $\Delta\chi = 0.05$, accordingly, $\chi_1 = 0.05$, $\chi_2 = 0.1$, etc. For the data measured at 8:15 and 14:00 (Figs. 2.76 and 2.78), the selection of the next level $\chi_1 = 0.05$ does not change the initial value of the height, that is, $h_{\text{up}}(\chi_1) = h_{\text{up}}(\chi_0) = h_{\text{max}}$. In other words, the level $\chi_1 = 0.05$ selected in these two cases yields the value of $R_Y(h)$, which remains below the interfering noise. For the data, obtained at 11:00 (Fig. 2.77), the level $\chi_1 = 0.05$ shifts the height $h_{\text{up}}(\chi_1)$ down, so that $h_{\text{up}}(\chi_1) = 3828$ m. Selecting the levels, $\chi_2 = 0.1$, $\chi_3 = 0.15$, etc., we can determine for each discrete χ, the corresponding maximum heights, $h_{\text{up}}(\chi_2)$, $h_{\text{up}}(\chi_3)$, etc., that is, the heights where the function $R_Y(h) = \chi_i$. The objective of the above operations is to establish when the noise-induced false fluctuations

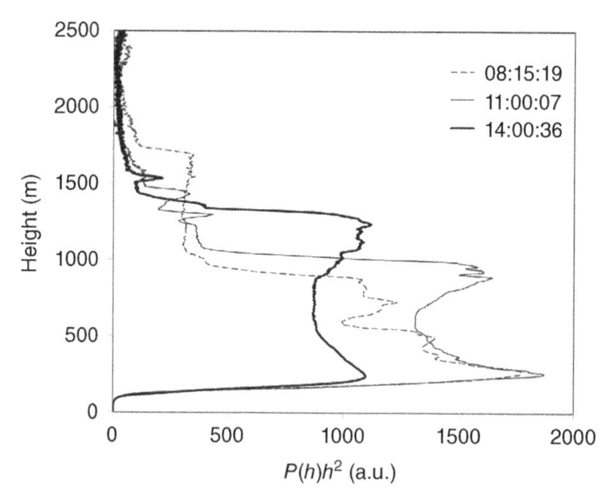

Fig. 2.74 Examples of the square-range-corrected lidar signals obtained in Pasadena, CA, on May 25, 2010.

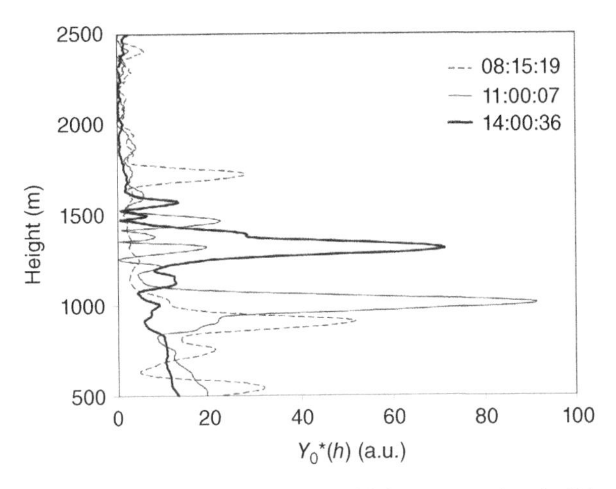

Fig. 2.75 Vertical profiles of the function $Y_0^*(h)$, which correspond to the lidar signals shown in Fig. 2.74.

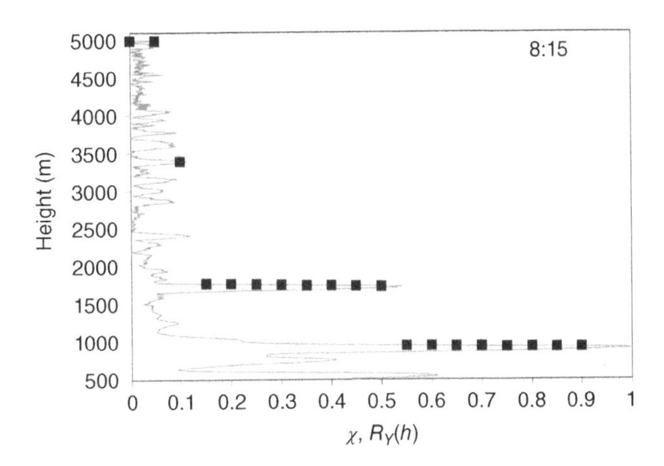

Fig. 2.76 Vertical profile of the ratio function, $R_Y(h)$ (thin solid curve), and the discrete heights $h_{up}(\chi)$ (filled squares), calculated using the signals of the lidar on May 25, 2012 at 8:15 local time.

of the function $R_Y(h)$ are below the level where reliable discrimination of layering occurs, so that the actual h_{up} can be found. In Figs. 2.76–2.78 the height $h_{\text{up}}(\chi_i)$ for each consecutive discrete χ_i is shown as a filled square.

The key question that should be answered is how to determine the optimal level χ_{opt} which would determine the most likely maximum height of the polluted layer of interest. As proposed in the study by Kovalev et al. (2011a), the determination of such an optimal level χ_{opt} and the corresponding height of interest $h_{\text{up}}(\chi_{\text{opt}})$ can be based on the calculation of the differences between the adjacent heights $h_{\text{up}}(\chi_i)$ and

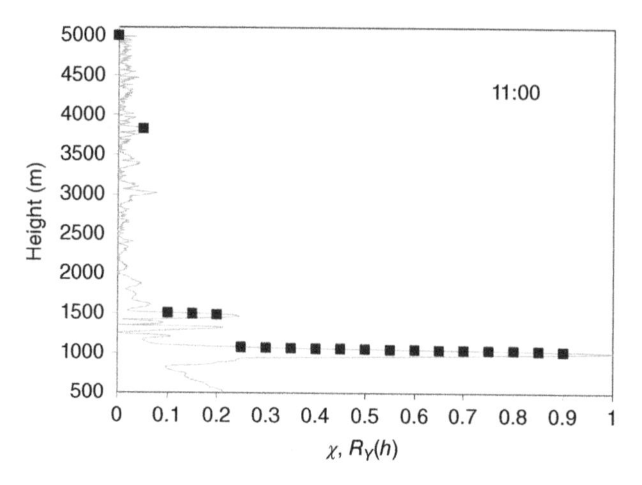

Fig. 2.77 Same as in Fig. 2.76 but measured at 11:00 local time.

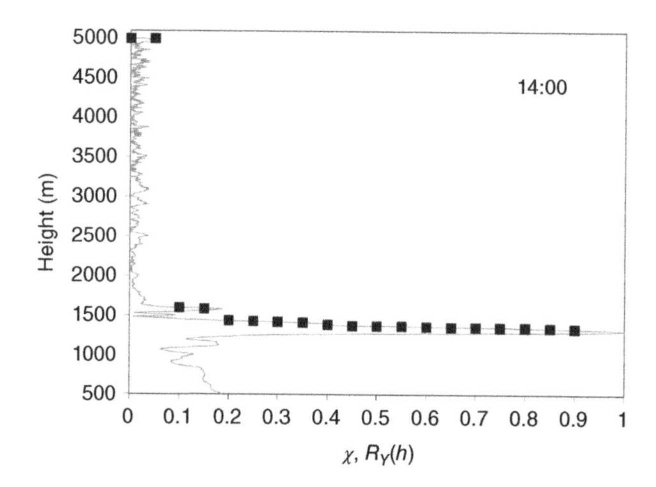

Fig. 2.78 Same as in Fig. 2.76 but measured at 14:00 local time.

$h_{up}(\chi_{i+1})$. The established value of χ_{opt} should meet two conditions. First, the difference between the adjacent heights $h_{up}(\chi_{opt} - \Delta\chi)$ and $h_{up}(\chi_{opt})$ should be maximal. Second, the next increase of χ, that is, the selection of the next consecutive level, equal to $\chi_{opt} + \Delta\chi$, then $\chi_{opt} + 2\Delta\chi$, etc., should result in only a slow decrease or invariability in the corresponding sequence of the heights $h_{up}(\chi)$. It is also worth mentioning that in order to avoid significantly underestimating the smoke plume maximum height, the scope of the level χ_{opt} should be restricted; it should not exceed, presumably, the range, $0.1 - 0.2$. This requirement puts reasonable restrictions on the level of the interfering noise in the lidar signal and accordingly, in the function $R_Y(h)$.

Analysis of experimental data showed that two typical situations can be encountered when determining χ_{opt}. If the smoke plume has an upper boundary with no local layering, the systematic difference between the heights determined using the consecutive levels, χ_{opt}, $\chi_{opt} + \Delta\chi$, $\chi_{opt} + 2\Delta\chi \ldots$, is small, whereas, the difference between the heights determined with the level χ_{opt} and the previous level $\chi_{opt} - \Delta\chi$ is large. The situation may be different when different levels of backscattering exist in the area of the upper boundary of interest. In such a case, the second condition may not be met.

Let us consider how the above requirements are met in the profiles in Figs. 2.76–2.78. In the data obtained at 8:15 (Fig. 2.76), the increase from $\chi = 0.05$ to 0.1 results in the decrease of $h_{up}(\chi)$ from 5000 to ~3400 m. The next consecutive increase from $\chi = 0.1$ to $\chi = 0.15$ results in the decrease of $h_{up}(\chi)$ from 3400 to 1775 m. After that, the consecutive increases of χ from 0.15 to 0.2, then from 0.2 to 0.25, etc. do not significantly reduce the extracted heights $h_{up}(\chi)$; the difference between any pair of consecutive heights, $h_{up}(\chi_i)$ and $h_{up}(\chi_{i+1})$, is less than 10 m. Thus, the application of this principle to the lidar data obtained at 8:15 yields the optimal level $\chi_{opt} = 0.15$ with the corresponding $h_{up}(\chi_{opt}) = 1775$ m.

In the data obtained at 11:00 (Fig. 2.77), the increase from $\chi = 0.05$ to 0.1 results in a sharp decrease of $h(\chi)$ from 3828 to 1510 m. The next two increases of χ from $\chi = 0.1$ to 0.15 and from 0.15 to 0.2 result in the heights 1500 and 1488 m. Thus, the data obtained at 11:00 yields the optimal level $\chi_{opt} = 0.1$ and the corresponding $h_{up}(\chi_{opt}) = 1510$ m.

The situation where the second condition is not properly met is shown in Fig. 2.78. Here, the maximum difference between the heights occurs when χ increases from 0.05 to 0.1. The heights $h_{up}(\chi)$, obtained with $\chi = 0.1$ and $\chi = 0.15$, are quite close to each other (1600 and 1585 m); however, the next increase of χ, that is, the selection of $\chi = 0.2$, decreases the height down to 1434 m. After the next increases of χ, the heights $h_{up}(\chi_i)$ change slightly and monotonically. In most cases, such behavior of $h_{up}(\chi)$ is caused by the presence of a detached aerosol layer close to the top of the boundary layer (see Fig. 2.80). The specifics in the variations of $h_{up}(\chi)$ may provide useful information about the boundaries of the examined aerosol layers, in particular, whether or not they are well defined.

B Determining the Boundaries of the Increased Backscattering Layers in the Troposphere Using Isoclinic Lines Let us discuss the variant when χ-isoclinic lines of the heights $h_{up}(\chi_i)$, extracted with fixed χ are used for determining the layering heights. Such a technique allows analyzing temporal behavior of the χ-isoclinic lines in different atmospheric conditions. In Fig. 2.79, the maximal heights of the polluted air, measured during the entire day on 24 May 2010 are shown. The empty diamonds show the heights $h_{up}(\chi)$ obtained with $\chi = 0.1$, whereas the filled diamonds show the heights obtained with $\chi = 0.15$. The increased difference between the corresponding heights $h_{up}(\chi = 0.1)$ and $h_{up}(\chi = 0.15)$ was caused by the poorly defined upper boundary between the polluted and clear air. Such an optical situation occurred mostly in the morning hours. Starting from 9:15, both heights tend to increase during the day, except for the short period from 13:30 to 15:00. Similarly, the heights

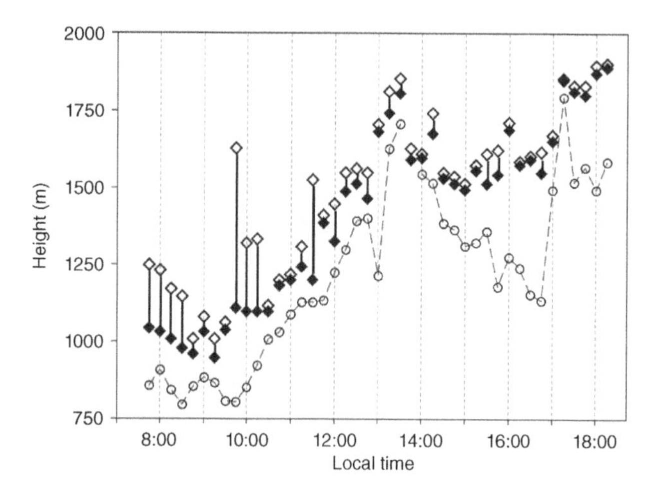

Fig. 2.79 The heights $h_{up}(\chi)$, obtained using different χ during the experiment in Pasadena, CA, on May 24, 2010. The empty diamonds show the heights obtained with $\chi = 0.1$, the filled diamonds are the heights obtained with $\chi = 0.15$, and the heights of maximum heterogeneity ($\chi = 0.9$) are shown as the empty circles. The vertical lines show the cases of the increased difference between the heights, $h_{up}(\chi = 0.1)$ and $h_{up}(\chi = 0.15)$ (Kovalev et al., 2011a).

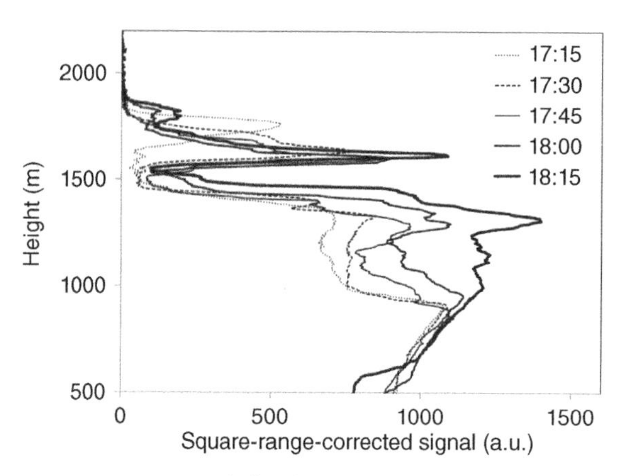

Fig. 2.80 The square-range-corrected signals versus height obtained on May 24, 2010 from 17:15 to 18:15. The detached aerosol layers close to the top of the boundary layer, at the heights ~1550–1850 m are clearly seen (Kovalev et al., 2011a).

of maximum heterogeneity $h_{up}(\chi = 0.9)$ (the empty circles) also show a typical daytime increase of height of the polluted area from approximately 10:00 till 13:30. Then from 14:00 to 16:45, these heights decrease, however, starting at 17:00, they increase again. This effect is caused by the appearance of detached aerosol layers close to the top of the planetary boundary layer, which are clearly seen in the corresponding

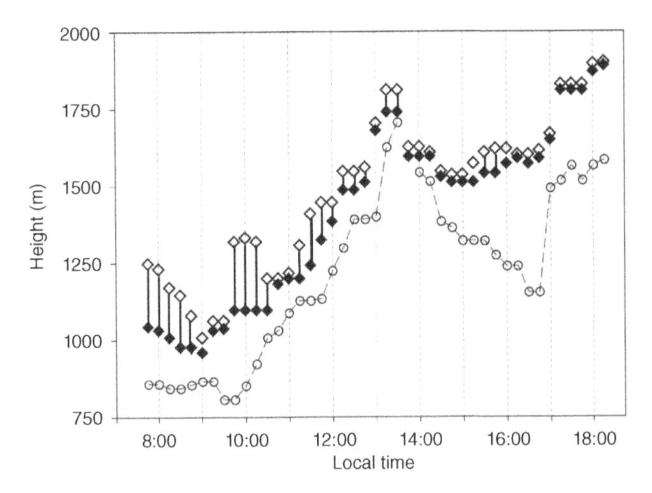

Fig. 2.81 Same as in Fig. 2.79 but here three-point median smoothing of the heights is done.

square-range-corrected signals (Fig. 2.80). Such an effect, which occurs typically during the afternoon hours, is well known (Davis et al., 2000; Hennemuth and Lammert, 2006; Mitev et al., 2011). When determining the boundary-layer height, the detached layers lead to ambiguity in the choice of the "relevant" minimum in the gradient that corresponds to this height.

Different methods have been proposed to avoid this challenge. For example, one can either combine the variance and the gradient methods (Lammert and Bösenberg, 2006), or apply the requirement of minimum continuity in the gradient method (Mitev et al., 2011). Obviously, the use of isoclinic lines can provide all the necessary information required for the analysis of such processes in the boundary layer.

To reduce the scattering effect caused by the appearance of detached layers, three-point median smoothing can be applied. Such a smoothing is performed for each consecutive height, $h^{(j-1)}(\chi_i)$, $h^{(j)}(\chi_i)$, and $h^{(j+1)}(\chi_i)$, using the formula

$$\begin{aligned} h^{(j,\,\mathrm{med})}(\chi_i) = \sum & \left\lfloor h^{(j-1)}(\chi_i),\ h^{(j)}(\chi_i),\ h^{(j+1)}(\chi_i)\right\rfloor \\ & -\min\left\lfloor h^{(j-1)}(\chi_i),\ h^{(j)}(\chi_i),\ h^{(j+1)}(\chi_i)\right\rfloor \\ & -\max\left\lfloor h^{(j-1)}(\chi_i),\ h^{(j)}(\chi_i),\ h^{(j+1)}(\chi_i)\right\rfloor. \end{aligned} \quad (2.138)$$

Such medians $h^{(j,\,\mathrm{med})}(\chi_i)$, calculated for the data points shown in Fig 2.79, are given in Fig. 2.81. One can see that three-point smoothing weakens the "noise effect" caused by the detached layers, and accordingly, allows better visualization of the processes in the boundary layer.

On May 24, 2010, the fluctuation between the heights $h_{\mathrm{up}}(\chi_i)$ determined with different χ_i varied slightly during the day. Another atmospheric situation was monitored the following day. The atmospheric conditions during that day were extremely variable, so that χ_{opt} varied from 0.05 to 0.5, making it difficult to compare the

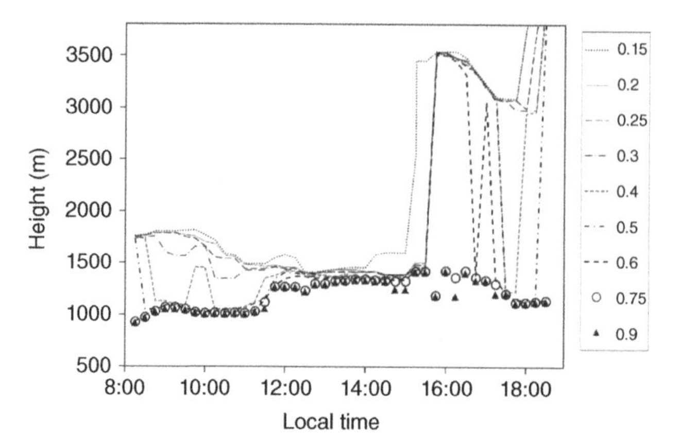

Fig. 2.82 The χ-isoclinic lines obtained on May 25, 2010 for fixed χ_i. The values of χ_i are shown in the legend (Kovalev et al., 2011a).

derived heights. For such a situation, the analysis of the temporal behavior of the heights extracted with different fixed χ is extremely fruitful. Obviously, in the areas of well-defined boundaries, adjacent values of χ_i create isoclinic lines that are close to each other. Contrastingly, large differences in heights obtained for adjacent χ_i indicate layers with poorly defined boundaries.

In Fig. 2.82, the isoclinic lines calculated for May 25, 2010 are shown. During that day, relatively well-defined boundaries existed: (i) for the period from 8:45 to 11:15 - at the heights 1000-1100 m and between 1450 and 1800 m; (ii) for the period from 12:00 to 15:00 - at the heights 1350-1450 m; and (iii) for the period from 15:45 to 17:30 - at the heights 3100-3500 m (for χ_i from 0.2 to 0.5). Two transition periods took place from 11:15 to 12:00 and from 15:00 to 15:45. Note that during the day, the heights of maximal heterogeneity, obtained with $\chi_i = 0.75$ and $\chi_i = 0.9$, change within the relatively restricted range, from, approximately, 1000 to 1400 m, having maximal values from 12:00 to 17:00, and minimal values in the morning, before noon, and in the afternoon, after 17:00.

Let us summarize. As is known, zones of increased backscattering, such as in the planetary boundary layer, in dispersed smoke plumes originated by wildfires, in dust clouds, or in the aerosol clouds created by volcano eruptions, often have poorly defined boundaries and extremely large ranging backscatter coefficients within the polluted area. This situation significantly impedes the determination of the temporal and spatial changes in these formations when using remote sensing instrumentation, such as lidar. No commonly accepted methodology of such investigations exists for both one-directional and scanning lidar. The result of the lidar study of such an important parameter as the maximum height of the layers with increased backscatter will depend on (i) the user's definition of the boundary, (ii) the of backscattering intensity within and outside the formation of interest, and (iii) on the relative level of noise in the measured lidar signals.

The methodology discussed in this section significantly simplifies the determination of the location and boundaries of the areas of increased backscatter. The proposed definition of the upper boundary of the area with increased backscatter as the maximal height where the aerosol heterogeneity is detectable is most practical. It provides maximal sensitivity in the detection of increased backscatter zones in the presence of unavoidable signal noise in the analyzed data. The use of different levels of χ_i for determining the maximum heights of the layers with increased backscatter allows discriminating of both well- and poorly defined boundaries. The extracted parameters of interest for the regions of increased backscatter are much more informative if analyzed in two-dimensional form, for example, as the height-time dependence for zenith lidar.

When investigating urban atmospheres, the use of the lidar scanning mode is generally impossible owing to eye safety regulations and the absence of open space for scanning. In other cases, such as investigating the smoke plume dynamics in the vicinity of wildfires, the multiangle mode is generally unavoidable. The data processing technique considered in this section can be applied to monitoring temporal changes in layers of increased backscattering using either one-directional or multiangle mode. The methodology is simple and robust.

3

PROFILING OF THE ATMOSPHERE WITH SCANNING LIDAR

3.1 PROFILING OF THE ATMOSPHERE USING THE KANO–HAMILTON INVERSION TECHNIQUE

3.1.1 Basics

Splitting lidar, such as high spectral resolution lidar (HSRL) or a combined elastic–inelastic (Raman) lidar enables the separation of the profiles of the relative backscatter component $C\beta_{\pi}(r)$ and the two-way transmittance $T_{\Sigma}^{2}(0, r)$ in the lidar equation, when operating in any fixed direction, either vertical or slope. Elastic lidar can make such a separation only when it operates in a multiangle mode in a horizontally stratified atmosphere. Until now the method proposed by Kano and Hamilton (Kano, 1968; Hamilton, 1969) remains the most common method used for processing the data of the scanning elastic lidar in such atmospheres.

The schematic of multiangle sounding of the atmosphere is shown in Fig. 3.1. The slope lines show the slope directions φ within the selected angular sector, from $\varphi_{\min}$ to $\varphi_{\max}$, in which the lidar is scanning. The thin solid arcs show the minimum and maximum measurement ranges of the lidar, $r_{\min}$ and $r_{\max}$. For simplicity, these ranges are shown in the figure as being equal for all slope directions.

The Kano–Hamilton multiangle method is the only method that allows the extraction of the vertical particulate extinction coefficient profile from elastic lidar data without using an *a priori* assumption about the vertical profile of the lidar ratio. However, the method works only in horizontally stratified atmosphere and requires two stringent conditions to be fulfilled:

Solutions in LIDAR Profiling of the Atmosphere, First Edition. Vladimir A. Kovalev.

First, the total backscatter coefficient, that is, the sum of the molecular and particulate components at any height should be invariable along the horizontal direction, that is,

$$\beta_\pi(h) = \text{const.} \tag{3.1}$$

The requirement in Eq. (3.1) means that the total of both components at any height should be the same for any slope direction used during scanning, at least, within the operative range of the scanning lidar restricted by the large arc in Fig. 3.1.

The second condition in the Kano–Hamilton multiangle method is the validity of the unanimous dependence between the total optical depth $\tau_j(0,h)$ measured along any elevation angle φ and the total vertical optical depth. This dependence can be written in the form

$$\tau_j(0,h)\sin\varphi = \tau_{\text{vert}}(0,h). \tag{3.2}$$

Here $\tau_{\text{vert}}(0,h)$ is the total optical depth of the layer from ground level ($h = 0$) to h in the vertical direction. Note that the subscript "vert" in $\tau_{\text{vert}}(0,h)$ means that this vertical optical depth profile was obtained through calculations using a set of measured slope signals, rather than being determined from the signal of the lidar directed toward the zenith. The optical depth that is obtained directly from the signal of the zenith-directed lidar will be marked with the subscript "90," that is, it will be denoted as $\tau_{90}(0,h)$. The same notation "90" will also be used for any test vertical profile in the numerical simulations.

To extract the optical depth $\tau_{\text{vert}}(0,h)$ the classic Kano–Hamilton inversion method uses the set of functions $y_j(h)$, calculated from the lidar signals $P_j(h)$

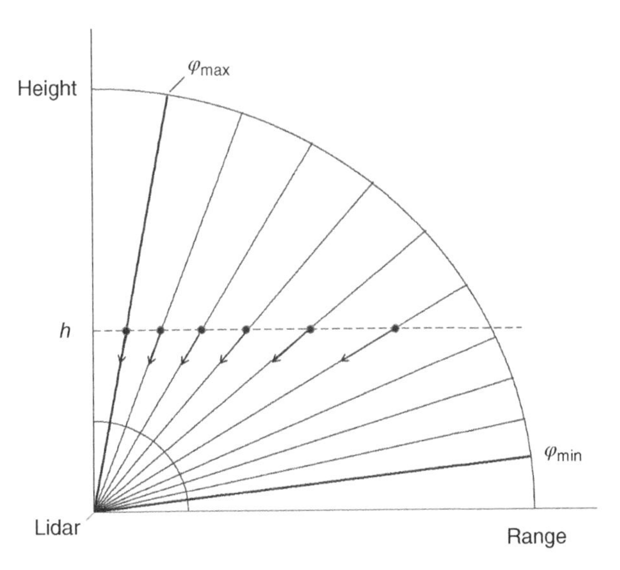

Fig. 3.1 Schematic of multiangle sounding of the atmosphere.

measured under different elevation angles. The functions are defined as

$$y_j(h) = \ln\lfloor P_j(h)(h/\sin\varphi)^2\rfloor = \ln[C\beta_{\pi,j}(h)] - 2\tau_j(0,h), \tag{3.3}$$

where the term $\lfloor P_j(h)(h/\sin\varphi)^2\rfloor$ is the square-range-corrected signal at the height, h, measured along the elevation angle φ; $\beta_{\pi,j}(h)$ is the total backscatter coefficient versus height for the same slope direction, and C is the lidar equation constant, which includes the laser emitting power with compensated variations (if such occur during scanning). The term $\tau_j(0,h)$ is the total optical depth from ground level to the height h measured along the elevation angle φ.

If the requirement in Eq. (3.2) is met, the function $y_j(h)$ can be rewritten in the form

$$y_j(h) = A_j(h) - \frac{2\tau_{\mathrm{vert}}(0,h)}{\sin\varphi}, \tag{3.4}$$

where $A_j(h) = \ln\lfloor C\beta_{\pi,j}(h)\rfloor$. In the atmosphere in which the requirement in Eq. (3.1) for all slope directions is also met, that is, $\beta_{\pi,j}(h) = \beta_\pi(h) = \text{const.}$, and C does not vary during the measurement, the function $A_j(h)$ for any discrete height has the same value for all slope directions, that is,

$$A_j(h) = A(h) = \ln[C\beta_\pi(h)]. \tag{3.5}$$

Accordingly, the linear fit $Y(x,h)$ obtained for the set of the discrete data points $y_j(h)$ can be written versus the independent $x = [\sin\varphi]^{-1}$ as

$$Y(x,h) = A(h) - 2\tau_{\mathrm{vert}}(0,h)x. \tag{3.6}$$

After performing lidar scanning, the data points $y_j(h)$ from the signals measured along different slope angles are initially found. This procedure is performed for a set of the discrete heights within the lidar operative zone, from the minimum height h_{min} up to the maximum height h_{max}. (The principles for determining these heights, h_{min} and h_{max}, will be clarified in Section 3.1.2). The data points $y_j(h)$ are calculated for the discrete heights h_{min}, $h_{\mathrm{min}} + \Delta h$, $h_{\mathrm{min}} + 2\Delta h$, $h_{\mathrm{min}} + 3\Delta h$, $\cdots h_{\mathrm{min}} + n\Delta h$, $\ldots$ h_{max}, where Δh is the selected height resolution. In the next step, the linear fit $Y(x,h)$ is determined. In this procedure, the least-square technique is applied to the data points $y_j(h)$ grouped separately for each discrete height. It follows from Eq. (3.6), that the functions $\tau_{\mathrm{vert}}(0,h)$ and $A(h)$ can be found from the slope of the linear fit $Y(x,h)$ and its point of intersection with the vertical axis, respectively. The intercept of the linear fit is found by extrapolating $Y(x,h)$ to $x = 0$. This intercept point determines the function $A(h)$ in Eq. (3.6), from which the product $[C\beta_\pi(h)]$ versus height can be found. Until recently, the latter function, which shows the relative behavior of the backscatter coefficient profile, was not the subject of researchers' interest. The main focus was on determining the slope of the linear fit $Y(x,h)$ which provides the profiles of the optical depth and the extinction coefficient.

The schematic of the conventional inversion procedure for the extraction of the extinction coefficient from the scanning lidar data is shown in Fig. 3.2. Here the

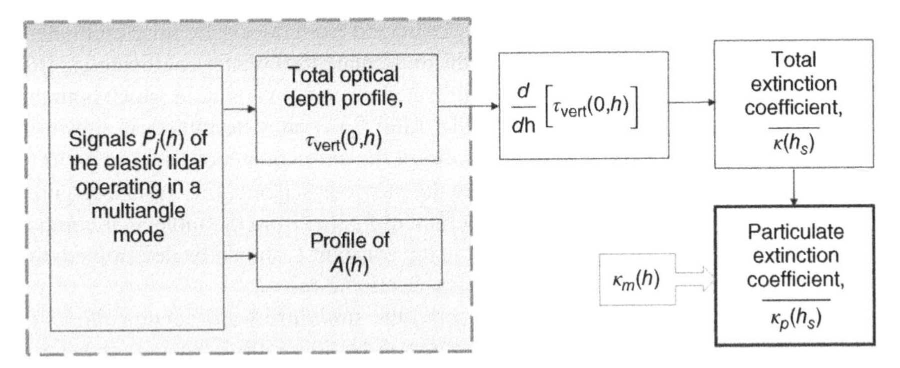

Fig. 3.2 Schematic of profiling of the particulate extinction coefficient through numerical differentiation of the optical depth profile. The blocks with gradient fill, located within the dashed square, show the functions $[C\beta_\pi(h)]$ and $\tau_{vert}(0, h)$, which can be directly extracted from Eq. (3.6).

blocks located within the left square with gradient fill show the data extracted from the linear fit $Y(x, h)$, the optical depth $\tau_{vert}(0, h)$ and the intercept $A(h)$. These functions are obtained under the assumption that the conditions in Eqs. (3.1) and (3.2) in the searched atmosphere are met. Therefore, as with one-directional profiling of the atmosphere, the extracted profiles cannot be considered as a measurement result, but only as a result of simulations based on the measured lidar signals and the applied model of the horizontally stratified atmosphere.

Using the simulated optical depth $\tau_{vert}(0, h)$, one can then get the vertical profile of the corresponding extinction coefficient. These operations are shown in blocks on the right side of the figure. The molecular profile $\kappa_m(h)$ used to separate and extract the particulate component, is shown as the arrow callout. Each slope optical depth $\tau_j(0, h)$, as with the related optical depth $\tau_{vert}(0, h)$, includes both the aerosol and the molecular components, that is, $\tau_j(0, h) = \tau_{m,j}(0, h) + \tau_{p,j}(0, h)$. As with any inversion method, extraction of the particulate component using multiangle lidar data requires knowledge of the corresponding molecular component, this time determined in the required slope directions φ.

The total extinction coefficient $\kappa^{(dif)}(h)$ derived through the numerical differentiation of the optical depth $\tau_{vert}(0, h)$ with the range resolution s is some mean extinction coefficient over the height interval s. As in Section 2.4.1, the extinction coefficient can be assigned to the middle point of the height interval from $h = h_s - 0.5\,s$ to $h = h_s + 0.5\,s$ (see Fig. 2.10). Note that the particulate extinction coefficient, the key function of researchers' interest, can be extracted either as shown in the schematic, by differentiation of the optical depth $\tau_{vert}(0, h)$ with the next subtraction the molecular component $\kappa_m(h)$; or by initially determining the particulate component $\tau_{p,vert}(0, h)$ as the difference between the total and molecular optical depths, and its numerical differentiation.

In the inversion scheme shown in Fig. 3.2, only the profile of the optical depth $\tau_{vert}(0, h)$ is retrieved from the Kano–Hamilton solution. Meanwhile, the particulate lidar ratio is another function that may be extracted from the elastic lidar multiangle data. This can be achieved by extracting the particulate backscatter coefficient $\beta_{\pi,p}(h)$ from the product $C\beta_{\pi}(h)$. This more complicated inversion variant, in which both the extinction coefficient $\kappa^{(\mathrm{dif})}(h)$ and the lidar ratio $S_p(h)$ are determined, is shown in Fig. 3.3. The lidar ratio is found as the ratio of the extinction coefficient $\kappa^{(\mathrm{dif})}(h)$ to the backscatter coefficient $\beta_{\pi,p}(h)$. Unlike the previous scheme, this variant requires separating the particulate backscatter coefficient $\beta_{\pi,p}(h)$ from the molecular component $\beta_{\pi,m}(h)$. To perform such a separation, the constant C should be determined; the details of such an operation are discussed in Section 2.6.2.

The determination of the lidar ratio using the straightforward version shown in Fig. 3.3 may result in extremely noisy profiles of the lidar ratio, especially in heterogeneous areas with sharp changes in aerosol loading. The reason for this hardship lies in the specifics of the above methodology. Instead of directly determining the lidar ratio as the ratio of two data points,

$$S_p(h) = \frac{\kappa_p(h)}{\beta_{\pi,p}(h)}, \tag{3.7}$$

it is determined as the ratio of the numerical derivative of the particulate optical depth over the range resolution s to the point datum of the backscatter coefficient $\beta_p(h)$, that is,

$$S_p(h) = \frac{\kappa_p^{(\mathrm{dif})}(h)}{\beta_{\pi,p}(h)} = \frac{1}{\beta_{\pi,p}(h)}\frac{d}{dh}[\tau_p(0,h)]. \tag{3.8}$$

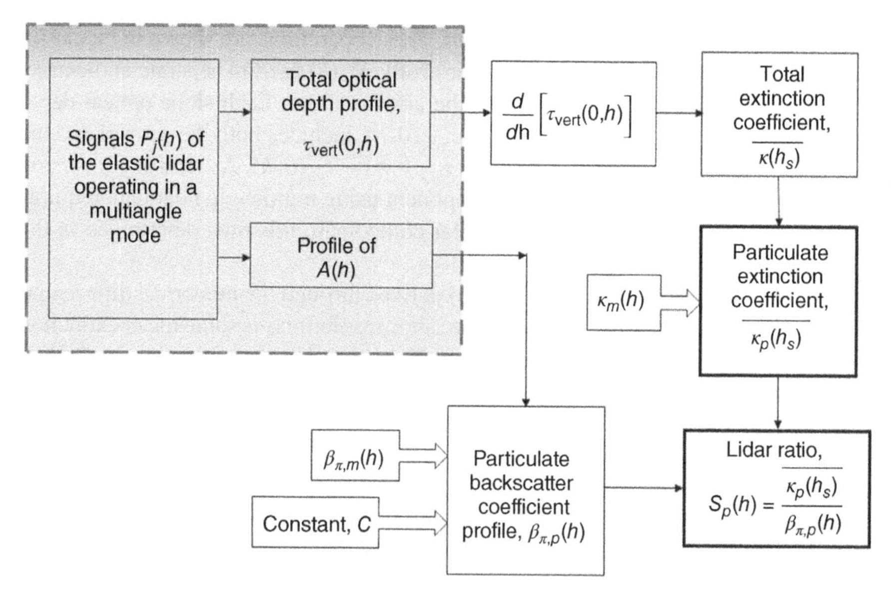

Fig. 3.3 Schematic for determining the particulate extinction coefficient $\kappa^{(\mathrm{dif})}(h)$ and the lidar ratio $S_p(h)$.

Because the extinction coefficient $\kappa_p^{(\mathrm{dif})}(h)$ is found as some mean within the range resolution s it may significantly differ from the local data point value at the height h especially in the areas of sharp change of the particulate extinction coefficient.

In some cases, the accuracy of the derived function $S_p(h)$ can be improved using a simple alternative methodology. To achieve such an improvement, the column-integrated particulate backscatter coefficient $\overline{\beta_{\pi,p}}(h)$ at the point h_s should be used instead the local data point $\beta_p(h)$. The integrated $\overline{\beta_{\pi,p}}(h)$ should be determined within the same sliding interval from $[h_s - 0.5\,\mathrm{s}]$ to $[h_s + 0.5\,\mathrm{s}]$, as the extinction coefficient $\kappa_p^{(\mathrm{dif})}(h)$, that is,

$$\overline{\beta_{\pi,p}(h_s)} = \frac{1}{s}\int_{h_s-0.5s}^{h_s+0.5s} \beta_{\pi,p}(h')dh'. \tag{3.9}$$

The alternative lidar ratio $\overline{S_p}(h_s)$ averaged over the same interval is then found as

$$\overline{S_p(h_s)} = \frac{\kappa_p^{(\mathrm{dif})}(h_s)}{\overline{\beta_{\pi,p}(h_s)}}. \tag{3.10}$$

Note that the sliding step of the numerical differentiation and the integration is the same for the functions in the denominator and the numerator of Eq. (3.10). This principle can produce a significant decrease in the scattering of the data points in the extracted lidar ratio. The schematic for the determination of the lidar ratio $\overline{S_p}(h_s)$ is shown in Fig. 3.4.

Let us demonstrate the difference in the results of determining the lidar ratio when using the data processing techniques shown in Figs. 3.3 and 3.4. Let a virtual elastic lidar scan the atmosphere, which contains sharp spatial changes in the aerosol loading. The synthetic profiles of the particulate backscatter coefficient $\beta_{\pi,p}(h)$ and the particulate optical depth $\tau_{p,\mathrm{vert}}(0, h)$ extracted from multiangle data of the virtual lidar, spoiled by quasi-random noise, are shown in the left and right panels of Fig. 3.5. The sharp changes of the particulate loading can be clearly visualized in Figs 3.5–3.7 at the heights 1500, 2500, and 3500 m. The inversion results of the synthetic profiles $\beta_{\pi,p}(h)$ and $\tau_{p,\mathrm{vert}}(0, h)$ are shown in Figs. 3.6 and 3.7. The thick solid curve in Fig. 3.6 is the lidar ratio extracted with Eq. (3.10), and the gray empty circles show the profile of the lidar ratio $S_p(h)$ obtained with Eq. (3.8). As could be expected, the latter data points are extremely scattered in the areas of sharp changes in the test profile, whereas the profiles extracted with Eq. (3.10) do not show such excessive scattering. In Fig. 3.7, the test and the retrieved profiles of the particulate extinction coefficient are shown. The thin solid curve is the test profile, and the thick solid curve is the particulate extinction coefficient retrieved with the formula

$$\kappa_p(h) = \overline{S_p(h_s)}\beta_{\pi,p}(h). \tag{3.11}$$

One can see that in the areas of sharp changes of the aerosol loading, the use of Eq. (3.11) provides the retrieval of the accurate boundaries of the zones with different backscattering.

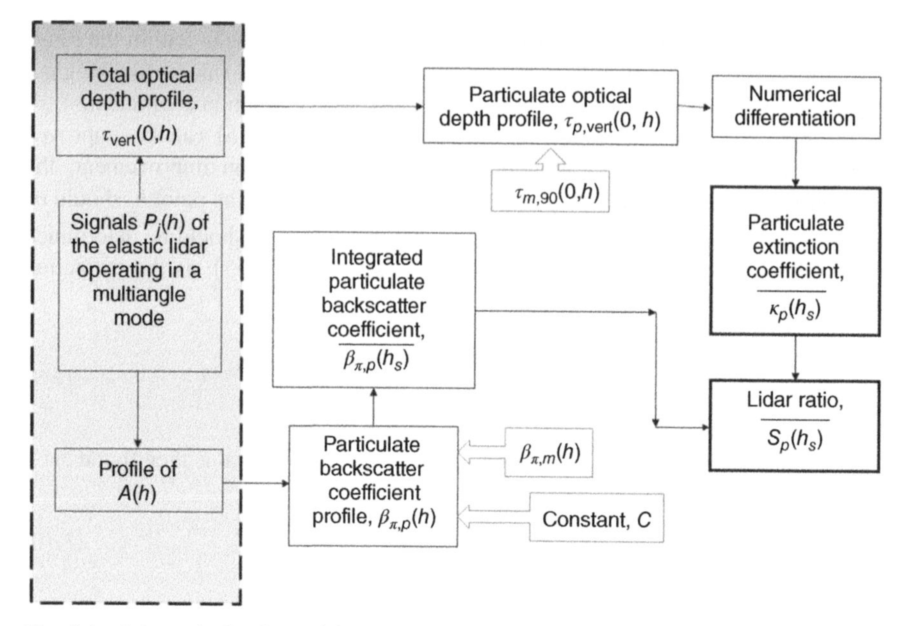

Fig. 3.4 Schematic for determining the particulate extinction coefficient and the lidar ratio using the integrated backscatter coefficient.

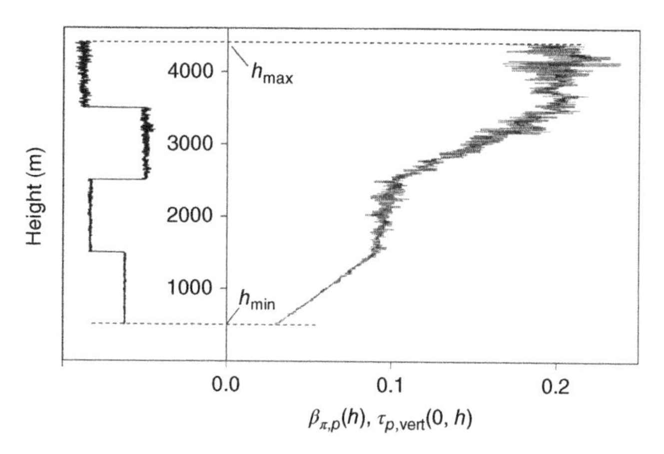

Fig. 3.5 Particulate backscatter coefficient $\beta_{\pi,p}(h)$ (left panel), and the optical depth $\tau_{p,vert}(0, h)$ (right panel), extracted from virtual lidar data and used for the determination of the lidar ratio in the numerical simulation (Kovalev and Kolgotin, 2008).

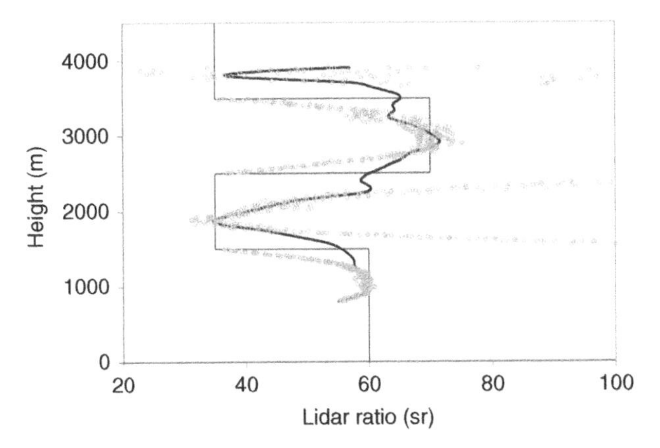

Fig. 3.6 The thick solid curve is the lidar ratio obtained with Eq. (3.10), whereas the gray empty circles show the profile obtained with Eq. (3.8). The thin solid curve is the test profile of the lidar ratio used in the numerical simulation (Adapted from Kovalev and Kolgotin, 2008).

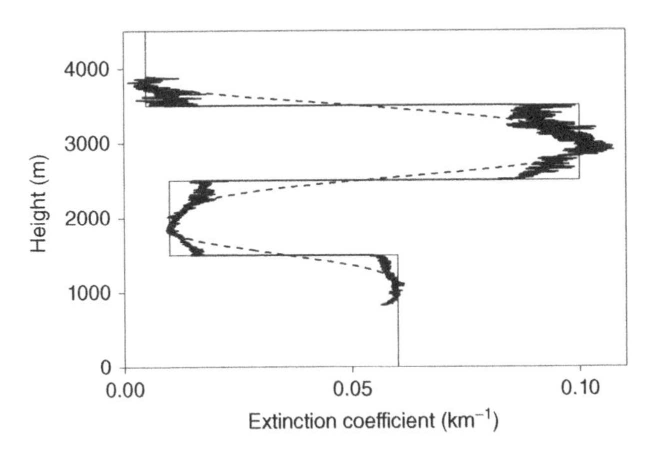

Fig. 3.7 The thin and thick solid curves are the profiles of the test extinction coefficient and that retrieved with Eq. (3.11), respectively. The dashed curve is the extinction coefficient extracted with numerical differentiation. (Adapted from Kovalev and Kolgotin, 2008).

3.1.2 Essentials and Specifics of the Methodology for Profiling of the Atmosphere with Scanning Lidar

As stated above, the conditions in Eqs. (3.1) and (3.2) should be met for both the particulate and molecular components. These requirements, especially, the key requirement in Eq. (3.1), are more satisfactorily met for the molecular than for the particulate component. The molecular backscatter coefficient $\beta_{\pi,m}(h)$ generally meets the condition of horizontal homogeneity significantly better than the particulate backscatter coefficient, which can often be heterogeneous in the horizontal directions. It follows from this observation that in clear and even in moderately polluted atmospheres, the Kano–Hamilton multiangle method works better when the elastic lidar operates at shorter wavelengths, for example, at 355 nm, where the atmospheric molecular component is relatively large. In clear atmospheres, the molecular component may be comparable or even larger than the particulate component. Such an increased molecular component can significantly stabilize the Kano–Hamilton solution. However, on the other hand, the presence of an increased molecular component, comparable with the particulate one, may significantly worsen the accuracy of the retrieved particulate extinction component (Kovalev and Eichinger, 2004). Indeed, the relative error of the particulate extinction coefficient $\delta\kappa_p(h)$ is related to the error of the total extinction coefficient $\delta\kappa(h)$ in the form

$$\delta\kappa_p(h) = \delta\kappa(h)\left[1 - \frac{\kappa_m(h)}{\kappa(h)}\right]^{-1}, \tag{3.12}$$

that is, the error $\delta\kappa_p(h)$ becomes significantly larger than the error $\delta\kappa(h)$ when the molecular extinction coefficient is large or even comparable with the particulate component. This unpleasant reality is a mandatory payment for the use of the elastic lidar in multiangle mode in clear atmospheres.

The experimental data, obtained in such clear atmospheres at different wavelengths in spring 2005 in Montana, confirmed that the elastic lidar signals measured at 355 nm yielded better inversion results than the backscatter signals at longer wavelengths (Adam et al., 2007). The signals at the wavelengths 532 and 1064 nm, not stabilized by the increased molecular component, are more sensitive to the horizontal fluctuations of the particulate backscatter coefficient. In addition, the optical depths at these wavelengths are significantly less than that at 355 nm. Another issue is that under the same atmospheric conditions, the signals at 1064 nm are significantly weaker than the signals at 355 nm; their signal-to-noise ratio is worse, and accordingly, the total measurement range at this wavelength is shorter than at 355 nm. The application of the multiangle solution to the elastic signals at 1064 nm measured during the experiment yielded discouraging results. The signals measured at 532 nm yielded more promising results. The authors' conclusion is that in favorable conditions, this wavelength can be used for multiangle profiling of the atmosphere.

It is worthwhile to comment on the combined azimuthal–vertical searching used in the above experiment. The practice shows that in many cases, the use of azimuthal averaging may be preferable as compared to temporal averaging over a fixed azimuthal direction. The real aerosol atmosphere is generally heterogeneous in both slope and horizontal directions. However, in most cases, the vertical heterogeneity is larger than the horizontal heterogeneity. In the vertical directions, systematic changes of the particulate concentration commonly prevail, so that even averaging of these data within any extended vertical interval should be done extremely cautiously. To reduce the influence of the horizontal heterogeneity in stable atmospheres, one can increase the time of vertical scanning or use combined azimuthal–vertical scanning. In the latter case, a sequence of lidar azimuthal scans at a fixed elevation angle is done and a mean azimuthal signal for this elevation angle is determined and used for the inversion. In practice, such azimuthal scanning within the selected azimuthal sector can be started using the smallest elevation angle. Then the next elevation angle is set and another set of the azimuthal scans is taken, etc., until the maximum elevation angle has been scanned. Relatively small (1–5°) angular resolution within the selected azimuthal sector are generally optimal. However, the smaller the angular resolution, the more sounding time is required.

One should keep in mind that fluctuations of the particulate matter in horizontal directions might be random, but unfortunately, not necessarily in a statistical sense. Therefore in real atmospheres, even in horizontal directions, signal spatial averaging should be done extremely cautiously. Initial examination of the scans recorded within the azimuthal sector by means of common statistical analysis (Adam et al., 2007) is not enough rigid, but it permits locating of local areas of heterogeneity, such as clouds and smoke plumes. Such an analysis allows the exclusion of the obvious heterogeneity areas before the determination of the mean azimuthal scan $\overline{y_j(h)} = \ln[\overline{P_j(h)}/(h/\sin\varphi)^2]$ along the selected slope direction φ. Spatial averaging might often be more preferable than temporal averaging over a single azimuthal direction. When single azimuthal scanning is made, the only option is to stop the scanning until the local cloud goes away (under the condition that the wind blows "in the right direction").

Another issue in the Kano-Hamilton solution is the selection of the minimum and maximum heights. Proper estimation of these heights, h_{min} and h_{max}, is required in order to restrict the lidar operative range to the heights within which the linear fit $Y(x, h)$ provides satisfactory accuracy for the extracted functions $\tau_{vert}(0, h)$ and $[C\beta_{\pi,p}(h)]$. To provide the tolerable accuracy for these functions, one should have at least some minimal number of data points $P_j(h)$ measured at the same height under different elevation angles φ. This condition is met if the minimal height h_{min} at which the profiles $\tau_{vert}(0, h)$ and $[C\beta_{\pi,p}(h)]$ are estimated is larger than the height $h_{dz} = r_{min}\sin\varphi_{min}$ (Fig. 3.8). The height h_{dz} which can be defined as the "dead zone" height, is proportional to the sine of the minimal slope angle φ_{min} used during scanning, and the length of the lidar incomplete overlap zone r_0, that is,

$$h_{dz} = r_0 \sin\varphi_{min}. \tag{3.13}$$

The fulfillment of the requirement $h_{min} > h_{dz}$ allows one to avoid inversion of the lidar data using points from the incomplete overlap zone, which would require knowledge of the exact profile of the overlap function $q(r)$ and could meet other significant difficulties (see Section 1.8.2).

A similar requirement of the availability of some minimum number of data points $y_j(h)$ is also valid for the maximum height h_{max}. In order to determine the linear fit $Y(x, h)$ at the maximal height, it should be less than the height $h_{\varphi m} = r_{max}\sin\varphi_{max}$ (Fig. 3.5). The maximum height h_{max} can be determined as

$$h_{max} = v_\varphi r_{max} \sin\varphi_{max},$$

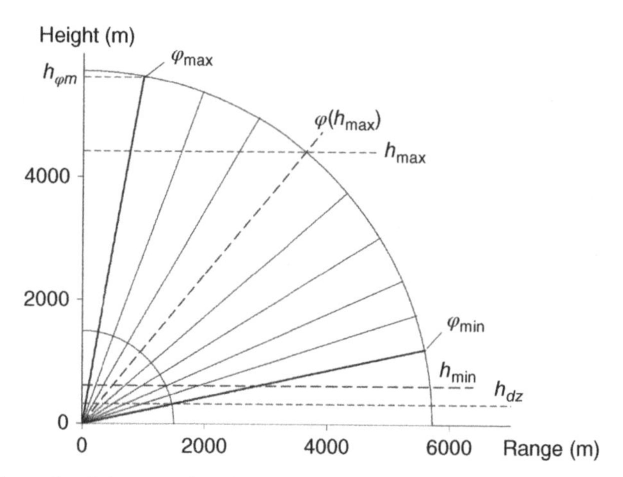

Fig. 3.8 Schematic of the scanning lidar operation and symbols used in the analysis of multiangle profiling of the atmosphere.

where the quantity $\upsilon_\varphi < 1$ defined as the ratio

$$\upsilon_\varphi = \frac{\sin\lfloor \varphi(h_{max})\rfloor}{\sin \varphi_{max}}, \tag{3.14}$$

should be initially established through the analysis of the available multiangle data. Unfortunately, there are neither rigid criteria for determining the optimum heights, h_{min} and h_{max}, nor sufficiently reliable algorithms to estimate the actual accuracy of such an *a posteriori* simulation. The stringent reality is that only some concepts, based on simplified statistical assumptions, mostly not properly grounded, are available to researcher.

Finally, it is interesting to compare the basic issues of the elastic lidar when it operates in the one-directional and multiangle modes. As stated in Section 1.1, the elastically backscattered signal contains insufficient information to separate the attenuation and the backscattering components in the lidar equation. For the inversion of the signal, the most critical issue is the *a priori* selection of the lidar ratio. The solution for the extinction coefficient profile is generally achieved by assuming the range-independent lidar ratio and selecting its value *a priori*. As has been stated in Section 1.2, the need for such a selection whose validity cannot be established precludes the determination of the extinction coefficient through lidar signal inversion as a measurement, transforming it into the a simulation based on lidar observations.

Let us summarize. As with the splitting lidar, the separation of the backscatter and the transmission terms is the main goal when utilizing the elastic lidar in the multiangle mode. After such a separation, the two-way transmission term generally becomes the basic term for the next analyses, as it allows determining the vertical optical depth and then the corresponding particulate extinction coefficient profile. The significant advantage of the Kano–Hamilton multiangle method is that unlike the one-directional method, it does not require the assumption of a range-independent lidar ratio and *a priori* selection of its numerical value. The main issue of the multiangle method is the stringent requirement of a horizontally stratified atmosphere. When selecting between the two above issues of the elastic signal inversion, the assumption of horizontal homogeneity in layered atmosphere looks more realistic than the assumption of vertical invariance of the lidar ratio. Moreover, the validity of the assumption of horizontal homogeneity can be explored by the analysis of the functions $y_j(h)$ obtained from the signals of the scanning lidar. However, the confirmation of the above assumption validity is not a guarantee of reliable quantitative estimation of the accuracy of the extracted profiles.

3.2 ISSUES IN PRACTICAL APPLICATION OF THE KANO–HAMILTON MULTIANGLE INVERSION TECHNIQUE

3.2.1 Multiplicative and Additive Distortions of the Backscatter Signal and Their Influence on the Inverted Optical Depth Profile

It is generally agreed that the main issue in multiangle profiling of the atmosphere stems from applying the method that assumes atmospheric homogeneity, whereas the atmosphere is heterogeneous. However, thorough analysis reveals that, in fact, the problem is more cumbersome. It shows that even minor errors that stem from instrumental uncertainties, inherent to real lidar data, may have dramatic consequences for the results of the inversion made using the multiangle method. The unavoidable signal distortions significantly impede the use of this method even in an ideally homogeneous atmosphere.

The influence of random noise when determining the slope of the inverted functions was highlighted in detail in literature (Whiteman, 1999; Kunz and Leeuw, 1993; Rocadenbosch et al., 2000, 2004; Volkov et al., 2002; etc). Therefore, in this section, the author will focus on the most influential systematic distortions in the optical profiles derived from the signals of elastic lidar operating in the multiangle mode. Let us consider the case when an elastic lidar works in an ideal horizontally stratified atmosphere. As follows from Eq. (1.7), the backscatter signal $\langle P_j(r)\rangle$, along the elevation angle φ, obtained after subtracting the estimated offset, $\langle B_j\rangle$ from the recorded signal, can be written as

$$\langle P_j(r)\rangle = P_j(r)\lfloor 1 + \delta_{P,j}(r)\rfloor + \sum w_{j,\mathrm{low}}(r) + \Delta B_j, \qquad (3.15)$$

where the uncorrupted backscatter signal $P_j(r)$ is

$$P_j(r) = Cq(r)\beta_{\pi,j}(r)r^{-2}\exp[-2\tau_j(0,r)]. \qquad (3.16)$$

Let us assume for simplicity that the noise component term $\sum w_{j,\mathrm{low}}(r)$ is minor and can be ignored. Then the distorted square-range-corrected signal versus height, defined as $\langle Z_j(h)\rangle = \langle P_j(h)\rangle/[h/\sin\varphi]^2$, can be written in the form

$$\langle Z_j(h)\rangle = Z_j(h)[1 + \delta_{P,j}(h)] + \left[\frac{h}{\sin\varphi}\right]^2 \Delta B_j, \qquad (3.17)$$

where $Z_j(h) = P_j(h)[h/\sin\varphi]^2$ is the undistorted square-range-corrected signal in the slope direction φ given as a function of height. Eq. (3.17), can be rewritten as the product of three factors, that is,

$$\langle Z_j(h)\rangle = Z_j(h)[1+\delta_{P,j}(h)]\left[1+\frac{\Delta B_j}{P_j(h)\,[1+\delta_{P,j}(h)]}\right]. \tag{3.18}$$

Taking into consideration Eqs. (3.1), (3.2), and (3.4), the distorted function $\langle y_j(h)\rangle = \ln\langle Z_j(h)\rangle$ can be written as the sum

$$\langle y_j(h)\rangle = y_i(h) + \ln[q(h)] + \ln[1+\delta_{P,j}(h)] + \ln\left[1+\frac{\Delta B_j}{P_j(h)\,[1+\delta_{P,j}(h)]}\right], \tag{3.19}$$

where

$$y_j(h) = \ln[C\beta_\pi(h)] - \frac{2\tau_{\text{vert}}(0,h)}{\sin\varphi},$$

is the undistorted Kano–Hamilton function in the complete overlap area, where $q(r) = 1$.

As shown in the previous chapters, different sources of systematic distortion are responsible for the corruption of the lidar signals over the near and distant ranges. The near-end distortions may occur as a result of receiving optics aberrations (Dho et al., 1997), the restricted frequency range of the photoreceiver (Tomine et al., 1989), induced low-frequency noise components (Sassen and Dodd, 1982), inaccurate determining of the minimal range of the complete overlap zone, and overlap changes caused by thermal effects (Sasano et al., 1979). Systematic distortions also occur because of the nonzero offset (invariant or range dependent) that remains in the backscatter signal after subtraction of the background component. These distortions are extremely influential over the distant ranges, where the small backscatter signal is found as a difference of two large quantities (see Section 1.5).

Thus, when $\sum w_{j,\text{low}}(r) \approx 0$, three remaining distortion components may be present in the real function $\langle y_j(h)\rangle$. The term $\ln[q(h)]$ effects the data points only within the zone where the assumed $q(r) = 1$, whereas the actual $q(r) < 1$. This term, as with the second one $\ln\lfloor 1+\delta_{P,j}(h)\rfloor$, is commonly most destructive when inverting the signals in the near zone. The third distortion term $\ln\left[1+\frac{\Delta B_j}{P_j(h)[1+\delta_{P,j}(h)]}\right]$ commonly increases with range due to decreasing the measured signal $P_j(h)$ over the distant ranges. Thus, the vertical optical depth $\tau_{\text{vert}}(0,h)$ determined from the slope of the linear fit of the data points $\langle y_j(h)\rangle$ will be corrupted over both the near-end and far-end ranges.

For a better understanding of the issue of multiangle profiling of the atmosphere related to the presence of systematic distortions, simple numerical simulations can be helpful. Let us consider the results of the numerical simulations for the case when a virtual scanning lidar operates in a synthetic atmosphere, ideally horizontally stratified. It is assumed that the particulate extinction coefficient in this atmosphere at the lidar wavelength 532 nm decreases linearly with height, from 0.1 km^{-1} at ground level to 0.04 km^{-1} at a height of 6000 m. The virtual elastic lidar measures the backscatter signals along discrete slope directions of 10°, 15°, 20°, 25°, 30°, 40°, 50°, and 60°.

Let us focus initially on the near-end distortions of the profile of the optical depth. In Fig. 3.9, the test optical depth $\tau_{90}(0, h)$ is shown as the thin slope line, whereas the optical depth profile $\tau_{\mathrm{vert}}(0, h)$ retrieved from the virtual scanning lidar data is shown as the filled squares. The overlap function of the virtual lidar $q(r)$ is shown as the bold curve in the bottom part of the plot. The incomplete overlap zone r_0 is defined here as the range where $q(r) < 0.98$, it extends up to $r_0 \approx 340$ m. To show how the incomplete overlap, when it is not taken into account, influences the retrieved optical depth, the profile $\tau_{\mathrm{vert}}(0, h)$ is calculated starting from the height ~100 m, and in the near zone, the inequality $q(r) < 1$ is ignored. The noticeable divergence of $\tau_{\mathrm{vert}}(0, h)$, extracted with the Kano–Hamilton solution from the test profile, $\tau_{90}(0, h)$ over the heights $h <$ 320 m, is entirely due to the presence of the non-corrected incomplete overlap zone. Note that near-end distortions may occur owing to different reasons: ignoring the true profile $q(r)$ in the incomplete overlap zone, distortions originating from the restricted frequency band of the photoreceiver and receiving optics aberrations, the presence of the low-frequency noise component in the signal, etc. Such distortions often result in erroneous negative optical depth values in the near-end range. The simulation results shown in Fig. 3.9 are obtained assuming that the lidar signals are not corrupted by random noise and no remaining offsets exist in the backscatter signals.

In Fig. 3.10, the optical depth profiles $\tau_{\mathrm{vert}}(0, h)$ are obtained from the signals of the same virtual scanning lidar, but this time the true overlap function $q(r)$ is taken into account. However, now the constant offsets B_j in the signals, originating from the background component and digitizer electronic shifts, are not precisely determined. Accordingly, after the subtraction of the estimated $\langle B_j \rangle$ from the recorded signal,

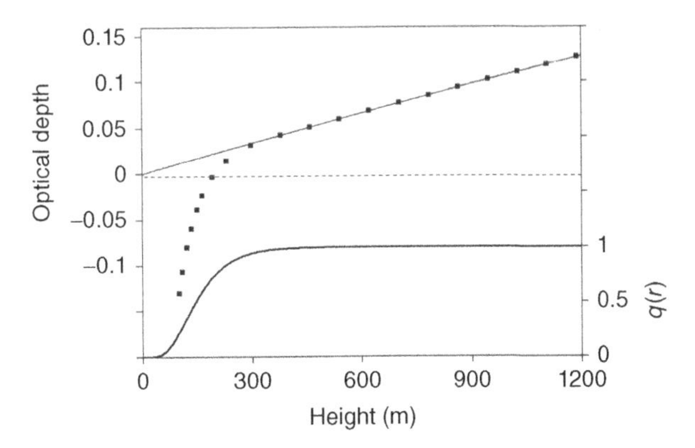

Fig. 3.9 The sloped solid line is the total optical depth $\tau_{90}(0, h)$ in the synthetic atmosphere used for the simulations. The profile $\tau_{\mathrm{vert}}(0, h)$ retrieved with the Kano–Hamilton solution is shown by the filled squares. The overlap function of the virtual lidar is shown as the thick bold curve at the bottom part of the plot.

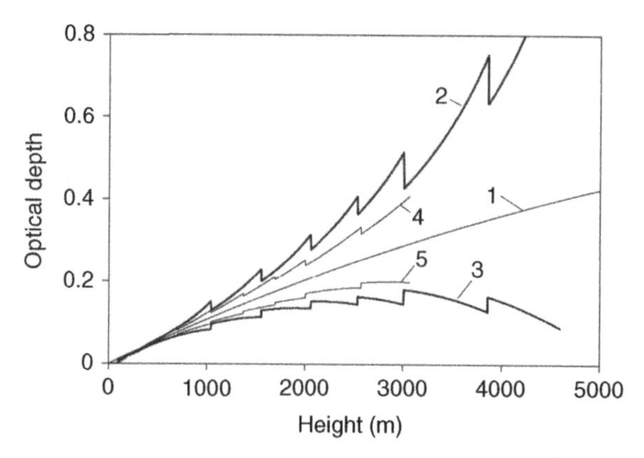

Fig. 3.10 Test optical depth $\tau_{90}(0, h)$ versus height (Curve 1) and the profiles $\tau_{vert}(0, h)$, retrieved when $\Delta B = \pm 0.5$ count. The profiles shown as Curves (2) and (3) are obtained from the signals of the virtual lidar when using $r_{max} = 6000$ m, the profiles shown as Curves (4) and (5) when $r_{max} = 4000$ m.

the offsets $\Delta B_j \neq 0$ are present in the backscatter signals $\langle P_j(r) \rangle$. For simplicity, it is assumed that no random noise is present in the signals, and that the other distortion components in Eq. (3.15), except ΔB_j, are absent. It is also assumed that the components B_j, $\langle B_j \rangle$, and ΔB_j are the same in all slope directions; therefore henceforth, the subscript "j" in these terms is omitted. The maximal lidar signal in the simulated data is ~4000 counts, and the true background component in each recorded lidar signal is $B = 200$ counts. The same maximal range $r_{max} = 6000$ m is initially used when inverting the lidar signals measured in the different slope directions. Curve 1 in the figure shows the initial, that is, the test optical depth $\tau_{90}(0, h)$; Curves 2 and 3 show the retrieved optical depth profiles $\tau_{vert}(0, h)$ obtained with the inaccurately estimated background levels $\langle B \rangle$ equal to 200.5 and 199.5 counts, respectively. One can see how such a seemingly insignificant systematic offset ±0.5 count, can dramatically distort the derived profiles $\tau_{vert}(0, h)$ over distant areas. In some instances, the distortion in the retrieved profile of the optical depth caused by the presence of the offset $\Delta B \neq 0$ can be reduced by decreasing the maximum range for the inverted backscatter signals. Curves 4 and 5 in Fig. 3.10 show the profiles of the optical depth derived from the same synthetic signals with the same offsets ΔB but when using the decreased maximum range $r_{max} = 4000$ m. Such a decrease from the initial $r_{max} = 6000$ m significantly reduces the distortions in the retrieved profiles of the optical depth. However, this way for reducing these distortions requires the absence of significant distortions of the signals in the nearest zone, close to r_{min}. Another problem is that the reduction of the maximal measurement range of the signals measured in slope directions decreases the number of data points $y_j(h)$ available for the regression, and accordingly, the maximum height of lidar profiling.

The periodic jumps in the curves in Fig. 3.10 are another specific of the derived optical depth profiles $\tau_{vert}(0, h)$. Such jumps appear when the optical depth is obtained from the backscatter signals in which the offset $\Delta B \neq 0$. As one can see in Fig.

3.1.8 of the previous section, the number of the data points $y_j(h)$ that are available for determining the linear fit $Y(x, h)$ changes with height. The jumps appear at the heights where the number of the points used for determining the linear fit changes. The magnitude of these nonphysical jumps depends on the relative level of the offsets ΔB in the backscatter signals. As one can see in Fig. 3.10, the jumps shift the profile of the derived optical depth $\tau_{\text{vert}}(0, h)$ toward the true profile $\tau_{90}(0, h)$. This is because at the jump point, the most distorted data point $y_j(h)$ from the preceding data is excluded. The jumps can be significantly decreased or even removed by using weighted data points $y_j(h)$ when determining the linear fit $Y(x, h)$. Such a decrease of the jump magnitude makes the profiles, $\tau_{\text{vert}}(0, h)$, much smoother, but unfortunately, not necessarily accurate enough. As usual, the devil is hiding in details. The commonly used least-squares method, which the weighted functions generally use, is based on statistical estimates of the random noise (Taylor, 1997; Barlow, 1999; Whiteman, 1999). In other words, when using weighted data points for the regression, the assumption is that the standard deviation or the variance is the dominant characteristic which determines the accuracy of the lidar signals. Meanwhile, this assertion, which ignores the systematic distortion components in the inverted signal, might be wrong. Furthermore, the use of these weighted data points makes the inverted optical depth smoother and nicer, but it does not remove completely the systematic divergence between the true profile $\tau_{90}(0, h)$ and the derived profile $\tau_{\text{vert}}(0, h)$. At best, it only reduces the difference between these profiles.

Let us focus on this specific issue using the study by Adam et al. (2007) and the related personal communication between the authors of this study. In the study, the numerical simulation was made for a virtual elastic scanning lidar, which operates in a synthetic atmosphere in which the particulate extinction coefficient decreases exponentially from 0.1 km^{-1} at ground level to 0.001 km^{-1} at a height of 4600 m. It was assumed that in addition to the quasi-random noise, the lidar signals were also systematically distorted in the near-lidar zone. These distortions were produced by an incorrect estimate of the incomplete overlap zone. As mentioned before, these distortions can be considered as the presence of the corresponding distortion component $\delta_{P,j}(r) \neq 0$. In Fig. 3.11 (a) and (b), the particulate optical depth profiles $\tau_{p,\text{vert}}(0, h)$ are shown, retrieved from the virtual lidar signals measured in this synthetic atmosphere. The optical depth profile shown in Fig. 3.11 (a) is retrieved using the weighted data points, where the weights were given by the ratio of STD to the signal $P_j(h)$; the profile in Fig. 3.11 (b) is retrieved without using the variable weights.

Comparing these profiles, one can see that the near-end distortions of the profiles in Fig. 3.11 (a) are significantly larger than those in Fig. 3.11 (b). In other words, the use of weights chosen as the ratio of the STD to the signal $P_j(h)$ is not compatible with the fact that the larger signals may also be corrupted by systematic distortions.

In Figs. 3.12 (a) and (b), the particulate optical depth profiles obtained from the same synthetic atmosphere are shown when $\delta_{P,j}(r) = 0$, but in the inverted backscatter signal, the nonzero offset ΔB is present. The simulation is made for the case where the offset is equal to ± 1 count. Such a relatively large offset ($\sim$3% from the maximal signal) is selected to allow better visualization of the details in the optical depths derived with and without using variable weights.

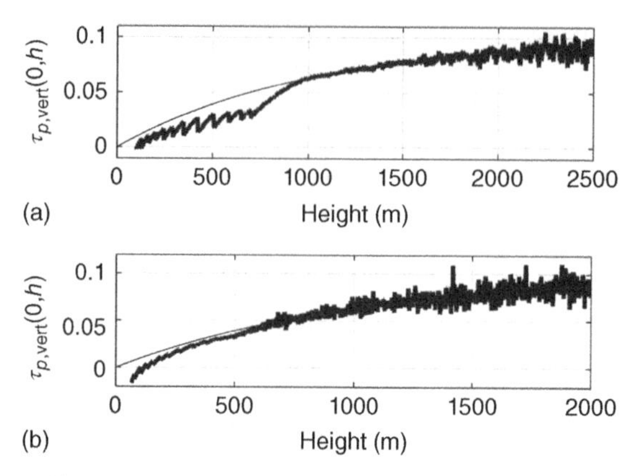

Fig. 3.11 (a) The thin solid line is the test profile $\tau_{90}(0, h)$ used in the numerical simulation, and the thick solid curve is the particulate optical depth $\tau_{p,\text{vert}}(0, h)$, calculated with the Kano–Hamilton solution using variable weights of the inverted function (Adapted from Adam et al., 2007). (b) As in (a) but where the particulate optical depth $\tau_{p,\text{vert}}(0, h)$ is calculated without using variable weights (Adam M., Institute for Environment and Sustainability, Joint Research Centre, Italy. Personal communication, 2006).

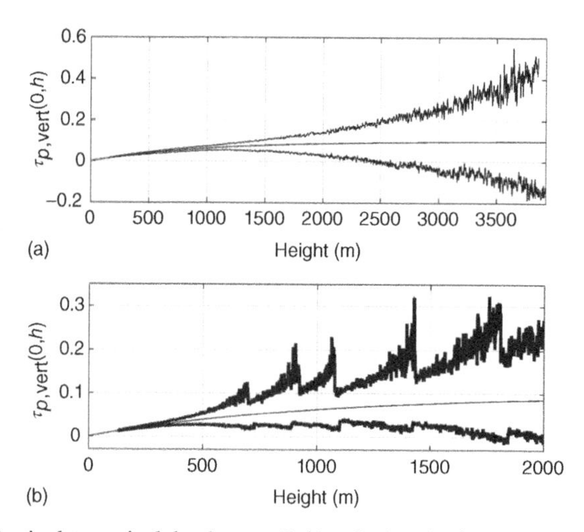

Fig. 3.12 (a) Particulate optical depth $\tau_{p,\text{vert}}(0, h)$, calculated using variable weights when the remaining offset ΔB is equal to -1 (upper curve) and $+1$ (bottom curve). The thin solid curve is the actual (test) profile (Adapted from Adam et al., 2007). (b) As in (a) but where $\tau_{p,\text{vert}}(0, h)$ is calculated without using the variable weights (Adam M., Institute for Environment and Sustainability, Joint Research Centre, Italy. Personal communication, 2006).

One can clearly see the relatively smooth optical depth profiles retrieved when variable weights are used [Fig. 3.12 (a)] and the large jumps on the profiles when variable weights are not used [Fig. 3.12 (b)]. From an initial glance, the use of variable weights is preferable. However, one should be cautious when improving the inversion results in this way. The use of weights removes the jumps but does not remove the systematic distortions in the derived optical depth. Actually, the upper profile in Fig. 3.12 (a) looks "too nice"; it gives no hint that it is actually distorted, creating an illusion of the absence of systematic shift in the retrieved profile. The profiles in Fig. 3.12 (b) immediately alert the researcher, giving him/her information about possible distortions in the inverted signals, and accordingly, in the corresponding inversion results. The smooth profiles obtained with weighted functions may result in overestimating the accuracy of the outputs of atmospheric profiling. From this point of view, the backscatter signals with the remaining negative offset are often preferable: the nonphysical decrease optical depth will immediately alert the researcher.

Let us summarize. The signal near-end distortions generally result in erroneous optical depth values in the near-end range. In Fig. 3.13 (a), such distortion is shown by the dotted curve at the bottom of the figure. Accordingly, the minimal lidar ranges r_{min} and corresponding heights h_{min} below which the inversion of the multiangle data points will produce inaccurate results should be thoroughly established. The imprecise estimate of the offset B in the recorded signal can substantially affect the quality of the retrieved optical depth $\tau_{p,\text{vert}}(0, h)$ over distant ranges. The use of a weighted function does not remove completely the distortions in the inversion result, but may mask them. Depending on the sign of the offset, the optical depth derived over distant ranges may be either larger or less than the actual value. In the latter case, the retrieved optical depth can show a nonphysical decrease with the range, and the corresponding extinction coefficient profile will yield erroneous negative values [Fig. 3.13 (b)].

Currently, no reliable method exists that would allow us to estimate the level of the true distortion in the retrieved optical depth caused by unknown systematic distortions in the lidar signal; the uncertainty estimates based on purely statistical methods generally yield underestimated distortions in both retrieved optical parameters – the optical depth and the extinction coefficient. As a temporary solution, one can perform the inversion with and without the weighted function, compare the results, and correct the inversion results by correcting the initial estimated offsets $\langle B_j \rangle$. Note that the far-end distortion caused by an inaccurate estimate of the constant component B can take place even if the recorded lidar signal $P_{\Sigma}(r)$ has no distortions. The distortion in the backscatter component is produced by the researcher, rather than by an imperfection of the lidar hardware; an imperfection of the hardware can only aggravate the situation. Unfortunately, directly detecting the presence of the systematic offset $\Delta B \neq 0$ that remains in the backscatter signal after subtracting the background component from $P_{\Sigma}(r)$ is hardly a solvable problem, even in the simplest case when the offset is range independent, that is, no low-frequency distorting components $\Sigma w_{j,\text{low}}(r)$ are present in the recorded lidar signal.

The random systematic errors and corresponding distortions are the important issues of multiangle profiling of the atmosphere. Numerical simulations show that

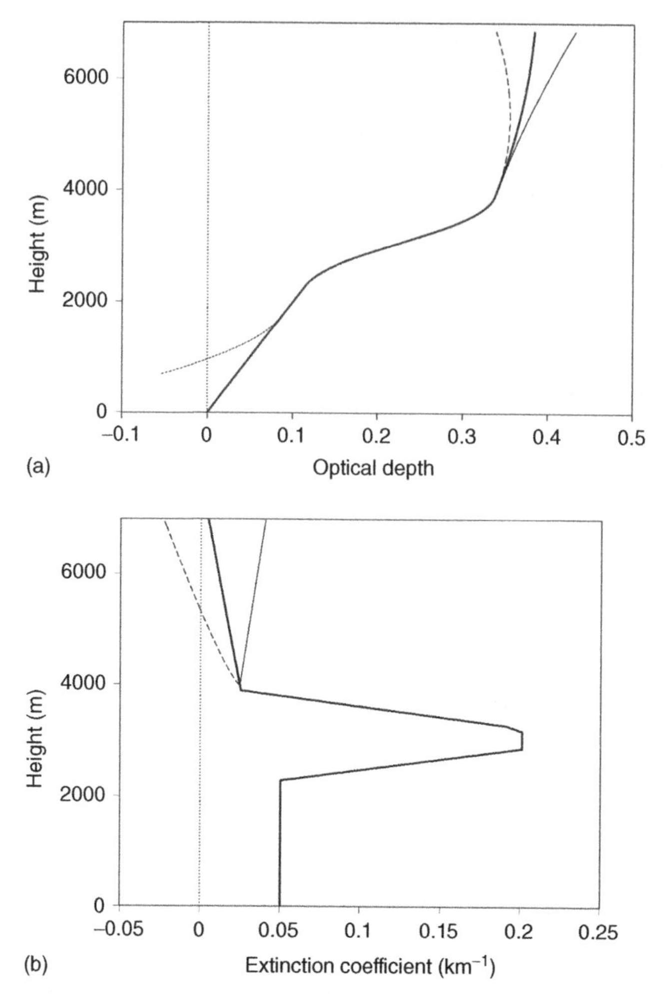

Fig. 3.13 (a) The true optical depth (thick solid curve) and its near-end distortion (dotted curve). The thin solid and dashed curves show typical far-end distortions. (b) The thin solid and dashed curves show the far-end distortions of the extinction coefficient profile corresponding to distortions of the optical depth in (a). The test extinction coefficient profile is shown as the thick solid curve.

such an effect would take place even in profiling of an ideal homogeneous atmosphere. However, as shown below, the primary issue for practical utilization of multiangle measurements is atmospheric heterogeneity.

3.2.2 Issues and Deficiencies in the Multiangle Inversion Technique

As indicated in Section 3.1, the Kano–Hamilton solution is the only model that permits extraction of the particulate extinction coefficient from elastic lidar data without an *a priori* assumption about the vertical profile of the lidar ratio. The main issue of this solution is that it works only in horizontally stratified atmosphere and requires fulfilling two rigid conditions. The more important and more susceptible of these conditions is that the total backscatter coefficient along the horizontal direction is invariable at all heights [Eq. (3.1)]. The second condition, which requires the slope optical depth within any fixed altitude range to be proportional to the reversed sine of the lidar elevation angle [Eq. (3.2)], is less stringent. In real atmospheres, these requirements are met, at best, only approximately. However even if these requirements are met and the signal distortions are minimized, the profile of the optical depth $\tau_{vert}(0, h)$ derived from multiangle data in clear atmospheres may significantly differ from the actual optical depth in the vertical direction. In such atmospheres, the difference in the optical depths, measured under adjacent slope directions, is small, so that the solution is ill-conditioned (see Section 1.4). Accordingly, the error factor and the corresponding error in the derived $\tau_{vert}(0, h)$ may be extremely large (Spinhirne et al., 1980, 1984; Kovalev and Eichinger, 2004). Meanwhile, the optical depth $\tau_{vert}(0, h)$ obtained using the Kano–Hamilton solution is generally only an intermediate inversion result; in most cases, the vertical profile of the particulate extinction coefficient, rather than the optical depth, is the subject of researchers' interest. Thus, to extract the extinction coefficient profile from the signals of scanning lidar, two consecutive procedures of slope determination are required. The first one is used to obtain the optical depth $\tau_{vert}(0, h)$ through determining the slope of the linear fit $Y(x, h)$ [Eq. (3.6)]. The second one, the extraction of the extinction coefficient, is commonly made through numerical differentiation of the derived $\tau_{vert}(0, h)$. The numerical differentiation result is extremely sensitive to any distortion in the inverted profile. This specific often yields a distorted profile of the vertical extinction coefficient, extracted from optical depth $\tau_{vert}(0, h)$. The error can reach tens and hundreds of percent (Chapter 2).

There are two more shortcomings with the classic Kano–Hamilton solution. The first one is the extremely poor inversion efficiency. To obtain a single vertical optical depth profile $\tau_{vert}(0, h)$ one should record and process at least a dozen lidar signals measured at different elevation angles. The second drawback of the classical multiangle inversion technique is that only the optical depth profile is generally extracted. The valuable information about particulate loading contained in the backscattering term, which can put additional constraints on the derived extracted extinction coefficient, is commonly not used. In a study by Kovalev (2006), an alternative technique for the extraction of the extinction coefficient was proposed, in which both the optical depth and the backscatter coefficient profiles were used in the inversion procedure. However, this technique still required utilizing numerical differentiation. One year later, Shcherbakov (2007) proposed a regularized algorithm for Raman lidar processing, in which the inversion was made without numerical differentiation. However, in that study, explicit relationships between the backscatter and the optical depth, and an *a priori* smoothness constraint of the retrieved characteristics, were required. The more advanced solutions, in which the most challenging operation of extracting the profile of $\tau_{vert}(0, h)$ through determining the slope of the data points $y_i(h)$ is omitted, will be considered in the next sections.

Owing to the significant issues discussed above, the multiangle method based on the assumption of atmospheric horizontal homogeneity is rarely used in practice (Spinhirne et al., 1980, 1984; Rothermal and Jones, 1985; Sicard et al., 2002; Takamura et al., 1994; Sasano, 1996). Different ways for improving the Kano–Hamilton profiling methodology have been proposed in the studies by Takamura et al. (1994), Sasano (1996), Sicard et al., 2002, etc., but the issues of multiangle methods are still far from being satisfactorily resolved.

The difficulties related to the practical use of the elastic lidar for separating two variable unknowns in the lidar equation has forced the researchers to look for other technical solutions to overcome the problem. They shifted their attention to splitting lidars, in particular, to combined elastic/inelastic Raman lidars and HSRLs. Due to such a shift to the one-directional methods, multiangle methods have not been properly investigated, and accordingly, properly valued. Meanwhile, multiangle methods are invaluable when properly used in proper conditions. In order to successfully utilize these methods, one should comprehend all the issues related to these methods and use all the possible alternatives that overcome, or at least, mitigate these issues. One can maintain that to be successfully applied to real lidar data, the classic Kano–Hamilton methodology needs to be modernized. Probably, the principal breakthrough in regard to the issues of multiangle measurements, especially with fulfilling the key requirement in Eq. (3.1), can be achieved by using HSRL in scanning mode (see Section 3.6.3).

In Table 3.1, the basic assumptions and premises used when processing data of multiangle profiling of the atmosphere are briefly tabulated.

Finally, the valuable uniqueness of the multiangle scanning mode as compared to the one-directional mode needs to be pointed out. In the inversion results obtained with one-directional zenith lidar, even large distortions in the extracted extinction coefficient are generally "not visible." Owing to masking large errors, the optical profiles obtained with the one-directional method of profiling of the atmosphere may create an illusion of their high accuracy.

The value of the multiangle method is its veracity. Multiangle data can also be aggravated by significant distortions, but all these "are visible." The multiangle processing technique immediately reveals the presence of significant lidar signal distortions by yielding unrealistic profiles of the retrieved optical parameters; the same effect takes place when working in a poorly stratified atmosphere. Therefore, the multiangle mode produces significantly more realistic estimates of the quality of the inverted data, and accordingly, permits the practical selection of the truly best results of atmospheric profiling. One can expect that this specific can be properly exploited by applying HSRL in the multiangle mode.

TABLE 3.1 Basic Assumptions or Implicit Premises Used When Processing Multiangle Signals of the Elastic Scanning Lidar

Data-processing Operation	Assumptions/Premises
Signal recording	(a) The atmosphere is frozen during the entire period of scanning within the established angular sector (b) No low-frequency noise component is present in the recorded lidar signals measured in all slope directions
Determination of the backscatter component in the recorded signals	The range-independent offset in all the recorded lidar signals is precisely determined, so that the remaining offset ΔB in the backscatter signals is negligible and does not influence the result of the signal inversion
Determination of the vertical optical depth and the corresponding extinction coefficient	(a) Atmosphere within the area searched by lidar is horizontally stratified; the existing horizontal heterogeneity of the atmosphere does not distort the inversion results (b) Selected operative range of the inverted multiangle signals is in conformity with the remaining offset ΔB, and existing level of horizontal heterogeneity of the searched atmosphere

3.2.3 Profiling of the Atmosphere Using Alternative Estimates of the Constant Offset in the Multiangle Signals

In Section 1.6, two alternative methods for estimating the offset B were considered: (i) determining the offset as a mean value of the recorded signal $P_{\Sigma}(r)$ over distant ranges and (ii) determining the offset B as the slope of the linear fit $y^*(Z)$ over the restricted interval at the distant signal range [Eqs. (1.56)–(1.60)]. The variable z in the general form was defined as the ratio of the squared range to the attenuated molecular backscatter coefficient, that is,

$$z(r) = \frac{r^2}{\beta_{\pi,m}(r)T_m^2(0,r)};$$

however, it can be reduced to $z(r) = r^2$, if the molecular component $\beta_{\pi,m}(r)$ is small as compared to the particulate component $\beta_{\pi,p}(r)$.

The methods for determining the signal offset discussed in Section 1.6 can also be used for the signals measured in the multiangle mode. Such a method can provide some useful estimates of the uncertainty boundaries in the optical profiles derived from the signals of scanning lidar. To clarify these methods, let us consider some results of multiangle optical depth profiling near Missoula, Montana, in August 2008. Signals at the wavelength 355 nm were measured in a combined slope–azimuthal mode (Adam et al., 2007) over 12 fixed slopes within the angular sector from 7.5° to 68°. For each azimuthally averaged signal measured under the fixed elevation angle, two alternative methods of separating the backscatter signal from the offset B were used. The offset $\langle B_j(P_\Sigma)\rangle$ was determined by averaging the signals $P_\Sigma(r)$ within some interval at the distant range, where the backscatter component of the signal presumably had vanished; the offset $\langle B_j(s)\rangle$ was calculated by determining the numerical sliding derivative $dy^*(z)/dz$ (see details in Section 1.6). Figs. 3.14–3.15 show typical inversion results, which are obtained from the azimuthally averaged signals recorded under the arbitrarily selected elevation angle 32°. In Fig. 3.14, the solid curve shows the profile of the sliding derivative $dy^*(z)/dz$ determined through numerical differentiation using the running least-square mean with step size $s = 300$ m. The corresponding offset $\langle B_j(s)\rangle = 420.61$ counts was determined as the average of the derivative over the range from 4000 m to 5500 m. The alternative offset $\langle B_j(P_\Sigma)\rangle$ was determined by averaging the data points of the same signal $P_\Sigma(r)$ at the far-end of the measured range, over the range from $r = 5640$ m to 6140 m; it yielded $\langle B_j(P_\Sigma)\rangle = 421.26$ counts. The corresponding backscatter signals, $P_1(r) = P_\Sigma(r) - \langle B_j(s)\rangle$ and $P_2(r) = P_\Sigma(r) - \langle B_j(P_\Sigma)\rangle$, were calculated and square range corrected. In Fig. 3.15, these square-range-corrected signals are shown as black filled dots and gray empty squares, respectively. Note that in spite of the extremely minor difference between the constants $\langle B_j(s)\rangle$ and $\langle B_j(P_\Sigma)\rangle$ equal to 0.65 counts, the difference in the shape of the corresponding square-range-corrected signals for the distant ranges $r > 3500$ m is significant.

The above operations were repeated for the entire lidar scan over each fixed slope. The analysis of the multiangle scan confirmed that for each azimuthally averaged signal, $\langle B_j(s)\rangle < \langle B_j(P_\Sigma)\rangle$. The difference between these values is small, ranging from 0.65 to 1.11 counts, which is less than 0.3% from the mean signal offsets. Nevertheless, even such a minor difference significantly influenced the inversion result. In Fig. 3.16, the resulting vertical particulate optical depth $\tau_{p,\text{vert}}(0, h)$ versus height

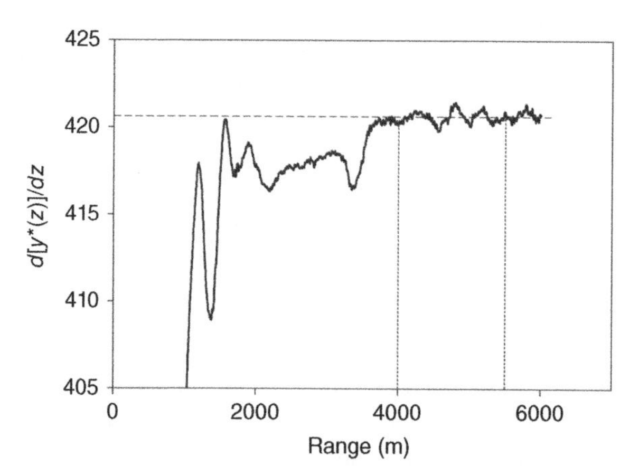

Fig. 3.14 The sliding derivative of $dy^*(z)/dz$ versus range determined numerically for the lidar signal azimuthally averaged over the elevation angle $\varphi = 32°$ (solid curve). The corresponding offset $\langle B_j(s)\rangle$, determined as the average of this function over the range from 4000 to 5500 m is shown as the horizontal dashed line (Kovalev et al., 2009a).

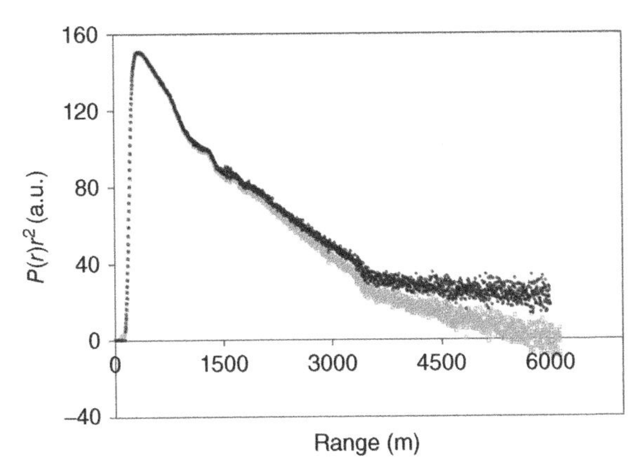

Fig. 3.15 Square-range-corrected signals versus range for the same elevation angle of 32°. The black filled dots show the signal calculated with the estimate $\langle B_j(s)\rangle = 420.61$ counts and the gray empty squares show that obtained with $\langle B_j(P_\Sigma)\rangle = 421.26$ counts (Kovalev et al., 2009a).

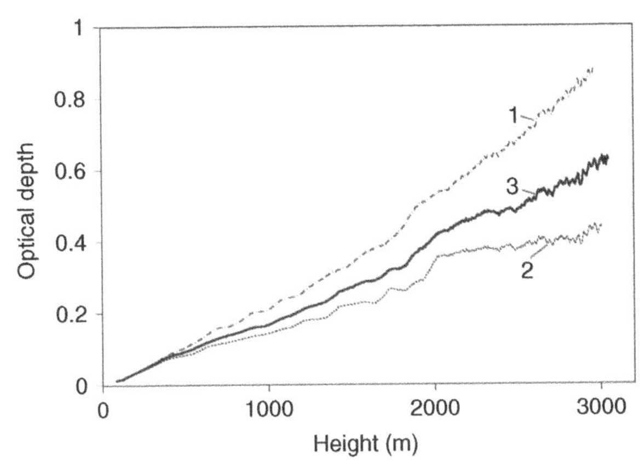

Fig. 3.16 Profiles of the particulate optical depth $\tau_{p,\mathrm{vert}}(0, h)$, calculated with alternative estimations of the signal offset. Curve 1 shows the profile of the optical depth obtained when the estimate $\langle B_j(P_\Sigma)\rangle$ was used. Curve 2 shows the optical depth profile obtained with the estimate $\langle B_j(s)\rangle$ and Curve 3 shows the optical depth profile calculated using the average of the offsets $\langle B_j(s)\rangle$ and $\langle B_j(P_\Sigma)\rangle$ (Kovalev et al., 2009a).

is shown calculated using the two alternative estimates of the signal offset. Curve 1 shows the optical depth obtained with the offset $\langle B_j(P_\Sigma)\rangle$, Curve 2 shows the optical depth obtained when using the offset $\langle B_j(s)\rangle$, and Curve 3 shows the mean profile, obtained using the average of $\langle B_j(s)\rangle$ and $\langle B_j(P_{\Sigma,j})\rangle$. In Fig. 3.17, the particulate extinction coefficient profiles $\kappa_p^{(\mathrm{dif})}(h)$ extracted from the optical depth profiles in Fig. 3.16 are shown. The profiles were extracted using numerical differentiation with the sliding range resolution $s = 200$ m. The systematic difference in these profiles originates in the systematic difference between the offsets $\langle B_j(s)\rangle$ and $\langle B_j(P_{\Sigma,j})\rangle$.

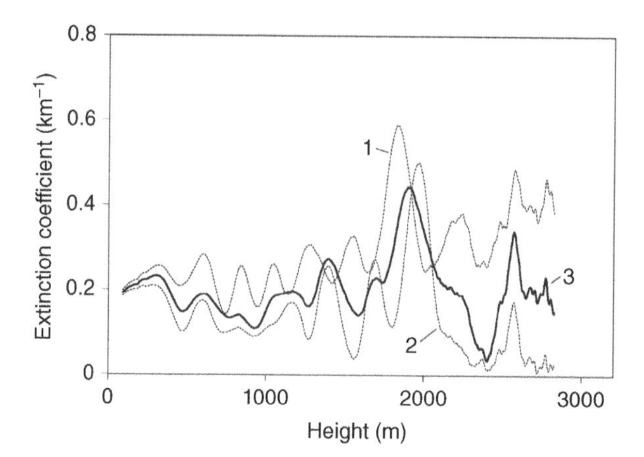

Fig. 3.17 Particulate extinction coefficient profiles $\kappa_p^{(\mathrm{dif})}(h)$ extracted from the optical depth profiles in Fig. 3.16.

No statistics works when analyzing the profiles $\kappa_p^{(\mathrm{dif})}(h)$ in Fig. 3.17, therefore, it is impossible to assert which of these is the profile with the largest likelihood. However, the above methodology gives some general estimation of the uncertainty limits in the retrieved profiles of the extinction coefficient. These three profiles show the general behavior of the particulate extinction coefficient versus height, at least up to the height ~2400 m and permit the estimation of the approximate scope of its uncertainty. One can discriminate a relatively clear zone with $\kappa_p^{(\mathrm{dif})}(h) \sim 0.2\,\mathrm{km}^{-1}$ up to the height ~1600 m and a layer of increased backscattering within the height interval from ~1600 to 2200 m.

When making inversion of the multiangle signals, one specific objective related to the distortions should be taken into consideration. The issue needing to be addressed is related to the selection of the optimal maximum range $r_{\max}$ up to which the lidar-signal distortions are acceptable. The inversion errors caused by signal systematic and low-frequency distortions is a matter that does not benefit from a universally accepted treatment protocol based on statistical estimates. Note also that in multiangle profiling of the atmosphere, signal distortions, such as the nonzero offset ΔB and the noise component $\sum w_{j,\mathrm{low}}(r)$ are not the only factors which restrict the maximum range $r_{\max}$. The heterogeneity of the searched atmosphere is also an extremely influencing factor, which restricts the acceptable maximum range. The unavoidable presence of these factors upholds the obvious fact that the arbitrarily selected signal-to-noise ratio as a criterion for the length of $r_{\max}$ is poorly grounded.

To illustrate the issue of determining optimal $r_{\max}$, let us consider the inversion results derived from the real signals of the scanning lidar measured in the vicinity of a wildfire in August 2009. The multiangle profiling was made using the azimuthally averaged signals at 12 fixed slope angles in the angular sector from 9° to 80°. The signals were inverted into the optical depth profiles $\tau_{\mathrm{vert}}(0, h)$ using different maximal ranges $r_{\max}$. In Fig. 3.18, these profiles, extracted with different maximum ranges

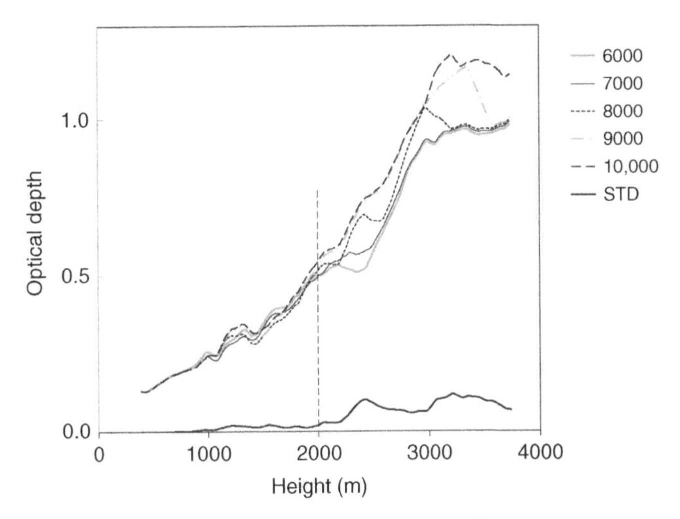

Fig. 3.18 Experimental profiles of the optical depth $\tau_{vert}(0, h)$ extracted from the signals of a scanning lidar using the different maximum ranges r_{max} shown in the legend. The thick curve at the bottom of the figure is the standard deviation for these optical depths.

from 6000 to 10,000 m are shown. The standard deviation of these optical depths is shown as the thick curve in the bottom of the figure. As one can see, the optical depth profiles obtained with different r_{max}, remain relatively close to each other only up to the height ~2000 m. For greater heights, the difference between the profiles sharply increases. Without additional analysis, which would clarify the optical situation at high altitudes, the extinction coefficient can be extracted from these optical depths only for heights below ~2000 m.

Thus, the presence of systematic distortions in the lidar signal, especially in the poorly stratified atmosphere, may significantly impede the selection of the optimal signal maximum range for the Kano–Hamilton solution. An arbitrarily selected r_{max} does not allow proper estimation of the actual maximum height up to which the optical depth and extinction coefficient profiles can be extracted. To overcome this issue, special algorithms that would allow estimation of an optimal maximum range for lidar searching are required. Such algorithms are considered in Section 3.6.2.

3.3 DETERMINATION OF THE EFFECTIVE OVERLAP USING THE SIGNALS OF THE SCANNING LIDAR

3.3.1 Effective Overlap: Definition and the Derivation Algorithm

The high sensitivity of the inversion results obtained in the multiangle mode to the distortions in the inverted signals forced researchers to look for a strategy to mitigate these errors. At the same time this objective has lead researchers to conclude that the multiangle method may be exploited not only for atmospheric profiling but also to examine the error sources inherent to lidar instrumentation. In other words, the multiangle methodology can be used not only for atmospheric profiling but also as a useful element of lidar tests and calibration. Such tests can include the determination of the lidar "effective"overlap, which in turn, allows experimental estimation of the minimal and maximal ranges, where the best accuracy of the signal inversion can be achieved. The test may even be performed using the lidar signals obtained during routine multiangle sounding, if it is made under favorable atmospheric conditions.

Let us clarify the notion of the effective overlap. Section 1.8 discusses the calibration of the overlap of the emitted laser light beam and the cone of the receiver telescope field of view using conventional methods. However, the real lidar signals, even when measured in a homogeneous atmosphere, do not guarantee the precise determination of the overlap, $q(r)$. This is especially true when the overlap calibration is made in a relatively turbid atmosphere, in which the atmospheric attenuation is comparable or larger than that in routine atmospheric profiling. The nonzero offset ΔB remaining in the backscatter signal can significantly influence the calibrated overlap. In other words, in real conditions, the experimentally extracted overlap function can be distorted relative to the actual overlap. When determining the overlap experimentally, we actually determine the shape of some "effective" overlap $q_{\text{eff}}(r)$ rather than the true overlap $q(r)$. Unfortunately, the offset ΔB may be different during the calibration and routine profiling of the atmosphere. For one-directional lidar, especially when the signals from the incomplete overlap zone need to be inverted, this is a practically insolvable issue.

When using scanning lidar, special methodology can be applied to analyze the shape of the effective lidar overlap. A simplified variant of such methodology was considered in the study by Adam et al. (2007). The important element in this straightforward technique is the identification and removal of the poor data points that "spoil" the linear dependence of the function $y_j(h)$ versus the sine of the searching angle [Eq. (3.4)]. The behavior of the overlap function can be determined only after this procedure.

The theoretical basis of such a multiangle lidar test is based on relatively simple formulas. As follows from Eq. (3.15), the distorted square-range-corrected signal under the slope direction φ, defined as $\langle Z_j(r)\rangle = \langle P_j(r)\rangle r^2$, can be rewritten as

$$\langle Z_j(r)\rangle = C\beta_{\pi,j}(r)q(r)[1+\delta_{P,j}(r)]\exp[-2\tau_j(0,r)] + \left[\Sigma w_{j,\text{low}}(r) + \Delta B_j\right] r^2, \quad (3.20)$$

where, the term $\Sigma w_{j,\text{low}}(r)$ implies that initial filtering of the high frequency noise that presumably obeys common statistics was performed, so that only low-frequency noise components remain in the square-range-corrected backscatter signal. Eq. (3.20) can be rewritten as

$$\langle Z_j(r)\rangle = C\beta_{\pi,j}(r)q(r)[1+\delta_{P,j}(r)]\exp[-2\tau_j(0,r)]\left[1+\frac{\Sigma w_{j,\text{low}}(r)+\Delta B_i}{P_j(r)[1+\delta_{P,j}(r)]}\right]. \quad (3.21)$$

where $P_j(r)$ is the undistorted backscatter signal. In turn, Eq. (3.21) can be rewritten in the form

$$\langle Z_j(r)\rangle = C\beta_{\pi,j}(r)q_{j,\mathrm{eff}}(r)\exp\lfloor -2\tau_j(0,r)\rfloor, \tag{3.22}$$

where the "effective" overlap $q_{j,\mathrm{eff}}(r)$ is defined as

$$q_{j,\mathrm{eff}}(r) = q(r)\left[1 + \delta_{P,j}(r) + \frac{\Sigma w_{j,\mathrm{low}}(r)}{P_j(r)} + \frac{\Delta B_j}{P_j(r)}\right]. \tag{3.23}$$

Thus, the effective overlap is the product of the true overlap function $q(r)$ and the combined distortion factor in the square brackets. In other words, it can be considered as the true overlap distorted by the combination of the multiplicative and additive components $\delta_{p,j}(r)$, $\sum w_{j,\mathrm{low}}(r)$, and ΔB_j. Because the backscatter signal $P_j(r)$ decreases with range, a larger discrepancy between $q(r)$ and $q_{j,\mathrm{eff}}(r)$ generally can be expected over distant ranges. In the ideal case when no backscatter signal distortion exists, the effective overlap does not differ from the true overlap, $q(r)$.

Experimental investigation of the effective overlap $q_{j,\mathrm{eff}}(r)$ in the multiangle mode can be made in a clear atmosphere. However, it can only be done if the atmospheric molecular scattering at the wavelength at which the elastic lidar operates is dominant. Under such a condition, the scanned atmosphere will better satisfy Eqs. (3.1) and (3.2).

The function $y_j(h) = \mathrm{In}\langle Z_j(h)\rangle$ for any slope direction φ can be written as

$$y_j(h) = \ln[C\beta_{\pi,m,j}(h)] + \ln\left[1 + \frac{\beta_{\pi,p,j}(h)}{\beta_{\pi,m,j}(h)}\right] + \ln[q_{j,\mathrm{eff}}(h)] - \left[\frac{2\tau_{\mathrm{vert}}(0,h)}{\sin\varphi}\right]. \tag{3.24}$$

Using the data points of the function $y_j(h)$ over the areas of presumably complete overlap, one can determine the linear fits $Y_j(x,h)$ of the function versus the independent $x = [\sin\varphi]^{-1}$, similar to that in Eq. (3.6), that is,

$$Y_j(x,h) = A(h) - 2\tau_{\mathrm{vert}}(0,h)x, \tag{3.25}$$

and accordingly determine the corresponding profiles of the functions $A(h)$ and $\tau_{\mathrm{vert}}(0,h)$. Note that in the area where $q_{j,\mathrm{eff}}(r) = 1$, the variations of the function $A(h)$ will originate only in the term $\ln\left[1 + \frac{\beta_{\pi,p,j}(h)}{\beta_{\pi,m,j}(h)}\right]$. If the molecular component is dominant, that is, $\beta_{\pi,p,j} < \beta_{\pi,m,j}$, the variations of $A(h)$ will be reduced. Using the profiles $A(h)$ and $\tau_{\mathrm{vert}}(0,h)$, determined as the intercept and the slope of the linear fit $Y_j(x,h)$, one can calculate the corresponding profile of a synthetic square-range-corrected vertical signal $Z^*_{\mathrm{vert}}(h)$ as (Adam et al., 2007)

$$Z^*_{\mathrm{vert}}(h) = \exp[A(h)]\exp[-2\tau_{\mathrm{vert}}(0,h)]. \tag{3.26}$$

The corresponding synthetic signals $Z_j^*(h)$ along the slope directions φ can be calculated as

$$Z_j^*(h) = \exp[A(h)] \exp\left[\frac{-2\tau_{\text{vert}}(0,h)}{\sin\varphi}\right]. \tag{3.27}$$

Using synthetic and real square-range-corrected signals for each elevation angle φ one can calculate the effective overlap function versus height along the slope direction. The effective overlap is determined as the ratio of the true square-range-corrected signal $\langle Z_j(h)\rangle$ to the synthetic profile $Z_j^*(h)$, that is,

$$q_{j,\text{eff}}(h) = \frac{\langle Z_j(h)\rangle}{Z_j^*(h)}. \tag{3.28}$$

For the next analysis, the height-dependent functions $q_{j,\text{eff}}(h)$ should be recalculated into functions of the slope range $q_{j,\text{eff}}(r)$. These range-dependent functions, determined under different slope directions, may prove to be different even over the ranges where the actual overlap $q(r) = 1$. Therefore, the effective overlap profiles should be investigated using all available signals measured in the slope directions. The goal of such an analysis of the individual functions $q_{j,\text{eff}}(r)$ is to determine (i) whether these functions differ from each other, (ii) if yes, over what range their distortions are minimal, and (iii) over what ranges the equality $q_{j,\text{eff}}(r) = q(r) = 1$ is fulfilled for individual profiles. Such an analysis establishes the presence of systematic differences in the functions $q_{j,\text{eff}}(r)$ and zones over which these differ from the assumed overlap function $q(r)$. Accordingly, it provides realistic estimates of the lidar minimum and maximum searching range, and the minimal value of the extinction coefficient below which the lidar cannot yield reliable inversion results because of its technical characteristics.

As stated above, when using the elastic lidar, the above method of determining the effective overlap has relatively stringent restrictions for its application. It can be used only under the condition that the molecular scattering at the wavelength at which the lidar operates is dominant as compared to the particulate scattering. This is why in the study by Adam et al. (2007), such a test was performed only at the wavelength 355 nm. However, this stringent condition is not required when using splitting lidar. The overlap tests described above can be extremely fruitful for investigation of the quality of data obtained with the HSRLs. In the molecular channel of HSRL, only the molecular backscatter component creates the output signal, so that the restrictive requirement in Eq. (3.1) is practically always met and only the requirement in Eq. (3.2) needs to be satisfied.

3.3.2 Divergence of $q_{\text{eff}}(h)$ from $q(h)$: Numerical Simulations and the Case Study

To illustrate a typical divergence of the effective $q_{\text{eff}}(h)$ from the overlap $q(h)$ when systematic distortions in the backscatter signal take place, let us consider some results of the numerical simulations made in the study by Adam et al. (2007). In these simulations, the data obtained by virtual elastic lidar in a synthetic horizontally homogeneous atmosphere were analyzed. The particulate extinction coefficient in this synthetic atmosphere decreased exponentially with height from 0.1 km^{-1} at ground level to 0.001 km^{-1} at a height of 4600 m. The virtual scanning lidar operated at the wavelength 355 nm along fourteen elevation angles from 6° to 80°. The lidar signals corrupted by quasi-random noise were recorded with a 12-bit digitizer. It was assumed that the background component B_j in the recorded signals was not precisely determined so that the backscatter signals obtained after subtracting the estimated $\langle B_j \rangle$ were distorted by the remaining nonzero offsets ΔB_j. The maximal ranges, at which the lidar data were considered as being still acceptable for inversion, were chosen as the ranges where the signal-to-noise ratio was unity. To focus on the influence of the constant offset ΔB_j on the effective overlap, two other distortions components $\delta_{P,j}(r)$ and $\sum w_{j,\text{low}}(r)$ were assumed insignificant.

The results of the numerical simulation revealed that the presence of moderate random signal noise that obeys conventional statistics generally does not mask the systematic divergence the functions $q(r)$ and $q_{j,\text{eff}}(r)$. In Figs. 3.19 (a) and 3.20 (a), the effective overlap functions obtained with Eq. (3.28) and recalculated as a function of range are shown for the case when the offsets ΔB_j in the backscatter signals, are equal to +1 count and −1 count, respectively. The average functions $\overline{q_{\text{eff}}(r)}$ and their boundaries, estimated as the average plus/minus STD, are shown in Figs. 3.19 (b) and 3.20 (b). One can see that in both cases, no significant difference between $q(r)$ and $q_{j,\text{eff}}(r)$ exist up to the ranges of about 3.5 km, but then the functions $q_{j,\text{eff}}(r)$ noticeably

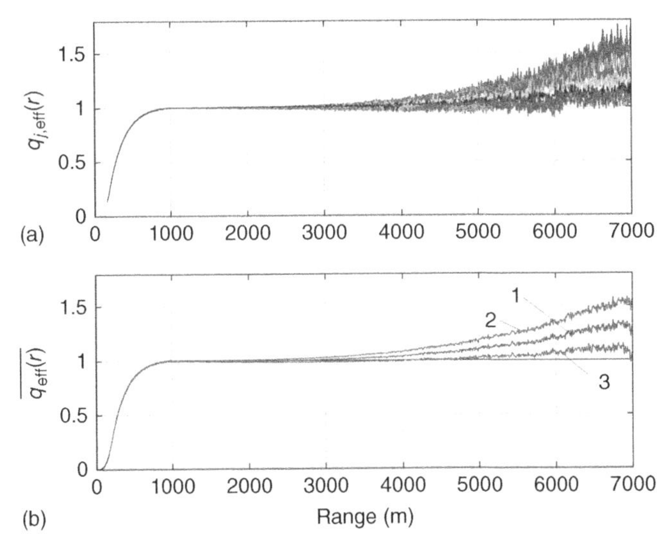

Fig. 3.19 (a) Effective overlap functions $q_{j,\text{eff}}(r)$ obtained from the numerical simulation for the case when $\Delta B_j = +1$ count. (b) The mean effective function $\overline{q_{\text{eff}}(r)}$ and its upper and lower boundaries, $\overline{q_{\text{eff}}(r)} \pm$ STD, shown as Curves 1, 2, and 3, respectively. The test overlap is shown as the thin black curve (Adam et al., 2007).

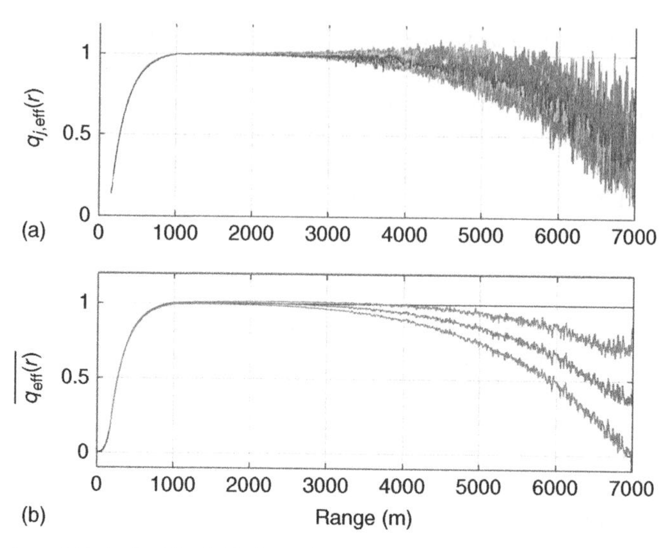

Fig. 3.20 (a and b) As in Fig. 3.19, but where the background component in the recorded signals is overestimated, and $\Delta B_j = -1$ count (Adam et al., 2007).

diverge. Thus, the complete overlap zone of the virtual lidar, where $q_{j,\mathrm{eff}}(r) = q(r)$, and the additive distortions will not worsen the routine inversion results, extends only from 1 to 3–3.5 km. As follows from the above simulations, the offset ΔB_j equal to ± 1 count, is unacceptable for lidar profiling of the atmosphere at higher altitudes. Note that the influence of the additive distortions ΔB_j is much more detrimental when the background component is overestimated.

Experimentally, the effective overlap function was investigated using signals of the 355-nm channel of the scanning lidar of the Montana Fire Sciences Laboratory (FSL). The lidar examination was performed in a stable and clear atmosphere, starting in the morning hours and finishing in the early afternoon, when scattered cloudiness prevented the continuation of the signal measurement. A typical result of such an examination is shown in Fig. 3.21. The plot represents the mean overlap profile and its boundaries. The behavior of the effective overlap function over distant zones, particularly, its deflection from unity, is presumably related to the remaining offsets ΔB_j in the inverted backscatter signals and the atmospheric heterogeneity. At a distance of 5000 m, the effective overlap $\overline{q_{\mathrm{eff}}(r)} = 0.96$; this means that in most cases, the constant offset in the lidar signals was overestimated. The assumption appears realistic as the estimates of the offsets were based on the recorded signals over distant ranges. As shown in Section 1.5, such a method may overestimate the actual offset in a recorded signal.

Finally, let us briefly discuss the specifics of effective overlaps $q_{j,\mathrm{eff}}(r)$, when the lidar multiangle signals are distorted by the multiplicative distortion component, $\delta_{P,j}(r)$. With the assumption that the distortion components $\Delta B_j = 0$ and

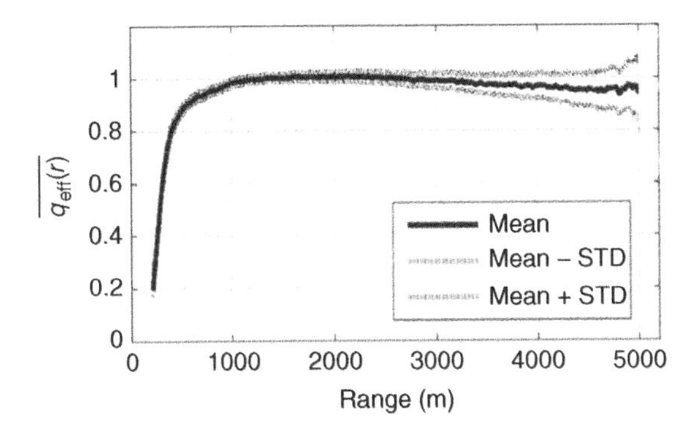

Fig. 3.21 Mean effective overlap function $\overline{q_{\text{eff}}(r)}$ for the 355-nm channel of the FSL lidar (thick solid line). The dotted curves represent the upper and lower boundaries of the optical depth, $\overline{q_{\text{eff}}(r)} \pm$ STD (Adam et al., 2007).

$\sum w_{j,\text{low}}(r) = 0$, Eq. (3.23) reduces to

$$q_{j,\text{eff}}(r) = q(r)\lfloor 1 + \delta_{P,j}(r)\rfloor, \tag{3.29}$$

that is, the effective overlap is directly proportional to the multiplicative distortion factor $[1 + \delta_{P,j}(r)]$. To illustrate the influence of the distortion factor on the effective overlap, we will exploit the fact that the use of an incorrect overlap is equivalent to the presence of the nonzero component $\delta_P(r)$ in the lidar signal. Such an observation makes it possible to confine our analysis of the influence of the factor $[1 + \delta_{P,j}(r)]$ by using the results of the numerical simulation made in the study by Adam et al. (2007), in which the consequences of inaccurately determining the overlap function are analyzed. In Fig. 3.22, the profiles $q_{j,\text{eff}}(r)$ are shown extracted from the data of a virtual scanning lidar operating in a synthetic atmosphere having an exponential change of the extinction coefficient. The simulated signals of the virtual lidar are corrupted by quasi-random noise and the offset ΔB_j is assumed to be zero. The actual length of the incomplete overlap zone was 1000 m, whereas the inaccurately estimated length was only 500 m. Obviously, the equivalent distortion component, which may be determined from Eq. (3.29) as $\lfloor 1 + \delta_{P,j}(r)\rfloor = \langle q(r)\rangle / q(r)$, differs from unity only within the restricted range from 500 to 1000 m. However, as can be seen in the figure, the deviations of the effective overlap function $q_{j,\text{eff}}(r)$, from unity take place over distant ranges significantly larger than 1000 m. These distortions become noticeable only when the effective overlap function is determined using the signals measured at small elevation angles; no distortions take place when the functions $q_{j,\text{eff}}(r)$ are determined from the signals measured at large elevation angles.

Thus, in clear and horizontally stratified atmosphere, the Kano–Hamilton method permits reliable determination of the effective overlap and exposes its divergence from unity in the complete overlap zone. This in turn, facilitates selection of the optimal ranges $r_{\min}$ and $r_{\max}$ for routine profiling of the atmosphere. One can maintain that the determination of the effective overlap $q_{\text{eff}}(r)$ in the multiangle mode and

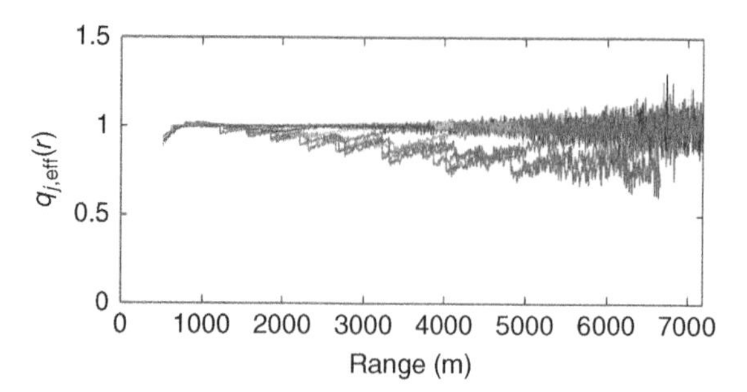

Fig. 3.22 The effective overlap functions $q_{j,\mathrm{eff}}(r)$ obtained from the numerical simulation when the synthetic signals of the virtual scanning lidar are corrupted with the multiplicative distortion factor $[1 + \delta_{P,j}(r)]$ (Adapted from Adam et al., 2007).

its comparison with the overlap function determined using traditional methods (e.g., Sasano et al., 1979; Sassen and Dodd, 1982; Ignatenko, 1985; Tomine et al., 1989; Dho et al., 1997) would allow significantly better estimation of quality of the lidar observations, especially, over ranges close to $r_{\max}$.

Let us summarize. Commonly, tests of the lidar instrumentation in the open atmosphere, including the determination of the incomplete overlap zone of the laser beam and the receiving telescope field of view, are performed in a fixed, generally in the horizontal, direction. First, such tests are difficult to perform, especially, when the incomplete overlap zone of the lidar is large. Second, such tests do not take into consideration all possible multiplicative and additive distortions in the measured lidar signal during overlap calibration. Failure to consider these can result in hidden distortions when routine one-directional atmospheric profiling is made.

The uniqueness of the multiangle Kano–Hamilton data processing technique is its extreme high sensitivity to any systematic distortion in the scanning lidar data, both in the near and the far end of the measurement range. This deficiency may be leveraged to perform comprehensive lidar instrumentation tests, which, under favorable conditions, can be made even in parallel to the routine profiling of the atmosphere. Inspection of the effective overlap shape made in the multiangle mode provides the researcher with invaluable information regarding the potential accuracy of the results obtained from the lidar instrumentation. Such a test may be extremely helpful when determining the optimal lidar measurement ranges and heights, minimal and maximal, over which the best inversion results can be obtained. One can maintain that tests and calibrations made in the multiangle mode may be extremely useful even for lidar that routinely operates in one-directional mode. Properly performed multiangle calibration can either confirm the declared accuracy of the one-directional lidar data or reveal the unpleasant fact that its accuracy is overestimated.

The test methodology discussed above is most worthwhile for splitting lidars, especially, for HSRL; elastic lidar, even when operated at wavelengths shorter than 532 nm, should be used for such a test extremely cautiously.

3.4 PROFILING OF THE ATMOSPHERE WITH SCANNING LIDAR USING THE ALTERNATIVE INVERSION TECHNIQUES

As discussed in Section 3.2, extraction of the extinction coefficient from scanning lidar data requires two consecutive slope determinations. The first one is used to obtain the optical depth, $\tau_{vert}(0, h)$, from the set of the measured lidar signals, and the second one, extraction of the extinction coefficient, is made through numerical differentiation of the optical depth. The use of numerical differentiation of the generally noisy lidar data yields unacceptably distorted profiles of the extinction coefficient. The additional drawback of such an inversion is that only the optical depth profile is used for extracting the extinction coefficient. Valuable information about particulate loading contained in the backscatter term is not used to put constraints on the profile of the extracted extinction coefficient. In this and the next sections, the substantiation and alternative variants for deriving the extinction coefficient profile from multiangle lidar data are considered.

3.4.1 Comparison of the Uncertainty in the Backscatter Coefficient and the Optical Depth Profiles Extracted from the Signals of the Scanning Lidar

In moderately clear atmosphere, the backscatter coefficient determined from the intercept of the linear fit can be obtained with better accuracy than the extinction coefficient (Kunz and Leeuw, 1993). It is sensible to take this specific of the slope method into account when processing multiangle data of scanning lidar. In clear and moderately clear atmospheres, the term $A(h)$ in Eq. (3.6) can be found with better accuracy than $\tau_{vert}(0, h)$, so that the profile of $A(h)$ can be used as an additional constraint for determining the profile of the vertical extinction coefficient.

Let us initially consider how violation of the requirements in Eqs. (3.1) and (3.2) influences the accuracy of the functions $\beta_\pi(h)$ and $\tau_{vert}(0, h)$, obtained with the Kano–Hamilton solution. For such estimates, the results of simple numerical simulations, in which the two-point multiangle solution is used, can be taken as evidence. Consider the data of the virtual lidar, which operates at the wavelength 355 nm along two elevation angles, $\varphi_1 = 90°$ and $\varphi_2 = 30°$, assuming that the systematic distortions in the derived optical parameters originate only from heterogeneity of the atmosphere.

It is sensible to analyze initially the case when the requirement of Eq. (3.2) is met but that of Eq. (3.1) is not satisfied. In the numerical experiment described below, it is assumed that the backscatter coefficients, $\beta_{\pi,30}(h)$ and $\beta_{\pi,90}(h)$, determined at the same height h along the slope directions 30° and 90° are not equal; in particular, the backscatter coefficient $\beta_{\pi,30}(h)$ differs by $\pm 30\%$ from that in the vertical direction. In Fig. 3.23, the corresponding data points $y_j(h)$ and the line $Y_j(h)$ versus $x = [\sin \varphi]^{-1}$ are shown. The line $Y(h)$ intersects these data points, $y_{30}(h)$ and $y_{90}(h)$, which are obtained with Eq. (3.6) if the condition, $\beta_{\pi,30}(h) = \beta_{\pi,90}(h)$, is satisfied; the lines $Y'(h)$ and $Y''(h)$ are obtained from the lidar data when negative and positive shifts in $\beta_{\pi,30}(h)$, presumably caused by the atmospheric heterogeneity, take place.

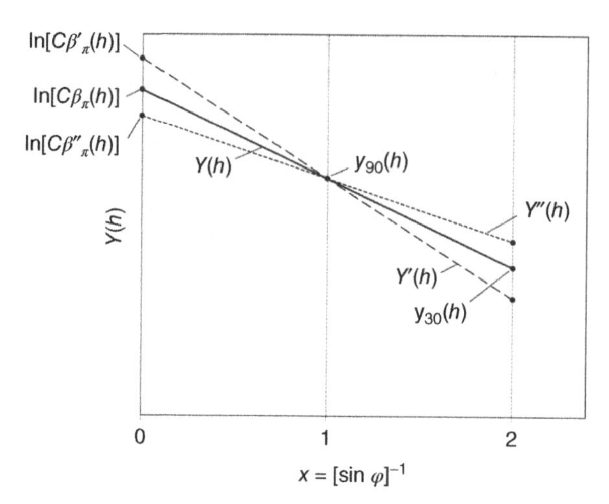

Fig. 3.23 The solid line $Y(h)$ corresponds to the condition $\beta_{\pi,30}(h) = \beta_{\pi,90}(h)$. The lines, $Y'(h)$ and $Y''(h)$, are obtained when the requirement in Eq. (3.1) is not met. The dashed line $Y'(h)$ is obtained when $\beta_{\pi,30}(h) = 0.7\,\beta_{\pi,90}(h)$, and the dotted line $Y'(h)$ is obtained when $\beta_{\pi,30}(h) = 1.3\,\beta_{\pi,90}(h)$ (Kovalev et al., 2011b).

Analysis of the retrieved optical profiles shows that the distortions in the extracted vertical optical depth $\tau_{\text{vert}}(0,h)$ and in the backscatter coefficient $\beta_\pi(h)$ will be different. This difference depends on the actual vertical optical depth $\tau_{90}(0,h)$ from ground level to the height of interest, which in our simulation is $h = 1000\,\text{m}$. In Fig. 3.24, the dashed curves show the relative errors in the calculated total optical depths $\tau_{\text{vert}}(0,h)$ versus true optical depth $\tau_{90}(0,h)$ for the above two cases: $\beta_{\pi,30}(h) = 0.7\beta_{\pi,90}(h)$ and $\beta_{\pi,30}(h) = 1.3\beta_{\pi,90}(h)$. The thick solid curves show the corresponding error in the retrieved backscatter coefficient $\beta_\pi(h)$. The simulation shows that in clear air conditions, when $\tau_{90}(0,h) \le 0.45 - 0.55$, the relative error in $\beta_\pi(h)$ is smaller than the error in the retrieved optical depth $\tau_{\text{vert}}(0,h)$. However, the range of optical depth $\tau_{90}(0,h)$ where the extinction coefficient $\kappa(h)$ can be extracted with good accuracy can be significantly larger than the above numbers. One should keep in mind that when the extinction coefficient $\kappa(h)$ is extracted without numerical differentiation, that is, using the auxiliary function $[C\beta_\pi(h)] = \exp[A(h)]$, its accuracy will generally be significantly better than in the alternative profile $\kappa^{(\text{dif})}(h)$ extracted by numerical differentiation.

Now let us see how the error in the retrieved $\beta_{\pi,90}(h)$ depends on the violation of the second requirement in Eq. (3.2), when the requirement in Eq. (3.1) is satisfied. Let the line $Y(h)$ in Fig. 3.23 correspond to the atmosphere where the requirement in Eq. (3.2) is met, whereas the line $Y'(h)$ is obtained when the requirement is not valid. Accordingly, the slope of the line $Y(h)$ yields the true optical depth $\tau_{\pi,90}(0,h)$, whereas the slope of the line $Y'(h)$ yields some shifted optical depth $\tau_{\text{vert}}(0,h)$. The Kano–Hamilton solution for these two cases may be written as $Y(h) = \ln[C\beta_\pi(h)] - [2\tau_{90}(0,h)]x$ and $Y'(h) = \ln[C\beta'_\pi(h)] - [2\tau_{\text{vert}}(0,h)]x$, respectively; here, $\ln[C\beta'_\pi(h)]$

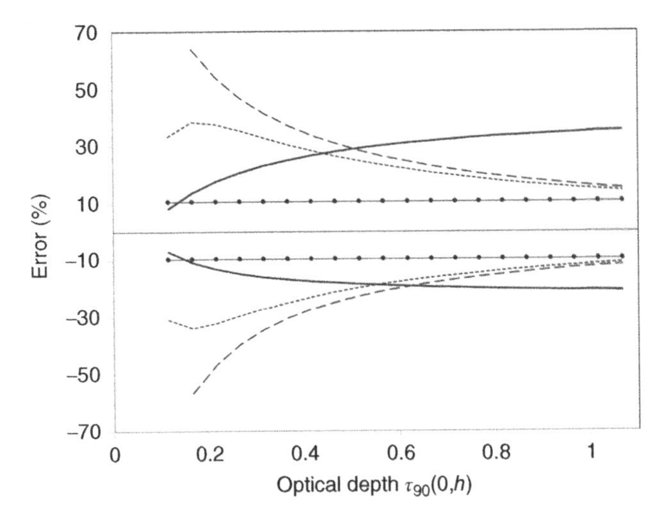

Fig. 3.24 Dependence of the relative errors in the total optical depth $\tau_{\text{vert}}(0, h)$ and its particulate component at 355 nm (dotted and dashed curves, respectively) and the corresponding errors in $\beta_\pi(h)$ (thick solid curves) caused by violation of the requirement in Eq. (3.1). The solid horizontal lines show the relative errors $\delta\beta_\pi(h)$, caused by violation of the requirement in Eq. (3.2) (Adapted from Kovalev et al., 2011b).

is the erroneous intercept, that corresponds to the shifted $Y'(h)$ (Fig. 3.23). Defining the difference between $\tau_{\text{vert}}(0, h)$ and $\tau_{90}(0, h)$ as $\Delta\tau_{\text{vert}}(0, h)$ and the corresponding shift in the logarithm of $[C\beta_\pi'(h)]$ as $\Delta \ln[C\beta_\pi(h)]$, one can obtain the formula

$$\ln\left[\frac{\beta_\pi'(h)}{\beta_\pi(h)}\right] = 2x_{\text{int}}\Delta\tau_{\text{vert}}(0, h), \tag{3.30}$$

where x_{int} is the interception point of the lines $Y(h)$ and $Y'(h)$. The corresponding relative error in the backscatter coefficient $\beta_\pi(h)$ is

$$\delta\beta_\pi(h) = \exp[2x_{\text{int}}\Delta\tau_{\text{vert}}(0, h)] - 1. \tag{3.31}$$

As follows from Eq. (3.31), the relative error $\delta\beta_\pi(h)$ depends on the difference between the actual and derived optical depths $\tau_{90}(0, h)$ and $\tau_{\text{vert}}(0, h)$ and on the location of the interception point x_{int}. In Fig. 3.24, the error $\delta\beta_\pi(h)$ calculated for $x_{\text{int}} = 1$ and $\Delta\tau_{\text{vert}}(0, h) = \pm 0.05$ is shown as two horizontal solid lines. If the intersection of the above linear fit occurs at $x = 2$, the error $\delta\beta_\pi(h)$ becomes larger; however, up to $\tau_{90}(0, h) \sim 0.6$, it still remains less than the error in the optical depth $\tau_{\text{vert}}(0, h)$. Note also that the error $\delta\beta_\pi(h)$ does not depend on the optical depth $\tau_{90}(0, h)$; it is proportional to the exponent of the absolute error $\Delta\tau_{\text{vert}}(0, h)$ and becomes noticeably larger for large optical depths.

3.4.2 Extraction of the Vertical Extinction Coefficient by Equalizing Alternative Transmittance Profiles in the Fixed Slope Direction: Basics

Let us consider the multiangle solution in which the most challenging operation of deriving the extinction coefficient profile through numerical differentiation is excluded. The backscatter signals measured under different slope directions are directly inverted into the optical parameters of interest. This alternative solution allows determination and comparison of the extinction coefficient profiles in different slope directions, providing better understanding of uncertainties in the retrieved profiles caused by atmospheric heterogeneity. Obviously, when a set of backscatter signals measured in different slope directions is analyzed separately, more accurate extinction coefficient profiles can be obtained as compared with the conventional way of extracting a single profile of the extinction coefficient $\kappa^{(\mathrm{dif})}(h)$ from the optical depth $\tau_{\mathrm{vert}}(0, h)$.

The modified Kano–Hamilton inversion method is focused on determining the profile of the function $[C\beta_\pi(h)]$ that is, the profile of the exponent of the intercept $A(h)$. Vertical profiles of $[C\beta_\pi(h)]$, transformed into the function of range $C\beta_\pi(r)$, and the range-corrected backscatter signal along the selected slope direction φ are the functions from which the optical depth and corresponding extinction coefficient are obtained. To use this method, the lidar-solution constant C needs to be known in order to determine the particulate backscatter coefficient $\beta_{\pi,p}(h)$ from the function, $[C\beta_\pi(h)]$. The determination of the slope of the linear fit $Y(x, h)$ with Eq. (3.6) is excluded. Instead, the two-way transmittance in the selected slope direction is found.

The square-range-corrected backscatter signal in the slope direction φ versus range r can be written as

$$P_j(r)r^2 = C\beta_\pi(r)T^2_{\Sigma,j}(0, r), \tag{3.32}$$

where $T^2_{\Sigma,j}(0, r)$ is the total two-way transmittance along the slope direction φ within the range from $r = 0$ to r, where $r_{\min} \le r \le r_{\max}$. The profile $T^2_{\Sigma,j}(0, r)$ within the corresponding height interval from $h_{j,\min} = r_{\min} \sin\varphi$ to $h_{j,\max} = r_{\max} \sin\varphi$ can be determined as

$$T^2_{\Sigma,j}(0, r) = \frac{P_j(r)r^2}{C\beta_\pi(r)}. \tag{3.33}$$

As with the zenith-directed lidar, the total range of the above signal measured in the slope direction should be divided into a number of smaller intervals $r' - r''$ within which the column-integrated lidar ratio $S^{(i)}_{p,j}$ can be taken as a constant. For each such interval, the profile of the two-way particulate transmittance is determined from the corresponding pieces of the total profile $T^2_{\Sigma,j}(0, r)$ and the molecular profile $T^2_{m,j}(0, r)$. The particulate transmission profile along the slope direction φ within the restricted interval from r' to r'' is found with a formula similar to Eq. (2.113), that is,

$$T^2_{p,j}(r', r) = \frac{T^2_{\Sigma,j}(0, r)T^2_{m,j}(0, r')}{T^2_{\Sigma,j}(0, r')T^2_{m,j}(0, r')}. \tag{3.34}$$

Using the assumption that the column-integrated lidar ratio $S^{(i)}_{p,j}$ within the segmented zone, $r' - r''$, can be taken as a constant, the vertical particulate extinction coefficient within the corresponding height interval from the height, $h' = r' \sin\varphi$ to $h'' = r'' \sin\varphi$ can be calculated. The extinction coefficient within the segmented zone is found as

$$\kappa^{(i)}_{p,j}(h) = S^{(i)}_{p,j}\beta_{\pi,p}(h). \tag{3.35}$$

As mentioned above, to perform this operation, the constant C should be somehow estimated, so that the profile of the backscatter coefficient $\beta_{\pi,p}(h)$ can be extracted from the product $C\beta_{\pi,p}(h)$; the details of such an operation are discussed in Section 2.6.2. After the extinction coefficient is found, the alternative two-way slope transmittance within the interval $r' - r''$ is found as

$$\langle T^2_{p,j}(r', r)\rangle = \exp\left[-2\int_{r'}^{r} \kappa^{(i)}_{p,j}(x)dx\right]. \tag{3.36}$$

Two transmission profiles, the initial, $T^2_{p,j}(r', r)$, and the alternative, $\langle T^2_{p,j}(r', r)\rangle$, are equalized by selecting the constant $S^{(i)}_{p,j}$ that minimizes the criterion $\Lambda(r', r'')$. After the criterion is minimized, the recalculated profile of $\kappa^{(i)}_{p,j}(h)$ in Eq. (3.35) is taken as the piecewise extinction coefficient profile of interest over the corresponding altitude interval from h' to h''. The above operations are repeated for all the segmented intervals (r', r'') within the corresponding altitude range.

3.4.3 Equalizing Alternative Transmittance Profiles along a Fixed Slope Direction: Numerical Simulations

To clarify the specifics of the modified multiangle technique, let us consider a virtual scanning lidar that operates in a synthetic atmosphere at the wavelength 355 nm. The lidar scans the atmosphere within the slope angular sector from 10° to 90° and its measurement range extends from $r_{min} = 500$ m to $r_{max} = 7000$ m. The particulate extinction coefficient in the synthetic atmosphere decreases with height, from $\kappa_p(h) = 0.15\,\text{km}^{-1}$ at ground level down to $\kappa_p(h) = 0.011\,\text{km}^{-1}$ at $h = 7000$ m. In the atmosphere, two horizontally stratified turbid layers exist with $\kappa_p(h) = 0.25\,\text{km}^{-1}$ at heights from 2500 to 3000 m and with $\kappa_p(h) = 0.1\,\text{km}^{-1}$ at heights from 3500 to 3800 m. The lidar ratio from ground level to the height 1000 m is equal to 20 sr, within both turbid layers it is 60 sr, and outside these layers, 30 sr. The corresponding profile of $C\beta_\pi(h)$, obtained from the synthetic multiangle signals with the Kano–Hamilton solution, is shown in Fig. 3.25 in the left side. To stress that this function is only an estimate of the true function $[C\beta_\pi(h)]$, it is symbolized below as $\langle C\beta_\pi(h)\rangle$.

The schematic of multiangle searching of the atmosphere, which clarifies the symbols used in the following formulas, is shown in the right side of Fig. 3.25. The slope lines show the directions of lidar scanning within the selected angular sector. To avoid superfluous details in the figure, only some slope directions are shown. The thin solid and dashed circular arcs show the minimum and maximum measurement

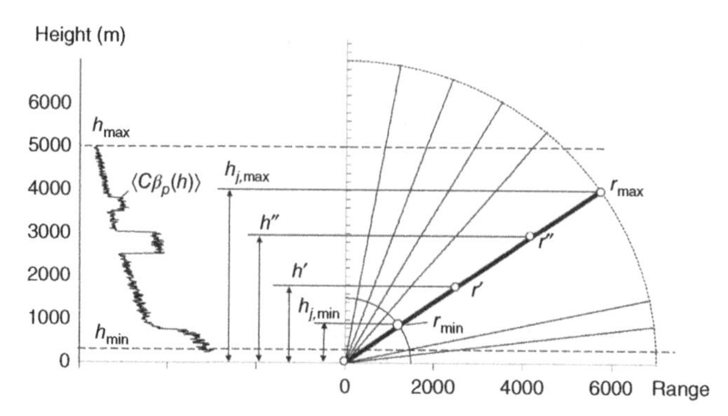

Fig. 3.25 Schematic of multiangle profiling of the atmosphere (right side) and the vertical profile of the function $\langle C\beta_\pi(h)\rangle$ extracted from the virtual lidar data (left side) (Kovalev et al., 2011b).

ranges of the lidar, r_{min} and r_{max}; for simplicity these ranges are selected the same for all slope directions. Accordingly, the square-range-corrected signal measured in any slope direction φ is determined within the height interval from $h_{j,min} = r_{min} \sin \varphi$ to $h_{j,max} = r_{max} \sin \varphi$. As made clear in Section 3.1, the maximum height h_{max} where the estimated profile $\langle C\beta_\pi(h)\rangle$ can be found with the Kano–Hamilton solution, is always less than the maximum lidar range r_{max}. In our case, the profile $\langle C\beta_\pi(h)\rangle$ is determined within the altitude range from $h_{min} = 180\,\text{m}$ to $h_{max} = 5000\,\text{m}$, whereas the lidar maximal range $r_{max} = 7000\,\text{m}$. The constant C required for determining the backscatter coefficient profile $\beta_\pi(h)$ from $\langle C\beta_\pi(h)\rangle$ was found using the assumption of purely molecular atmosphere at the maximum altitude, 5000 m. The constant proved to be as much as 8% higher than the constant C taken for the numerical simulations. Such an increased value is obtained because the condition of the purely molecular scattering at the maximum altitude, 5000 m, assumed during the retrieval procedure, was not precisely met.

Let us start inversions using data extracted from the signal of the virtual lidar measured at $\varphi = 90°$. In Fig. 3.26, the thick solid curve is the test profile of the two-way transmittance $T^2_{\Sigma,90}(0, h)$ and the noisy dashed curve is the derived profile, calculated Eq. (3.33), using the function $\langle C\beta_\pi(h)\rangle$ shown in the left side of Fig. 3.25. To calculate the piecewise extinction coefficient with Eq. (3.35), one should initially select the number and location of the fragmented intervals within the altitude heights from h_{min} to h_{max}, in which the column-integrated lidar ratio can be assumed a constant. In the following numerical simulations, overlapping intervals, whose length increases with range, are used; their consequent locations and lengths are the same as in Fig. 2.7.9.

After the intervals and their overlapping parameters are selected, the piecewise extinction coefficient $\kappa^{(i)}_{p,90}(h)$, with arbitrarily selected $S^{(i)}_{p,90}$ is calculated and the alternative profiles, $T^2_{p,90}(h', h)$ and $\langle T^2_{p,90}(h', h)\rangle$ for the first interval are compared and equalized. The latter is done by selecting the piecewise column-integrated lidar ratio $S^{(i)}_{p,90}$ that minimizes the criterion, $\Lambda(h', h'')$, as discussed in Section

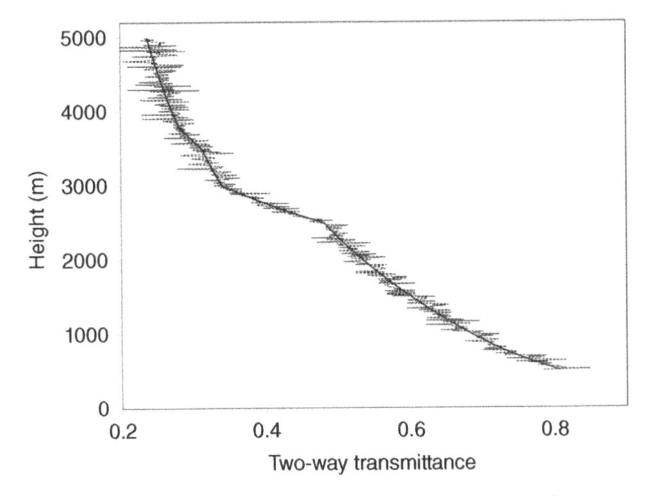

Fig. 3.26 Test profile of the vertical two-way total transmittance $T^2_{\Sigma,90}(0,h)$ (solid curve) and that retrieved from virtual lidar data using the Kano–Hamilton solution (dashed curve). (Adapted from Kovalev et al., 2011b).

2.6.3 [Eqs. (2.116)–(2.118)]. When minimization is achieved, the corresponding piecewise profile of $\kappa^{(i)}_{p,90}(h)$ is taken as final. After such operations are repeated for all selected intervals, the piecewise profiles of $\kappa^{(i)}_{p,90}(h)$ are "sewed together" as explained in Section 2.7.2. The initial test profile and the piecewise continuous extinction coefficient profile $\kappa_{p,90}(h)$ derived from the signal of the zenith-directed lidar are shown in Fig. 3.27 as dashed and solid curves, respectively.

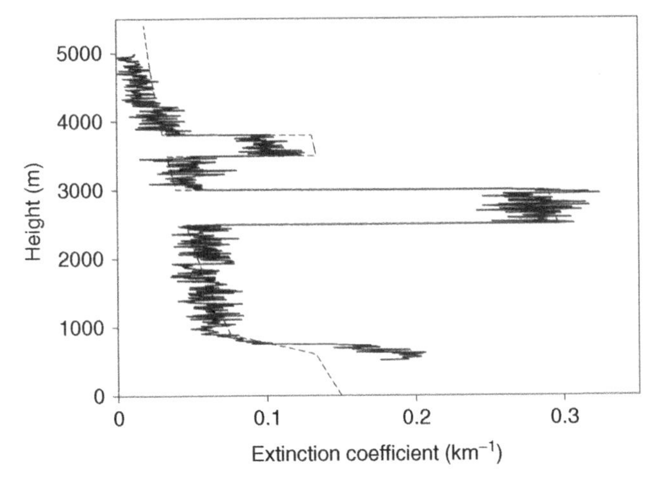

Fig. 3.27 The dashed curve is the test profile of the extinction coefficient $\kappa_p(h)$ used in the numerical simulation and the solid curve is the retrieved profile $\kappa_{p,90}(h)$. (Adapted from Kovalev et al., 2011b).

To obtain accurate extinction coefficient profiles over lower heights, signals measured under smaller elevation angles are preferable. This observation follows from the fact that sharp changes in the lidar ratio with height are generally softened along low angles. With the decrease in elevation angle, the discriminated height also decreases, whereas the slope range resolution remains the same. In Fig. 3.28, the inversion results are shown for the same synthetic atmosphere, but when the profile $T^2_{\Sigma,j}(0, h)$ is extracted from the signal measured in the slope direction $\varphi = 45°$. The solid curve is the piecewise continuous extinction coefficient $\kappa_{p,45}(h)$ "stitched" from the corresponding pieces $\kappa^{(i)}_{p,45}(h)$, whereas the thin solid curve is the test profile used in the simulations, and is the same as in Fig. 3.27.

It is always useful to analyze inversion results by using alternative inversion techniques to compare the extinction coefficient profiles obtained from the same lidar data. In our case, we can compare the profiles obtained by the above piecewise inversion technique and the numerical differentiation technique. The latter determines the sliding derivative of the logarithm of the ratio

$$\kappa_p^{(\text{dif})}(r) = -0.5\frac{d}{dr}\ln\left[\frac{T^2_{\Sigma,j}(0, r)}{T^2_{m,j}(0, r)}\right]. \tag{3.37}$$

The profile of the extinction coefficient $\kappa_p^{(\text{dif})}(h)$ obtained with Eq. (3.37) from the two-way transmittance profile at slope angle 45°, is shown in Fig. 3.28 as the dotted curve. The range resolution selected for the numerical differentiation $s = 510\,\text{m}$

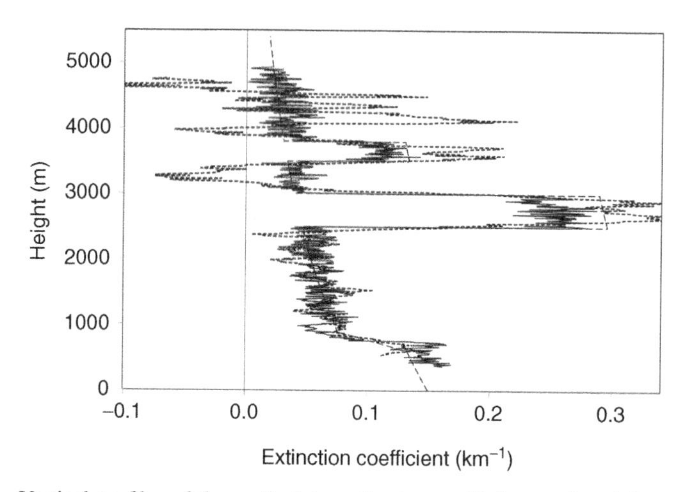

Fig. 3.28 Vertical profiles of the particulate extinction coefficient retrieved from the virtual lidar signal measured at $\varphi = 45°$. The dashed curve is the same test profile of the extinction coefficient as in Fig. 3.27, and the solid curve is the piecewise continuous profile $\kappa_{p,45}(h)$ retrieved using the technique under consideration. The dotted curve shows the profile of the extinction coefficient $\kappa_p^{(\text{dif})}(h)$ extracted through numerical differentiation (Kovalev et al., 2011b).

makes it possible to visualize both heterogeneous layers in the simulated atmosphere. However, the intense noise spikes in the retrieved profile mask the second layer. It is impossible to establish whether the spike originates in the actual aerosol layer with increased backscattering or it is due to noise.

In Fig. 3.29, the profiles of the extinction coefficient $\kappa_p^{(\text{dif})}(h)$ obtained with Eq. (3.37) at the angles 90°, 45°, and 30° are shown as the thick solid curve, the curve with the open circles, and the gray solid curve, respectively; the thin dashed curve shows the initial test profile of $\kappa_p(h)$ used in the simulation. All these profiles are obtained with the same range resolution, $s = 1000$ m, as a smaller resolution yields extremely noisy profiles. As can be seen in this figure, numerical differentiation provides acceptable measurement accuracy only over near zones; over distant zones, the distortion levels become unacceptable. The analysis shows that in the simulated case, neither large nor small range resolution allows discrimination of the local layer located at the heights 3500–3800 m. The profile of $\kappa_p^{(\text{dif})}(h)$ extracted from the signal measured at $\varphi = 45°$ shows some increase in the aerosol loading at these altitudes. However, it also shows the presence of a nonexisting aerosol layer in the height interval between the heights of 3400 and 4500 m. Meanwhile, both layers can be easily distinguished in the extinction coefficient profile, obtained by the piecewise technique (Figs. 3.27 and 3.28). The use of the backscatter coefficient profile, $\beta_{\pi,p}(h)$, for extracting the extinction coefficient appears to be the best way to examine such perturbations. Obviously, this is only achievable when such perturbations are well defined in the original profile of $C\beta_\pi(h)$, determined with the Kano–Hamilton solution.

Let us summarize the essentials of the piecewise inversion technique for data obtained in the multiangle mode. The basic idea of this technique is the same as that in

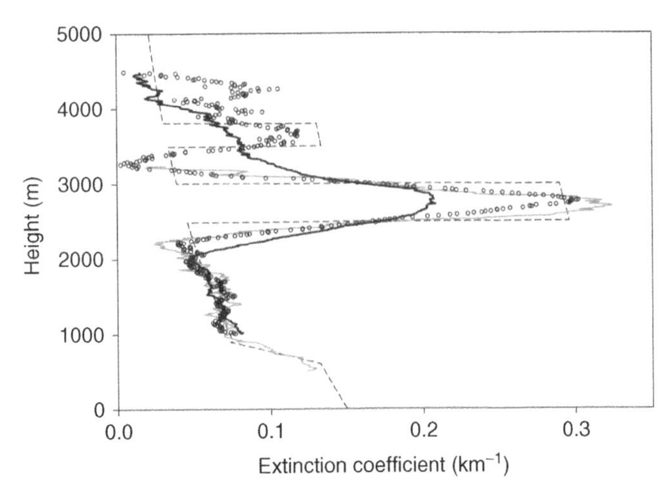

Fig. 3.29 Vertical profiles of the extinction coefficient obtained through numerical differentiation of the lidar signals measured at 90° (thick solid curve), 45° (open circles), and 30° (gray solid curve). The thin dashed curve is the test profile of the extinction coefficient, the same as in Fig. 3.28 (Kovalev et al., 2011b).

one-directional measurements, discussed in Section 2.6, that is, to extract the piecewise continuous extinction coefficient profile from the profile $C\beta_\pi(h)$ instead of using numerical differentiation of the optical depth profile. To exclude the latter procedure, two alternative profiles of the two-way transmittance, $T^2_{p,j}(h', h)$ and $\langle T^2_{p,j}(h', h)\rangle$, in the selected slope direction are determined over restricted intervals, in which the lidar ratio is assumed to be constant. Then the lidar ratio that minimizes the discrepancy between these alternative transmittances is found. After the ratio is found, the piecewise extinction coefficient for this segmented range is calculated. Such an approach allows extraction of the extinction coefficient profile both along slope elevation angles and in the vertical direction.

Using such a technique, one can obtain and compare the profiles of the vertical extinction coefficient derived under different slope directions and analyze the validity of the initial assumption of a horizontally stratified atmosphere. One can determine (i) whether the requirements in Eqs. (3.1) and (3.2) are satisfactorily met, (ii) how significant the discrepancies between profiles extracted under different elevation angles within the same altitude interval are, and (iii) under what elevation angles the above conditions are not valid and should be excluded for getting more reliable inversion results.

The piecewise data processing technique does not yield negative values for the extinction coefficient. This statement is valid even if lidar data are obtained from noisy signals or in a poorly stratified atmosphere. Unlike the common numerical differentiation technique, this technique enables discrimination of thin stratified layering with sharp boundaries. It also permits a more realistic estimation of possible distortions in the inversion results as compared to the methods based on pure statistics, when only random errors are taken into consideration.

The method was used to process experimental data obtained with the FSL lidar both in clear-air conditions and in the vicinity of wildfires, and has demonstrated its validity.

3.4.4 Essentials and Issues of the Practical Application of the Piecewise Inversion Technique

In this subsection, real experimental data obtained with the scanning lidar in smoke-polluted atmospheres in the vicinity of wildfires are discussed. These data were acquired during investigation of dispersed smoke plumes from the Kootenai Creek Fire in Montana in August 2009. The lidar signals at wavelength 355 nm were measured along 12 azimuthally averaged scans within the slope angular sector from 9° to 80°. For each slope direction, the slope profile $T^2_{\Sigma,j}(0, r)$ was found and recalculated into the profile versus height, $T^2_{\Sigma,j}(0, h)$. The piecewise particulate extinction coefficient, $\kappa^{(i)}_{p,j}(h)$, was then determined using the procedures described in Section 3.4.2. To obtain the piecewise profiles of the particulate extinction coefficient, eight overlapping intervals for each slope direction were used, the same as in the numerical experiment in Section 3.4.3. To reduce the error due to the possible existence of nonzero aerosol loading at the reference altitude, the constant,

$\langle C \rangle$, estimated with the condition $\langle C \rangle = \min \gamma(h)$, was reduced by 10%. [For details, see Eqs. (2.99)–(2.103)]. In addition to simple averaging of the piecewise extinction coefficient in the overlapping areas, as was done in the simulations in Section 3.4.3, here the weighted averaging of the extinction coefficient was also performed. The weight function for the extinction coefficient over the overlapping zone, (r_m, r_n), was calculated as

$$w(r_m, r_n) = \left\{ \frac{1}{n\left(r_m, r_n\right)} \sum_{r_m}^{r_n} \left[\left\langle T_{p,j}^2\left(r_m, r\right)\right\rangle - T_{p,j}^2(r_m, r) \right]^2 \right\}^{-1}, \tag{3.38}$$

where $n(r_m, r_n)$ is the number of data points within the overlapping interval, $r_m - r_n$.

The inversion results for two arbitrarily selected slope directions, $\varphi = 15°$ and $\varphi = 68°$ are shown in Figs. 3.30 and 3.31. In both figures, the solid gray curve is the average profile of the piecewise continuous extinction coefficient, whereas the solid black curve is the weighted average function. The analysis of the experimental data confirmed that the weighted and non-weighted averages yield close results only within intervals where no sharp changes in aerosol loading take place. In the zone where the sharp change occurs, the derived extinction coefficient profiles may significantly differ, indicating the areas, where the assumption of the invariable lidar ratio is not valid. In principle the weighted average may yield a more accurate inversion result. However, this statement requires an additional comment. As mentioned earlier the inversion of lidar data is a typical ill-posed problem where no rigid estima-

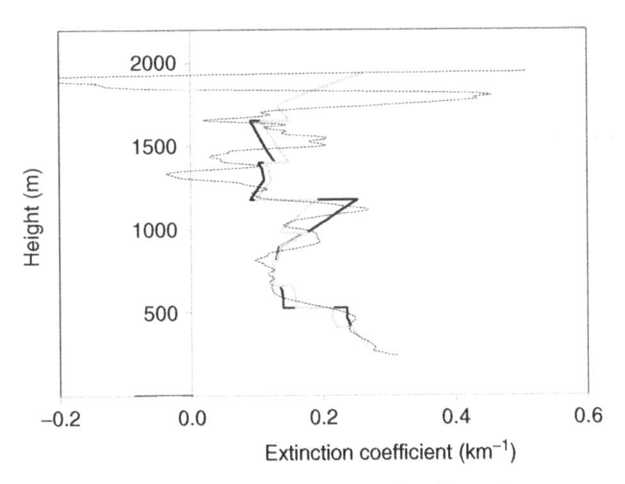

Fig. 3.30 Particulate extinction coefficient profiles derived from the two-way transmittance $T^2_{\Sigma,15}(0, h)$ for the slope direction $\varphi = 15°$. The solid gray curve shows the average profile of stepwise continuous extinction coefficient $\kappa_{p,15}(h)$ and the solid black curve is the profile obtained with the weighted average. The dotted curve is the profile $\kappa_p^{(\mathrm{dif})}(h)$ obtained through numerical differentiation (Kovalev et al., 2011b).

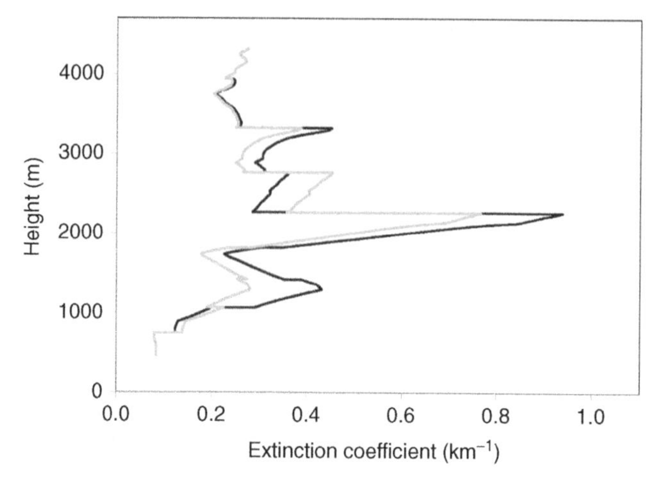

Fig. 3.31 Particulate extinction coefficient profiles derived from the two-way transmittance $T^2_{\Sigma,68}(0, h)$ for the slope direction $\varphi = 68°$. The solid gray curve shows the average profile of the extinction coefficient $\kappa_{p,68}(h)$ and the solid black curve is the profile obtained with the weighted average (Kovalev et al., 2011b).

tion of the accuracy of the retrieved optical characteristic is possible. The selection of the weight function for the inverted data points is also some sort of modeling. When using alternative weight functions, the inversion results of the same lidar signal may be significantly different, especially in regions of increased heterogeneity; the selection of weighted or non-weighted averages is the selection of different solution models.

Nevertheless, the use of different weights when profiling the atmospheric parameters may be extremely beneficial. Being applied to the same scanning lidar data, the different inversion variants allow determination of (i) the discrepancies that exist between the extracted profiles when different weights, models, or *a priori* assumptions are used; and (ii) the heights over which the different models or assumptions yield moderate divergence in the derived profiles. In other words, in many cases, it is useful to analyze the profiles of the extinction coefficient obtained using alternative methods, including the profile of the extinction coefficient obtained through numerical differentiation. In Fig. 3.30, such a profile of $\kappa_p^{(\mathrm{dif})}(h)$ obtained through numerical differentiation with the slope range resolution $s \approx 500$ m, is shown as the dotted curve. This profile agrees well with two others up to the height ~1200 m, but then large erroneous fluctuations take place. Nevertheless, these fluctuations are centered close to two other profiles, which in turn, are close to each other. One can maintain that certain likelihood exists that the extracted, or more correctly, the *a posteriori* simulated profiles of the extinction coefficient, are valid.

The area scanned by a typical elastic lidar extends, at least, up to five or more kilometers, and the atmospheric properties, including the horizontal homogeneity, can be significantly different within this area. To improve the multiangle inversion results, signals from heterogeneous "spots" should be revealed and removed before inversion is performed. Accordingly, two questions need to be answered when making multiangle lidar signal inversion. The first one is how to discriminate "good data" from "bad data" and separate these. The second question is related to selection of the best solution algorithm or algorithms. This question is directly related to selecting the assumptions that are valid for inversion in the existing atmospheric conditions.

Let us consider some principles for solving the above issues, starting with consideration of the solution, which allows excluding poor input data before determining the profile $C\beta_{\pi,p}(h)$ with the Kano–Hamilton solution. This inversion variant permits excluding poor data without initially determining the individual profiles of the extinction coefficient, $\kappa_{p,j}(h)$, at different slope directions. To clarify this variant, data obtained with the FSL lidar in clear atmosphere in September 2008 are used. During that lidar searching, 12 azimuthally averaged scans were made within the vertical sector from 9° to 80°. After the function $C\beta_{\pi,p}(h)$, derived from the signals measured in the slope directions was determined, the corresponding profiles of the two-way total transmittance, $T^2_{\Sigma,j}(0,r)$, in the slope directions were found. These slope profiles, transformed into functions of height, $T^2_{\Sigma,j}(0,h)$, were recalculated into vertical profiles versus height, $T^2_{\Sigma,j,\mathrm{vert}}(0,h)$, using the formula, $T^2_{\Sigma,j,\mathrm{vert}}(0,h) = [T^2_{\Sigma,j}(0,h)]^{\sin\varphi}$. In Fig. 3.32, the full set of these functions, determined in the angular sector from 9° to 80°, is shown as the cluster of gray and dark gray curves. Aside from the profile in the direction of 80°, all other profiles are close to each other. This means that the profile of the two-way transmittance for $\varphi = 80°$ has to be excluded from data processing; that is, for the inversion, the restricted angular sector from 9° to 68° should be used. The remaining functions, $T^2_{\Sigma,j,\mathrm{vert}}(0,h)$, which are close to each other, can be then used

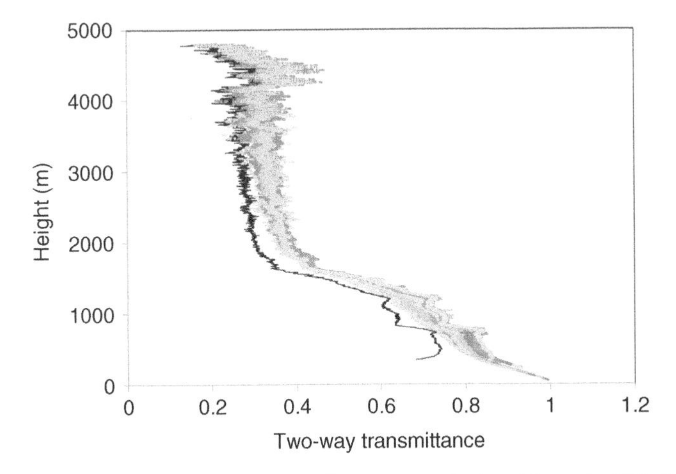

Fig. 3.32 Vertical two-way transmittance $T^2_{\Sigma,j,\mathrm{vert}}(0,h)$ calculated from the FSL lidar signals measured at different elevation angles φ. The black solid curve on the left is the profile derived at $\varphi = 80°$. (Kovalev et al., 2011b).

to determine their average, $\overline{T^2_{\Sigma,\text{vert}}(0,h)}$, and the corresponding vertical extinction coefficient for the whole lidar scan. Note that this can be done without determining the piecewise continuous extinction coefficient $\kappa_{p,j}(h)$ for each individual slope direction.

Analysis of the experimental data obtained in this case study revealed that even the use of the average profile does not necessarily improve the inversion result. Bulges in the individual curves $T^2_{\Sigma,j,\text{vert}}(0,h)$ create bulges in the averaged profile $\overline{T^2_{\Sigma,\text{vert}}(0,h)}$, causing erroneous spatial variations in the extracted extinction coefficient. Accordingly, some changes in the procedure were tested. The use of minimal values of the profiles $T^2_{\Sigma,j,\text{vert}}(0,h)$ resulted in a more smoothed two-way transmittance profile than the average $\overline{T^2_{\Sigma,\text{vert}}(0,h)}$. In Fig. 3.33, the gray curve shows the average particulate transmission profile $\overline{T^2_{p,\text{vert}}(0,h)}$ obtained from $\overline{T^2_{\Sigma,\text{vert}}(0,h)}$ after excluding the molecular component, and the black curve shows the minimal vertical profile, $T^2_{p,\text{vert,min}}(0,h)$. One can see that the latter has significantly fewer bulges and concavities over the altitude range from ~400 to 1300 m than the average profile. Accordingly, the two-way transmittance profile $T^2_{p,\text{vert,min}}(0,h)$ was taken as the best estimate of aerosol loading in the searched atmosphere. Having stating this, we point out that when different models of the lidar-signal inversion result in different output results, it is the researcher's right to select the result he or she believes is most valid.

Fig. 3.34, the particulate extinction coefficient profiles are shown, extracted from the profile $T^2_{p,\text{vert,min}}(0,h)$ in Fig. 3.33. The solid gray curve in the figure shows the

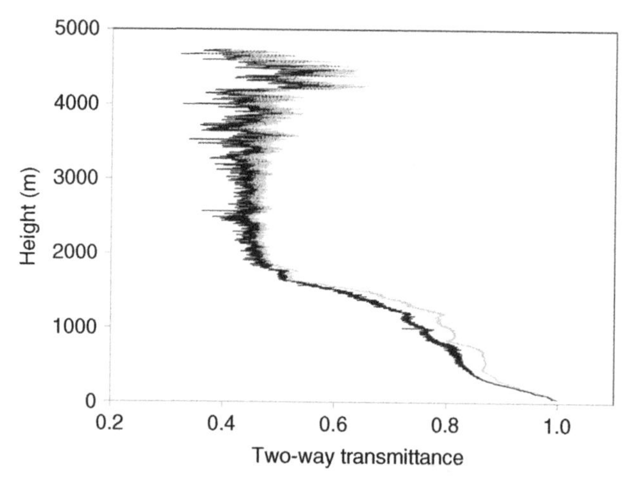

Fig. 3.33 Two-way particulate transmittance profiles extracted from the set of profiles shown in Fig. 3.32 after excluding the profile for the slope direction $\varphi = 80°$. The gray curve shows the average particulate profile $\overline{T^2_{p,\text{vert}}(0,h)}$ and the black curve shows the minimal vertical profile $T^2_{p,\text{vert,min}}(0,h)$ (Kovalev et al., 2011b).

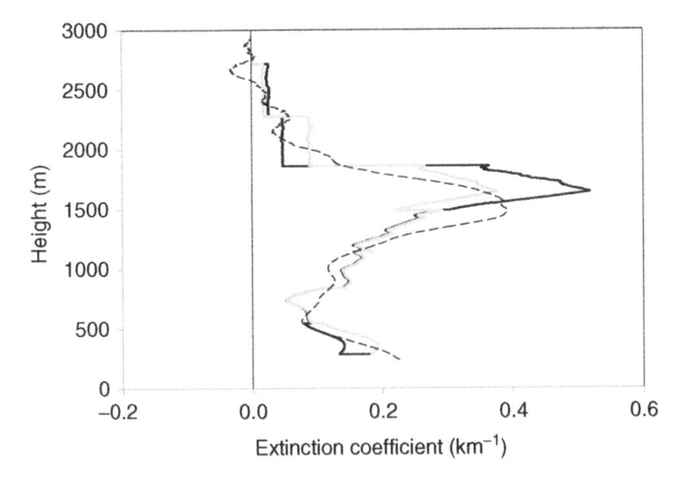

Fig. 3.34 Particulate extinction coefficient profiles retrieved from $T^2_{p,\text{vert,min}}(0, h)$ in Fig. 3–33. The solid gray curve shows the average profile of the stepwise continuous extinction coefficient $\kappa_p(h)$ and the solid black curve is that obtained using the weighted average. The dashed curve is the profile $\kappa_p^{(\text{dif})}(h)$ obtained through numerical differentiation (Kovalev et al., 2011b).

average profile of the piecewise continuous extinction coefficient, and the solid black curve is the weighted average obtained with the weight functions calculated with Eq. (3.38). The dashed curve shows the profile $\kappa_p^{(\text{dif})}(h)$ extracted using numerical differentiation with a height resolution of 500 m. The disagreement between these profiles is rather moderate, slightly increased in the heterogeneous area and at the far end, where numerical differentiation yields even negative values for the extinction coefficient.

The conclusions that follow from the above consideration are as follows. When analyzing the results of the lidar signal inversion, the basic question is whether or not the solution is sensitive to the applied assumptions. Depending on the answer to this question, one can estimate the level of reliability of results obtained from lidar profiling of the atmosphere. In areas where the atmospheric profiles extracted using different assumptions are close to each other, the assumptions used in the solutions can be considered as appropriate. In areas where alternative assumptions used for the inversion result in large discrepancies in the derived profiles, some assumption or assumptions are probably not valid.

One should always keep in mind that any algorithm selected for lidar signal inversion is valid only for the selected solution model, that is, the model which researcher considers as most trustworthy. No reliable criteria exist for selection of the model, which would ensure its applicability to the current conditions. This means that any simulation of the optical properties of the searched atmosphere based on the previously measured multiangle signals is some art, which requires common sense and certain creativity; simple pushing computer buttons generally does not provide reliable and trustworthy results.

3.5 DIRECT MULTIANGLE SOLUTION

The direct multiangle solution is based on a combination of the one-directional and multiangle inversion techniques of lidar data processing. The usual multiangle scanning in the searched atmosphere is performed; but when inverting the lidar data, the backscatter signal measured in zenith direction (or in the direction closest to the zenith) is taken as a basic function. Unlike the conventional one-directional method, the direct multiangle solution uses the multiangle signals. However, these signals are used as auxiliary data when inverting the zenith (or close to zenith) signal. As with other multiangle algorithms, the direct multiangle solution avoids the uncertainty related to the unknown lidar ratio, inherent to vertically pointed elastic lidar, but requires the estimation of the function $[C\beta_\pi(h)]$ and the assumption of an invariable column-integrated ratio within restricted height intervals.

3.5.1 Essentials of the Data Processing

The case below is considered when during multiangle scanning, the lidar signal in the direction of the zenith is also measured, The corresponding square-range-corrected backscatter signal $P_{90}(h)h^2$ at height h is the product of the lidar solution constant C, the total (molecular and particulate) vertical backscatter coefficient $\beta_{\pi,90}(h)$, and the total two-way vertical transmittance $T_{90}^2(0, h)$ from ground level to the height h, that is,

$$P_{90}(h)h^2 = C\beta_{\pi,90}(h)T_{90}^2(0, h), \tag{3.39}$$

where the two-way vertical transmittance

$$T_{90}^2(0, h) = \exp[-2\tau_{90}(0, h)], \tag{3.40}$$

and the corresponding vertical optical depth $\tau_{90}(0, h)$ are the profiles to be determined. Similar to Eq. (3.33), the two-way vertical transmittance $T_{90}^2(0, h)$ can be determined as the ratio of the square-range-corrected backscatter signal $P_{90}(h)h^2$ and the product $C\beta_{\pi,90}(h)$. For such a determination, the function $[C\beta_{\pi,90}(h)]$, should be someway estimated. The estimate of this function, symbolized below as $\langle C\beta_{\pi,\text{vert}}(h)\rangle$, may be found from the set of multiangle signals $P_j(h)$ using the conventional Kano–Hamilton solution. Accordingly, the direct multiangle solution is derived from Eq. (3.39) in the simple form

$$T_{\text{vert}}^2(0, h) = P_{90}(h)h^2/\langle C\beta_{\pi,\text{vert}}(h)\rangle. \tag{3.41}$$

Note that in Eq. (3.41), two different subscripts, "90" and "vert" are used. The subscript "90" denotes the functions obtained or related to direct profiling in the zenith direction, whereas the subscript "vert" denotes the same vertical profiles, but determined indirectly from the lidar signals measured in slope directions.

Eqs. (3.39) and (3.41) are similar to Eqs. (3.32) and (3.33) in Section 3.4. The only difference is that the formulas in Section 3.4 are valid for signals measured in any slope direction, whereas the profile of the two-way transmittance in Eq. (3.41) is obtained from the signal measured in the zenith direction. As in the Kano–Hamilton method, the data points $y_j(h)$ for signals measured along the set of slope directions φ are calculated from the logarithms of the square-range corrected signals, that is,

$$y_j(h) = \ln\lfloor P_j(h)(h/\sin\varphi)^2\rfloor.$$

The linear fit $Y(x, h)$ for the set of data points $y_j(h)$ at height h is found in a general form, which can be written as

$$Y(x, h) = \langle A(h)\rangle + b(h)x, \tag{3.42}$$

where, as in Eq. (3.6), $x = 1/\sin\varphi$ is the discrete variable, $b(h)$ is the slope of the linear fit, and the intercept, denoted here as $\langle A(h)\rangle$, is found by extrapolation of the linear fit $Y(x, h)$ to $x = 0$.

The important feature of the direct multiangle solution is the method of obtaining the estimate $\langle C\beta_{\pi,\mathrm{vert}}(h)\rangle$. To improve accuracy of the estimate, the linear fit $Y(x, h)$ is shifted laterally into the point $y_{90}(h)$, which is the logarithm of the zenith signal at the height h, that is,

$$y_{90}(h) = \ln\lfloor P_{90}(h)h^2\rfloor. \tag{3.43}$$

Accordingly, the initial intercept $\langle A(h)\rangle$ is shifted to the point $A'(h)$, which can be determined as

$$A'(h) = y_{90}(h) - b(h). \tag{3.44}$$

The parameter of interest $\langle C\beta_{\pi,\mathrm{vert}}(h)\rangle$, in Eq. (3.41) is determined as the exponent of the intercept, that is,

$$\langle C\beta_{\pi,\mathrm{vert}}(h)\rangle = \exp[A'(h)]. \tag{3.45}$$

To clarify the essence of the above technique, let us perform a simple numerical simulation for a virtual lidar, scanning a synthetic, poorly stratified atmosphere, at nine elevation angles, 12°, 15°, 18°, 24°, 30°, 40°, 55°, 70°, and 90°. The arbitrarily dispersed data points of $C\beta_{\pi,j}(h)$ and $\tau_\varphi(0, h)$ for the above nine angles φ at a fixed height h are shown in Fig. 3.35 as filled squares and filled triangles, respectively. The empty triangles are the corresponding optical depths recalculated in the zenith direction, that is, $\tau_{\mathrm{vert}}(0, h) = \tau_\varphi(0, h)\sin\varphi_j$. One can see that this synthetic atmosphere is not stratified horizontally. The corresponding data points $y_j(h)$ and their linear fit $Y(x, h)$ versus the independent $x = (\sin\varphi)^{-1}$ are shown in Fig. 3.36 as filled circles and the dashed line, respectively. The interception point of the linear fit $\langle A(h)\rangle$, obtained by extrapolation of the line into $x = 0$, is shown on the vertical axis as the filled triangle. The corresponding quantity $\langle C\beta_{\pi,\mathrm{vert}}(h)\rangle = \exp\langle A(h)\rangle$ is equal to 105.6 a.u. Meanwhile, the actual value used for the simulation is $[C\beta_{\pi,90}(h)] = 200$ a.u. Thus, when using the original intersection point $\langle A(h)\rangle$ the relative error of the estimated $\langle C\beta_{\pi,\mathrm{vert}}(h)\rangle$ is ~47%.

The determination of the intercept with Eq. (3.44) significantly decreases this error, shifting the new intersection point $A'(h)$ (the filled square) closer to the actual

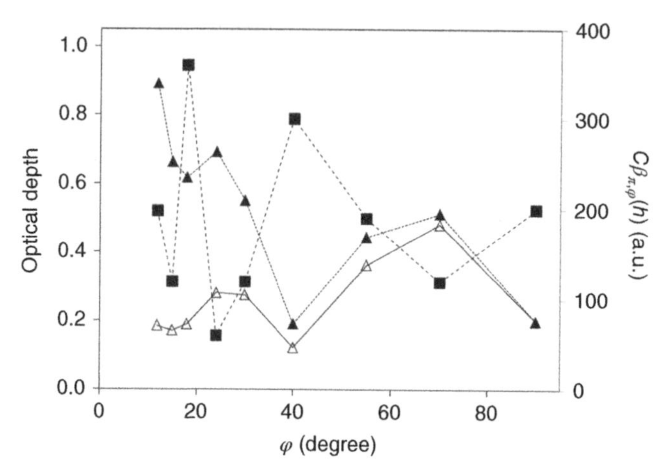

Fig. 3.35 Simulated data points $C\beta_{\pi,\varphi}(h)$ (filled squares), $\tau_{\varphi}(0,h)$ (solid triangles), and $\tau_{\varphi}(0,h)\sin\varphi$ (empty triangles) versus the elevation angle φ at the fixed height h (Kovalev et al., 2012a).

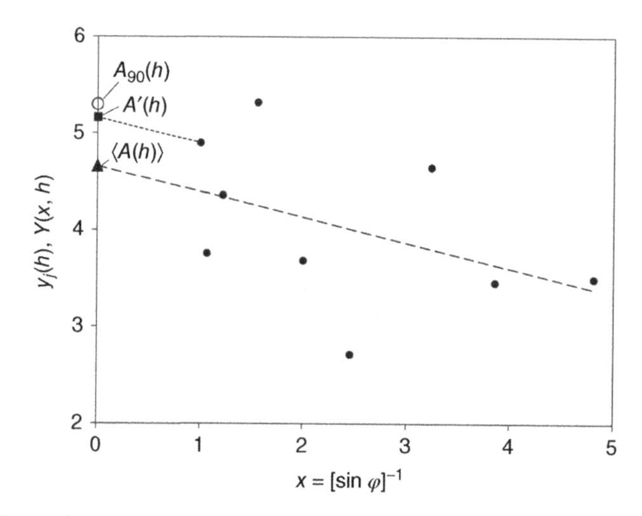

Fig. 3.36 Dependence of the data points $y_j(h)$ (filled circles) and the linear fit $Y(x,h)$ (dashed line) on $x = 1/\sin\varphi$ at the fixed height h obtained from the data points in Fig. 3.35 (Kovalev et al., 2012a).

point $A_{90}(h)$. The use of the intersection point $A'(h)$, instead of the initial $\langle A(h)\rangle$, reduces the corresponding error in the estimate $\langle C\beta_{\pi,\text{vert}}(h)\rangle$ down to ~13%. Note that the difference between the actual optical depth $\tau_{90}(0,h)$, and that retrieved from the slope $b(h)$ remains large. Indeed, $b(h) = -0.26$, that is, the corresponding estimate of the vertical optical depth is $\langle\tau_{90}(0,h)\rangle = 0.13$, whereas the true (test) value

$\tau_{90}(0, h) = 0.2$; the relative difference between these is ~35%. Obviously, the use of the estimated $\langle C\beta_{\pi,\text{vert}}(h)\rangle$ instead of $\langle \tau_{90}(0, h)\rangle$ for extracting the extinction coefficient is preferable.

Numerous simulations and experimental results showed that in many cases the shifted point $A'(h)$ proves to be significantly closer to the actual location of the point $A_{90}(h)$ than the initial intersection point $\langle A(h)\rangle$ found with Eq. (3.42). Such a lateral shift of the linear fit into the point $y_{90}(h)$ may be quite helpful when the data points $y_j(h)$ are strongly scattered, and the point $y_{90}(h)$ at $x = 1$ is significantly shifted from the linear fit $Y(x, h)$.

Let us consider a worse-case situation. In spite of the strong inhomogeneity of the optical parameters given in Fig. 3.35, the corresponding slope b in Fig. 3.36 is still negative. In some cases of a poorly stratified aerosol atmosphere, the retrieved slope may be positive or zero, that is, obviously erroneous. Let us continue the numerical simulation, considering now the case when the optical depth derived from the slope b of the linear fit $Y(x, h)$ yields such an erroneous negative optical depth. This case is illustrated by Figs. 3.37 and 3.38. The optical depth extracted from the slope of the linear fit in Fig. 3.38 is negative, causing nonphysical negative extinction coefficients to be retrieved from it. The direct use of Eq. (3.44) would yield $A'(h) = 4.52$ instead of the true $A_{90}(h) = 4.87$ (Fig. 3.38), and the extracted $\langle C\beta_{\pi,\text{vert}}(h)\rangle$ would differ from the true $C\beta_{\pi,90}(h)$ by as much as ~30%.

When the derived slope $b(h)$ is obviously erroneous, a special correction procedure is required to reduce large shifts in the estimated $\langle C\beta_{\pi,\text{vert}}(h)\rangle$. There is no unique solution for such a case, but some recommendations can be given, at least, for clear atmospheres. The intercept can be determined more accurately if the vertical optical depth from ground level up to the height h can be someway estimated, at least, approximately; this would permit the replacement of the obviously erroneous positive slope b with a negative slope. When working in clear atmospheres at wavelengths close

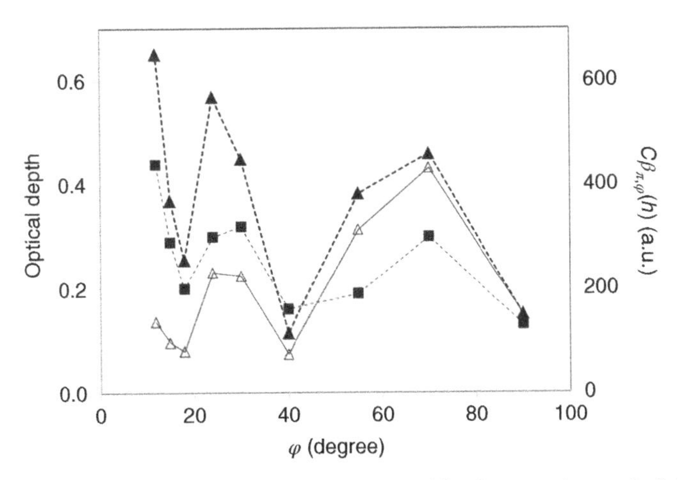

Fig. 3.37 The dependencies of the data points on x used for the second numerical simulation. The symbols are the same as in Fig. 3.35 (Kovalev et al., 2012a).

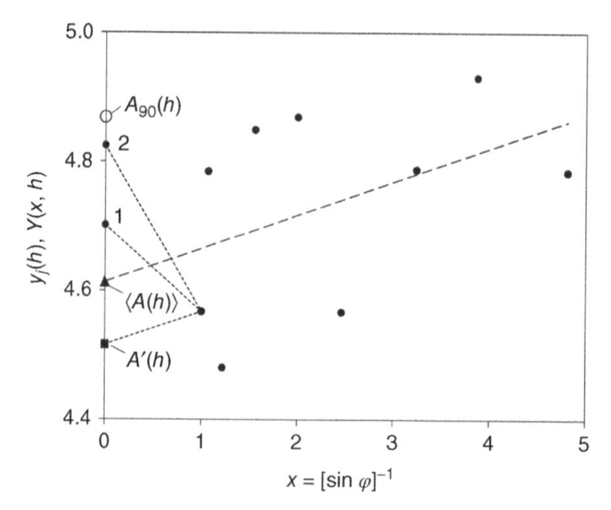

Fig. 3.38 Same as Fig. 3.36, but where the data points shown in Fig. 3.37 are used. The symbols are the same as in Fig. 3.36 (Kovalev et al., 2012a).

to the UV spectra, at least the attenuation of the purely molecular atmosphere can be estimated. The simplest correction may be done by replacing the erroneous positive $b(h)$ by the negative $b_{\text{mol}}(h)$, calculated using the molecular extinction coefficient profile $\kappa_m(h)$, that is,

$$b_{\text{mol}}(h) = -2 \int_0^h \kappa_m(h')dh'. \tag{3.46}$$

When $b(h)$ calculated with Eq. (3.42) is larger than $b_{\text{mol}}(h)$, the latter should be used instead of the nonphysical $b(h)$. In the case shown in Figs. 3.37 and 3.38, the replacement of the erroneous $b(h)$ by $b_{\text{mol}}(h)$ yields $A'(h) = 4.70$ for height $h = 1000$ m and $A'(h) = 4.82$ for $h = 2000$ m (points 1 and 2 in Fig 3-38), significantly closer to the true $A_{90}(h)$. Unfortunately, such a simple correction is only effective when aerosol loading is minor as compared with a molecular atmosphere. Otherwise, more sophisticated correction methods should be used, based on the consideration of either the shape of the square-range-corrected signals, or the numerical values of the optical depth $\tau_{\text{vert}}(0, h)$ in the vicinity of the height of interest.

Simple transformations yield the following relationships between the estimated value of $\langle C\beta_{\pi,\text{vert}}(h) \rangle$, and the actual product $C\beta_{\pi,90}(h)$ in the zenith direction:

$$\frac{\langle C\beta_{\pi,\text{vert}}(h) \rangle}{C\beta_{\pi,90}(h)} = \exp[-2\tau_{90}(0, h) - b(h)]. \tag{3.47}$$

The corresponding relative error in the derived two-way vertical transmittance profile obeys the formula

$$\delta[T_{90}^2(0, h)] = \exp[-2\tau_{90}(0, h) - b(h)] - 1. \tag{3.48}$$

As follows from Eqs. (3.47) and (3.48), the relative error in the estimated profile $\langle C\beta_{\pi,\mathrm{vert}}(h)\rangle$ and accordingly in the retrieved profile $T_{90}^2(0,h)$ depends only on the difference between the actual vertical optical depth in the zenith direction and that determined from the slope $b(h)$.

3.5.2 Selection of the Maximum Range for the Multiangle Lidar Signals

When processing multiangle data, the selection of maximum range for inverted signals of scanning lidar is a significant issue. Practice shows that heterogeneity of the real atmosphere exacerbated by systematic distortions of the backscatter signal over distant areas are extremely destructive factors. These are much more destructive than random noise, commonly used as the criterion for selection of the maximum operative range of scanning lidar.

To clarify possible ways to minimize this issue, let us analyze typical experimental data obtained in August 2009 with the FSL scanning lidar. The profiling of the atmosphere was made in the vicinity of wildfires where the atmospheric inversion created multiple smoke layers at different heights. To reduce the influence of local atmospheric horizontal heterogeneity, combined azimuthal–vertical sounding was made (Adam et al., 2007). Azimuthal scanning within the angular sector of 100° was made along 12 elevation angles from 9° to 80°. For each elevation angle, a mean azimuthal signal and the corresponding functions $y_j(h)$ were calculated and used for the inversion. In Fig. 3.39, the logarithms of the mean azimuthal signals for the selected elevation angles given in the legend, are shown. The multilayered atmospheric structure is clearly seen over three zones, 1, 2, and 3, located at the heights 1400–1800, 2400–3100 m, and between 4200 and 5000 m, respectively.

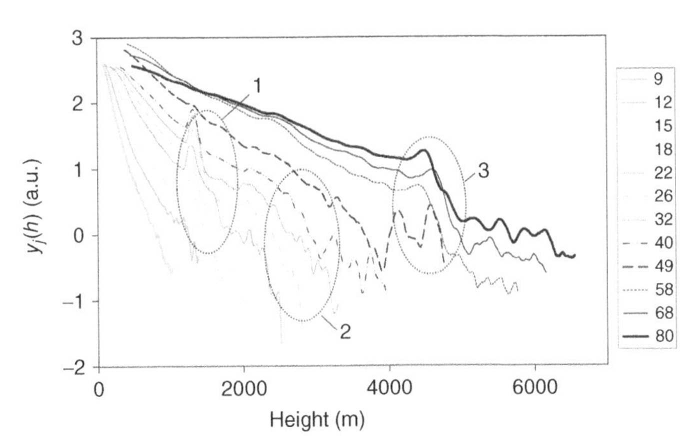

Fig. 3.39 Logarithms of the square-range corrected signals versus height for 12 elevation angles measured in a smoke polluted atmosphere. The minimum and maximum ranges for the functions $y_j(h)$ shown in the figure are chosen subjectively to provide better visualization of the specifics of these functions (Kovalev et al., 2012a).

Following the common inversion procedure, the linear fits $Y(x, h)$ for the data points $y_j(h)$ at discrete heights should be found initially. However, before determining the linear fits, the operative range $r_{min} - r_{max}$ for the inverted signals needs to be established. The minimal range r_{min} is generally selected somewhere outside and close to the incomplete overlap zone r_0. The selection of the maximum range r_{max} is not so straightforward. Typically, when selecting the maximum range, the random noise is considered the dominating factor (Whiteman, 1999; Rocadenbosch et al., 1998; Volkov et al., 2002; Adam et al., 2007). Practically, no other criteria for the selection of r_{max} exist. Moreover, the most common criterion for selecting r_{max}, the minimal level of the signal-to-noise ratio, is chosen arbitrarily. Such a method of estimating the maximal range is based on an implicit premise that no systematic distortions in the backscatter signal exist which will influence the inversion result close to the maximum range. Obviously, the use of the signal-to-noise ratio as an only criterion is far from being sensible. For a real backscatter signal with a nonzero additive distortion determined in real, not ideally stratified atmosphere, the slope of the linear fit $Y(x, h)$ may strongly depend on selected r_{max}. No theory exists that would give practical recommendations for selecting the best maximum range under such conditions. The only way to address this issue is the use of different r_{max} and the analysis of the obtained inversion results. However, straightforward use of this idea yields discouraging results. This is illustrated in Fig. 3.40, where the optical depth profiles extracted with different r_{max} are shown; the values of the maximum ranges in meters are given in the legend. The corresponding particulate extinction coefficient profiles $\kappa_p^{(dif)}(h)$ extracted from these optical depth profiles using the sliding derivative with the range resolution $s = 300$ m are shown in Fig. 3.41. Ignoring the presence of the bumps and concavities in these profiles, typical for results obtained when using numerical differentiation, one can see that these profiles significantly differ from each

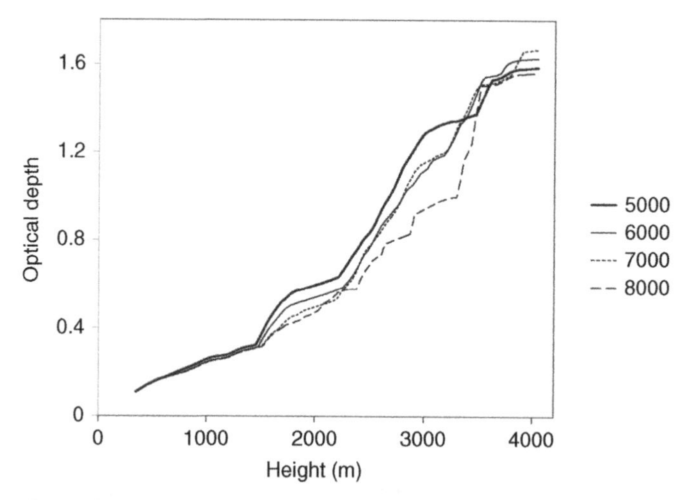

Fig. 3.40 The particulate optical depth profiles, extracted from the functions $y_j(h)$ in Fig. 3.39 with the conventional Kano–Hamilton retrieval procedure. The maximum ranges r_{max} (in meters) used for the inversion are shown in the legend.

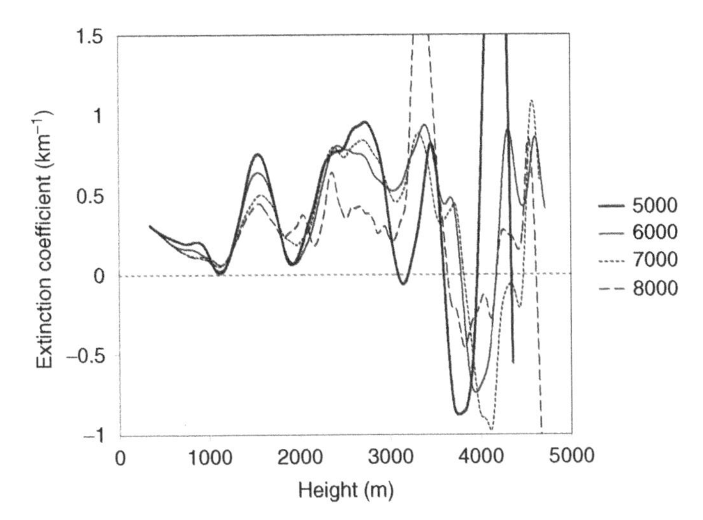

Fig. 3.41 The particulate extinction coefficient profiles $\kappa_p^{(\mathrm{dif})}(h)$ extracted from the optical depth profiles shown in Fig. 3.40 with the sliding range resolution of 300 m. The values of $r_{\max}$ (in meters) are shown in the legend (Kovalev et al., 2012a).

other at heights greater than ~ 1200 m. Meanwhile, the profiles are extracted from the same set of the functions $y_j(h)$ using the same range resolution s; the only difference is in the selected $r_{\max}$. Obviously, the issue of ambiguity in the selection of the optimal maximum range, should be somehow overcome, or at least, reduced. This goal can be achieved by using the direct multiangle solution.

Let us consider the procedures that provide solutions to this issue using the same experimental data given in Fig. 3.39. The first step in such a procedure is to establish a set of sensible maximum ranges $r_{\max,1}, r_{\max,2}, \ldots r_{\max,N}$, which will be used for the analysis. For the case under consideration, the maximum ranges can be selected from 4000 to 10,000 m with the interval between the discrete $r_{\max,j}$ equal to 1000 m; that is, $r_{\max,1} = 4000$ m, $r_{\max,2} = 5000$ m, etc. The same minimum range, $r_{\min} = 500$ m, is used in all cases. To find the optimum maximum range, the inversion is repeated N times, using the established intervals. In our case, these intervals are 500–4000 m; 500–5000 m; 500–6000 m, and the maximum interval under investigation extends from 500 to 10,000 m. For each range, $r_{\min} - r_{\max}$, the profiles of $A'(h)$, $\langle C\beta_{\pi,\mathrm{vert}}(h)\rangle$, and $T^2_{\mathrm{vert}}(0, h)$ are found as explained in Section 3.5.1.

In the case under consideration, the maximum slope angle was $\varphi = 80°$, so the corresponding $x_{\min} = 1/\sin 80°$. The basic profile of interest $T^2_{p,\mathrm{vert}}(0, h)$ is obtained using the square-range-corrected signal measured at the slope direction $\varphi = 80°$ and compensating the molecular component $T^2_{m,80}(0, h)$, that is,

$$T^2_{p,\mathrm{vert}}(0,h) = \left[\frac{P_{80}(h)(hx_{\min})^2}{T^2_{m,80}(0,h)\exp[A'(h)]}\right]^{\left(1/x_{\min}\right)}. \tag{3.49}$$

Such an operation is repeated using all the selected above operative ranges, yielding N profiles of the particulate transmittance profiles $T^2_{p,\text{vert}}(0, h)$. When all the profiles are determined, their average $\overline{T^2_{p,\text{vert}}(0, h)}$ and standard deviation $\Delta T^2_{p,\text{vert}}(h)$ are calculated. Under favorable conditions, that is, when the searched atmosphere is well stratified and the signal distortions are minor, all the profiles $T^2_{p,\text{vert}}(0, h)$ are close to each other, $\Delta T^2_{p,\text{vert}}(h) \ll \overline{T^2_{p,\text{vert}}(0, h)}$, and there is no problem selecting the proper $r_{\max}$. Otherwise, the marginal profiles that significantly differ from the average $\overline{T^2_{p,\text{vert}}(0, h)}$ should be excluded and the above procedures repeated. Once again, the researcher has the last say; he or she decides whether the inversion results are acceptable or not. The people who need this information should clearly understand that they acquire the result of *a posteriori* simulation, rather than the result of a measurement.

Thus, the profiles that significantly differ from the average $\overline{T^2_{p,\text{vert}}(0, h)}$ should be excluded. It is sensible to exclude the profiles which extend outside the area restricted by the boundaries, $\lfloor \overline{T^2_{p,\text{vert}}(0, h)} - \Delta T^2_{p,\text{vert}}(h) \rfloor$ and $\lfloor \overline{T^2_{p,\text{vert}}(0, h)} + \Delta T^2_{p,\text{vert}}(h) \rfloor$. In our case, the extreme profiles $T^2_{p,\text{vert}}(0, h)$ determined with $r_{\max}$ equal to 4000, 9000, and 10,000 m, were excluded. After that, the new average for the remaining profiles was calculated; these profiles and their average are shown in Fig. 3.42. Note that no statistics work here, therefore, one cannot claim 68% confidence in the retrieved profile. However, as Taylor (1997) stated, in such cases, this is the only way to provide at least some reasonable estimates.

In Fig. 3.43, the extinction coefficient profiles extracted from the average profile $\overline{T^2_{p,\text{vert}}(0, h)}$ are shown. The dashed curve is the averaged profile of the piecewise continuous particulate extinction coefficient, whereas the thick solid curve is the average

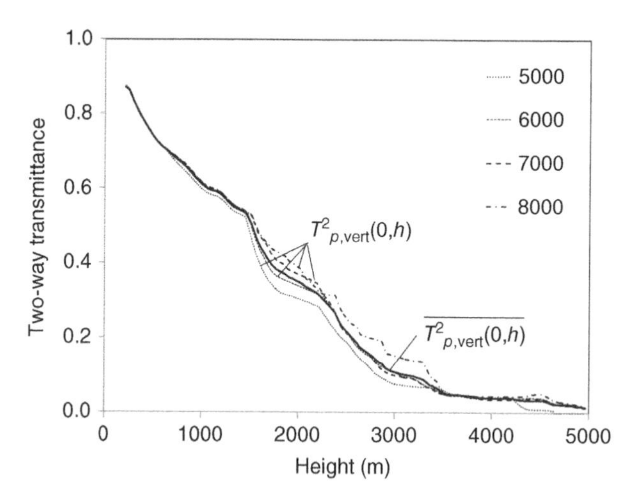

Fig. 3.42 Two-way vertical particulate transmission profiles $T^2_{p,\text{vert}}(0, h)$ extracted from the functions shown in Fig. 3.39 with different $r_{\max}$ and their average $\overline{T^2_{p,\text{vert}}(0, h)}$. The ranges of $r_{\max}$ are given in the legend (Kovalev et al., 2012a).

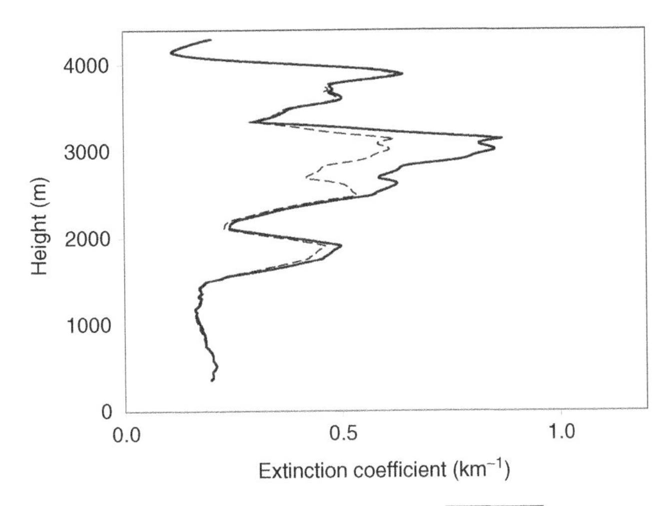

Fig. 3.43 Extinction coefficient profiles, extracted from $\overline{T^2_{p,\text{vert}}(0, h)}$ shown in Fig. 3.42. The thick solid curve is the piecewise continuous particulate extinction-coefficient profile determined with weighted averaging, and the thin dashed curve is the average profile determined without using variable weights.

profile determined using the weight function defined in Eq. (3.38). Both profiles are close to each other, excluding the zone within the height interval from ~2400 to 3200 m, where the weighted average yields significantly larger extinction coefficients. The profiles show that in the searched atmosphere, three layers can be discriminated.

The corresponding piecewise lidar ratios $S^{(i)}_{p,\text{vert}}$ and their average are shown in Fig. 3.44 as thick vertical lines and the dashed curve, respectively. To obtain some notion about how reliable the extracted piecewise profiles are, one can perform the analysis of the weights used for determining $\kappa_p^{(i)}(h)$. The weights $w(\Delta h_{m,n})$, calculated with Eq. (3.38), are shown in Fig. 3.45. They significantly differ for different height intervals, therefore, the logarithmic scale is used for the Y-axis. The sensible lidar ratios are present only within the height interval from 1000 to 2000 m. Then the weights dramatically decrease, so the lidar ratios $S^{(i)}_{p,\text{vert}}$ at the high altitudes are extremely unreliable. Such an effect is common and it occurs due to large signal distortions at distant areas.

The above estimates of the weights present more information about the profiling accuracy than commonly used estimates based on the evaluation of the random noise. On the basis of Fig. 3.45, one can definitely conclude that the profiles of the extinction coefficient and the lidar ratio at the heights above ~2000 – 3000 m are not sufficiently reliable. The lidar ratios at high altitudes are significantly underestimated, and their extremely small values are undoubtedly erroneous. In other words, one can assert that at heights 3000 – 4000 m, an aerosol layer with increased backscattering is probably present but its optical properties cannot be properly estimated.

It is interesting to compare the profiles $\kappa_p^{(\text{dif})}(h)$ obtained using the classic Kano–Hamilton retrieval technique with such a profile extracted from the two-way

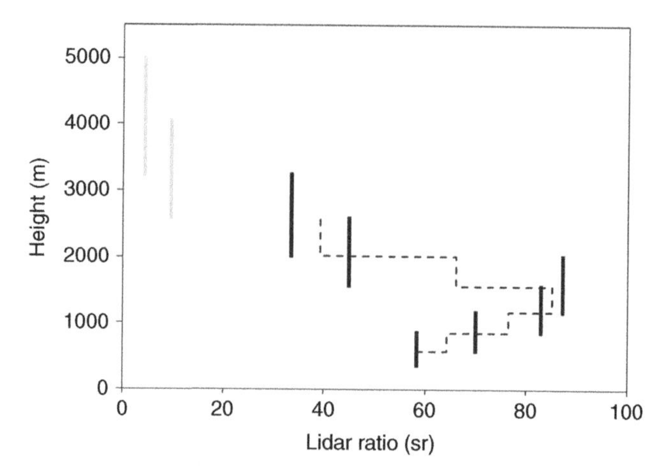

Fig. 3.44 The thick solid lines are the vertical profiles of the lidar ratios $S^{(i)}_{p,\mathrm{vert}}$ corresponding to the extinction coefficient profiles in Fig. 3.43. The dashed curve shows the average profile of the piecewise lidar ratios at the heights below 2600 m.

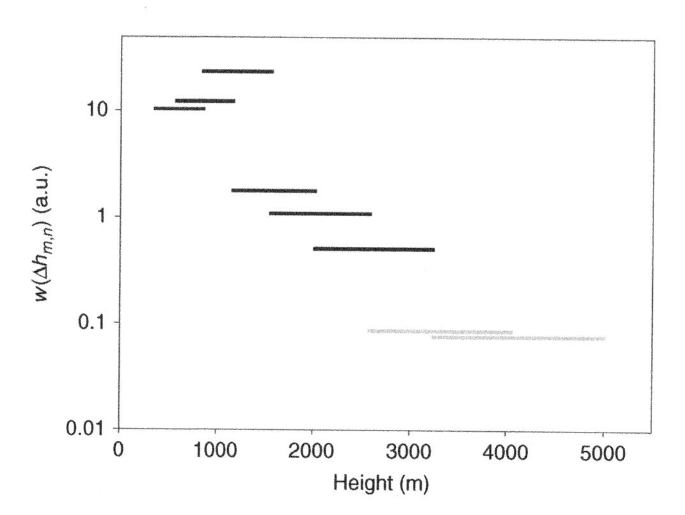

Fig. 3.45 Weight functions of the piecewise extinction coefficient profile $\kappa_p^{(i)}(h)$ shown in Fig. 3.43 as the thick solid curve.

transmittance $\overline{T^2_{p,\mathrm{vert}}(0,h)}$ when using numerical differentiation with the same range resolution of 300 m. The comparison shows that the best agreement between these profiles takes place when the extinction coefficient is extracted with the Kano–Hamilton method from the signals with $r_{\mathrm{max}} = 7000\,\mathrm{m}$. In Fig. 3.46, such a profile $\kappa_p^{(\mathrm{dif})}(h)$ is shown as the dashed curve; the bold curve is the profile of the extinction coefficient extracted from the logarithm of $\overline{T^2_{p,\mathrm{vert}}(0,h)}$. In the near zone up

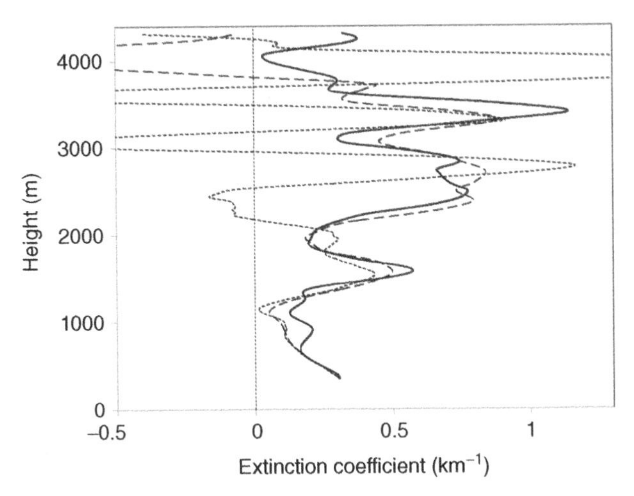

Fig. 3.46 Profile of the particulate extinction coefficient $\kappa_p^{(dif)}(h)$ extracted through numerical differentiation of the optical depth derived from the profile $\overline{T_{p,vert}^2(0,h)}$ (bod solid curve) and that extracted from the optical depth obtained with the Kano–Hamilton solution using $r_{max} = 7000$ m (dashed curve) and $r_{max} = 9000$ m (dotted curve) (Adapted from Kovalev et al., 2012a).

to the height ~3500 m, they agree with each other relatively well, but significantly diverge at higher altitudes. Thus, the classic Kano–Hamilton method, even used with the optimal range $r_{max} = 7000$ m, does not provide the proper inversion results at high altitudes. All other profiles $\kappa_p^{(dif)}(h)$ extracted with the Kano–Hamilton solution show much worse agreement, as for example, the profile obtained with $r_{max} = 9000$ m, shown in the figure as the dotted curve.

Let us briefly summarize. In the direct multiangle solution, the signal measured by the scanning lidar in the zenith direction (or close to zenith direction) is used as a core source of information about the aerosol vertical profile. The set of multiangle signals extracted from the same scanning lidar is used as the source of auxiliary information, which permits extraction of the profile of the vertical transmittance from the zenith signal. To avoid an arbitrary selection of the maximal range of inverted lidar signals, which may yield a suboptimal inversion result, the set of two-way transmittance profiles are determined using different maximum ranges of the inverted backscatter signals. The average of these profiles and the corresponding standard deviation allow excluding from the final analysis the profiles that are significantly shifted from the average.

3.5.3 Direct Solution for High Spectral Resolution Lidar Operating in Multiangle Mode

In the study by Adam (2012), the comparison of the backscatter and extinction coefficient profiles, derived from simulated signals of the virtual scanning and zenith-directed lidars, was made. The author concluded that, potentially, multiangle retrieval methods are more accurate than one-directional retrieval methods. This conclusion was based on simulated lidar profiling in an ideal horizontally stratified atmosphere. Unfortunately, no such ideal atmosphere exists. However, the level of horizontal stratification in the searched atmosphere depends on the hardware solutions used in the lidar instrumentation. In particular, for HSRL, the atmosphere is much better horizontally stratified than for elastic lidar. Therefore, the use of HSRL in multiangle mode may significantly improve the potential accuracy of multiangle solutions and give them "a new birth."

Let us consider the principal advantage of the direct multiangle solution when it is used for the inversion of the backscatter signals of scanning HSRL. Following the study by Liu et al. (2009), the backscatter signal from the molecular channel of HSRL measured in the slope direction φ can be defined as a function of height as

$$P_m(h) = C_m \Delta r_b \left(\frac{h}{\sin \varphi} \right)^{-2} f_m(h) \beta_{\pi,m}(h) \exp\{-2[\tau_{m,\varphi}(0,h) + \tau_{p,\varphi}(0,h)]\}, \quad (3.50)$$

where $P_m(h)$ is the signal in the molecular channel, C_m – a constant, Δr_b – the range resolution, $\beta_{\pi,m}(h)$ – the molecular backscatter coefficient, $f_m(h) = f_m(T,p)$ – the molecular temperature and pressure-dependent factor (see Section 2.1.2), and $\tau_{m,\varphi}(0,h)$ and $\tau_{p,\varphi}(0,h)$ are the molecular and particulate optical depths within the layer $(0,h)$ in the slope direction φ. For the horizontally stratified atmosphere, the logarithm of the function in Eq. (3.50) may be written in the form

$$y_j(h) = \ln \left[\frac{P_m(h) h^2}{\sin^2 \varphi} \right] = A_m(h) - \frac{2}{\sin \varphi} [\tau_{m,90}(0,h) + \tau_{p,\text{vert}}(0,h)], \quad (3.51)$$

where

$$A_m(h) = \text{In} \lfloor C_m \Delta r_b f_m(h) \beta_{\pi,m}(h) \rfloor, \quad (3.52)$$

and $\tau_{m,90}(0,h)$ and $\tau_{p,\text{vert}}(0,h)$ are the molecular and aerosol optical depths of the layer $(0,h)$ in the vertical direction.

As with multiangle lidar profiling with elastic lidar, the linear fit of the data points $y_j(h)$ is found and its slope and interception points determined. However, the aerosol backscatter component in the molecular channel of HSRL is filtered out, so the requirement in Eq. (3.1) is not as restrictive as for elastic lidar.

In Fig. 3.47, the data points $y_j(h)$ and the corresponding linear fit $Y(x,h)$ obtained from a virtual HSRL in the same atmospheric conditions as in Fig. 3.37 are shown. However, this time, these parameters are found from signals of the molecular channel of HSRL, where the term $A_m(h)$ does not include the particulate component. Because $A_m(h) = \text{const.}$, only variations in aerosol slope optical depths $\tau_{p,\text{vert}}(0,h)$ will distort the inversion results. The comparison of Figs. 3.47 and 3.38 nicely demonstrates the potential advantages of HSRL as compared to elastic lidar.

A specific advantage of scanning HSRL as compared to zenith-directed HSRL should be mentioned. As indicated in Section 2.1, when using zenith-directed

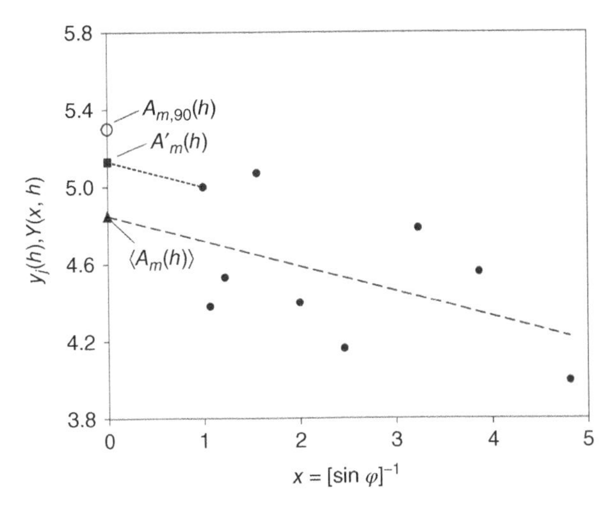

Fig. 3.47 Dependence of the data points $y_j(h)$ (filled circles) and the linear fit $Y(x, h)$ (dashed line) on x for the scanning HSRL, operating in the atmospheric conditions shown in Fig. 3.37. The symbols are the same as in Fig. 3.36 (Kovalev et al., 2012a).

HSRL, the function $f_m(h)$ is most critical for the solution. It depends on air temperature T, pressure p, and height h, and needs to be initially transformed into the height-dependent function. The use of HSRL in multiangle mode will minimize the issue related to estimation of the portion of the molecular scattering blocked by the iodine filter. The filtered backscatter spectrum, which has an extremely complicated spectral shape, can be determined directly from the intercept point of the linear fit $A_m(h)$.

One can expect that the direct multiangle solution will demonstrate its real value in the very near future when spectral resolution lidars are utilized in multiangle mode. Taking into consideration the progress in the HSRL technique and methodology (Esselborn et al., 2008; Hair et al., 2008; Liu et al., 2009; Rogers et al., 2011), one can expect significant progress in future multiangle profiling of the atmosphere.

3.6 MONITORING BOUNDARIES OF THE AREAS OF INCREASED BACKSCATTERING WITH SCANNING LIDAR

3.6.1 Images of Scanning Lidar Data and their Quantification

When using the inversion methodology based on the assumption of an invariable lidar ratio within segmented intervals, an inappropriate selection of these intervals can yield significant distortions in the derived data. The distortions generally occur when the backscattering has large variations within such intervals. To avoid these distortions, it is desirable to select the length and location of the segmented intervals that match the length and location of the atmospheric layers having approximately the same level of backscattering within these. Initial analysis of the lidar signals, which allows clarifying the locations of such layers, can be very helpful. In the following, a possible technique for the analysis of such scanning lidar data is considered.

In the case of multiangle profiling of the atmosphere, Eq. (2.132) should be presented as a function of the slope angle φ under which the lidar search is made, that is,

$$y(v_\varphi) = P_\Sigma(v_\varphi)v_\varphi = [P(v_\varphi) + B_\varphi]v_\varphi, \tag{3.53}$$

where $P(v_\varphi)$ is the backscatter signal measured in this direction and B_φ is the corresponding range-independent offset in the recorded lidar signal. For determining the boundaries of increased backscattering at the wavelength $\lambda = 1064\,\text{nm}$ where the molecular component is minor, the variable v_φ can be reduced to the squared range r^2 as was done in Section 2.9. In the multiangle mode, the variable is related to the slope angle φ, accordingly, $v_\varphi = [h/\sin\varphi]^2$.

After the sliding linear fit $Y(v_\varphi)$ of the data points $y(v_\varphi)$ is calculated and extrapolated to $v_\varphi = 0$, its intercept with the Y-axis can be found as

$$Y_0(v_\varphi) = Y(v_\varphi) - \frac{dY_\varphi}{dv_\varphi}v_\varphi, \tag{3.54}$$

where the sliding derivative dY_φ/dv_φ is

$$\frac{dY_\varphi}{dv_\varphi} = \frac{d}{dv_\varphi}[P(v_\varphi) + B_\varphi v_\varphi]. \tag{3.55}$$

As with vertical profiling (Section 2.9.1), the sliding numerical derivative is determined with the constant resolution Δr that is, with the variable Δv_φ. The function $Y_0(v_\varphi)$, obtained for each slope direction is regularized. The regularized function is defined as the absolute value of the ratio

$$Y_0^*(v_\varphi) = \left|\frac{Y_0(v_\varphi)}{v_\varphi + \varepsilon_v v_{\varphi,\max}}\right|, \tag{3.56}$$

where $v_{\varphi,\max}$ is the maximum value of the variable v_φ over the selected height interval $h_{\min} - h_{\max}$ and ε_v is a positive nonzero constant, whose value can be selected from ~0.02 to 0.05. As with zenith profiling, the goal of adding the component $\varepsilon_v v_{\varphi,\max}$ in the denominator of Eq. (3.56) is to avoid infinite increase of $Y_0^*(v_\varphi)$ when $v_\varphi \to 0$, and accordingly, to avoid obtaining excessively high values of $Y_0^*(v_\varphi)$ close to $h_{\min}$. The additive component $\varepsilon_v v_{\varphi,\max}$ has little practical influence on the function $Y_0^*(v_\varphi)$ at the distances of interest, where $v_\varphi \gg \varepsilon_v v_{\varphi,\max}$; therefore, the selection of the value of ε_v is not critical.

The regions where the function $Y_0^*(v_\varphi)$ increases are the regions with increased gradient of backscattering; these can be examined by using the conventional range-height indicator (RHI). It is interesting to compare the conventional RHI, which shows the spatial image of the square-range-corrected backscatter signal, with the similar image of $Y_0^*(\varphi, h)$, obtained by transforming the function $Y_0^*(\nu_\varphi)$ into a function of height. An example of such a comparison is shown in Fig. 3.48 (a) and (b). The signals were obtained with the FSL lidar when scanning smoke plumes in the vicinity of the Montana I-90 Fire in August, 2005. The lidar scanned vertically over 37 slope directions from $\varphi_{\min} = 7.5°$ to $\varphi_{\max} = 79.5°$, with angular separation $\Delta\varphi = 2°$. The typical RHI plot retrieved from the lidar scan at 1064 nm is shown in Fig. 3.48 (a); the corresponding image of the function $Y_0^*(\varphi, h)$ calculated for the same lidar data is shown in Fig. 3.48 (b). In both plots, one can see horizontally stratified smoke layers located between the heights ~1000 and 2600 – 3000 m. Note the difference between the maximum smoke-plume height on RHI (~2600 m) and that on the $Y_0^*(\varphi, h)$ scan (~3000 m).

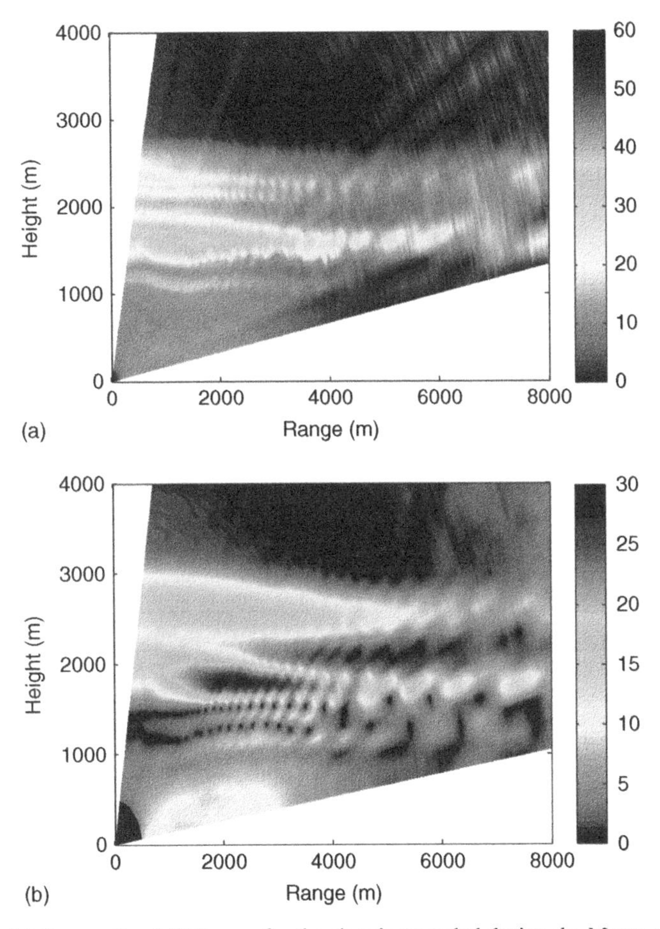

Fig. 3.48 (a) Conventional RHI scan for the signals recorded during the Montana I-90 Fire. The gray scale shows the attenuated backscatter intensity in arbitrary units. (b) The $Y_0^*(\nu_\varphi)$ scan extracted from the same signals. The gray scale to the right of the plots shows the magnitude of these functions in arbitrary units (Kovalev et al., 2009b).

For practical investigation of the location of aerosol layers, the images of aerosol layering should be transformed into numbers. For this purpose, the so-called atmospheric heterogeneity height indicator (AHHI), proposed in the study by Kovalev et al. (2009b), can be used. The transformation of the image of the function $Y_0^*(\varphi, h)$ into numbers makes it applicable for computer processing. The principle of determining the numerical values of the heights of the increased backscatter layers is clarified in Figs. 3.49 and 3.50. The virtual scanning lidar, located at point A, is scanning in a synthetic atmosphere, in which the horizontally stratified aerosol layer L between the heights h_3 and h_4 exists (Fig. 3.49). At lower altitudes of this synthetic atmosphere, within the height range from h_1 to h_2, a spatially restricted aerosol cloud C is also present. From the lidar signals measured in 17 slope directions, the corresponding profiles $Y_0^*(\varphi, h)$ are determined, from which the areas with an increased backscatter gradient are defined. When measuring signals from the smallest angles, along slopes 1 and 2, the laser beam propagates in clear atmosphere, outside the cloud C. Increased backscatter gradients are not observed in these signals. When the lidar is scanning over larger angles, along slopes 3–6, the cloud C is the source of increased backscatter signals. In these signals, backscatter gradients increase at the heights from h_1 to h_2. Along slope 7, no cloud is detected. Presumably, the horizontal layering L at higher altitudes $h_3 - h_4$ is still not detected along this slope because of restricted lidar measurement range. This layering is revealed by the lidar when searching occurs along slope directions from 8 to 17. The localities of increased backscatter are shown in Fig. 3.49 as the black rectangles.

The set of profiles $Y_0^*(\varphi, h)$ calculated with the height resolution Δh produces the AHHI profile, shown in Fig. 3.50. The AHHI is a histogram, which establishes (i) the

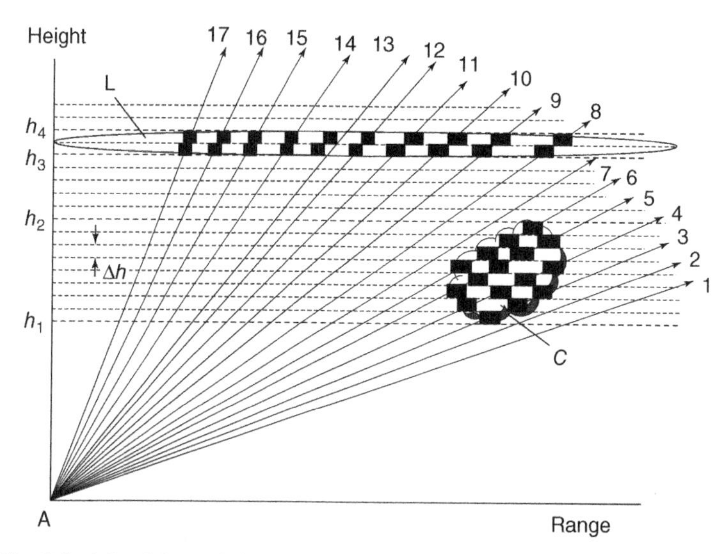

Fig. 3.49 Principle of determining the areas and layers with increased backscatter gradient. The lidar located at the point A scans in multiple slope directions. The black rectangles show localities with increased backscatter (Kovalev et al., 2009b).

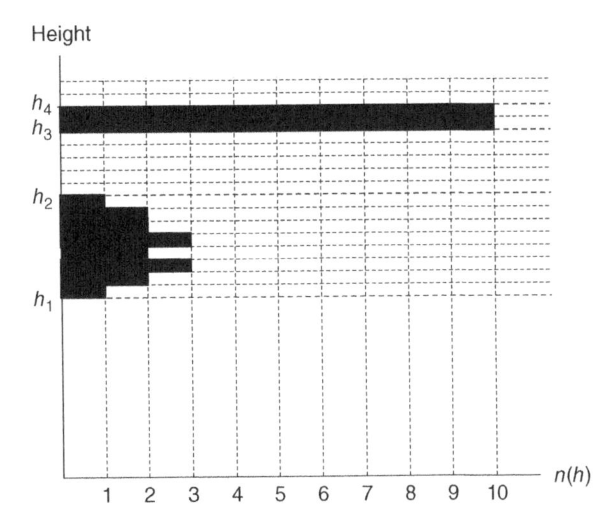

Fig. 3.50 AHHI plot showing the number of heterogeneity events $n(h)$ and heights at which the increased smoke-plume gradients were fixed (Kovalev et al., 2009b).

total number of heterogeneity events $n(h)$ defined by scanning lidar at consecutive height intervals, (ii) the number of slope directions where these events were observed, and (iii) the corresponding heights at which increased smoke-plume backscatter gradients take place. In short, the AHHI plot shows the number of heterogeneity events $n(h)$ fixed at different heights (see Section 2.9). The increased backscatter gradient in the layer located over the height interval from h_3 to h_4 is detected along 10 elevation angles (from 8 to 17), that is, the number of heterogeneity events at these altitudes is $n(h) = 10$. For the cloud C, within the height interval from h_1 to h_2, the increased backscatter gradient is visible only over two or three slope directions. Thus, the AHHI plot shows not only the smoke-plume altitude ranges but also discriminates local clouds from the extended aerosol layers. Note that one-directional lidar data do not allow direct determining of whether the signal decrease occurs because of decrease in the backscattering intensity or because of increase of the optical depth in the transmittance term. This yields an additional uncertainty when investigating the upper boundaries of dense smoke plumes.

3.6.2 Determination of the Upper Boundary of Increased Backscattering Area

The determination of the upper boundary of increased backscattering area is required not only for common investigations of the boundary layer, but also in many other cases, for example, in areas polluted by a volcanic eruption, in dusty clouds, in the vicinity of wildfires, etc. Here, we will focus on the evaluation of the maximal heights of smoke plumes originated in wildfires. Obviously, this specific task generally requires the use of scanning lidar. Let us consider the case when scanning is made in a fixed azimuthal direction θ along N slope directions within the angular sector from $\varphi_{\min}$ to $\varphi_{\max}$. The stepped angular resolution is $\Delta\varphi$, that is, $\varphi_1 = \varphi_{\min}$, $\varphi_2 = \varphi_{\min} + \Delta\varphi, \dots, \varphi_j = \varphi_{\min} + (j-1)\Delta\varphi, \dots$, and $\varphi_N = \varphi_{\max}$. The recorded signals are transformed into the corresponding set of functions $Y_0^*(\varphi, h)$; these functions are calculated within the altitude range from $h_{\min}$ to $h_{\max}$ with the selected height resolution Δh; that is, $h_1 = h_{\min}$, $h_2 = h_{\min} + \Delta h, \dots, h_i = h_{\min} + (i-1)\Delta h, \dots$, and $h_M = h_{\max}$; M is the number of height intervals Δh within the operative altitude range $(h_{\min}, h_{\max})$.

To locate the maximum height of the region of increased backscatter, the regularized function $Y_0^*(\varphi, h)$ for each slope direction φ_j within the altitude range from $h_{\min}$ to $h_{\max}$, is calculated, that is,

$$f_{j,i} = Y_0^*(\varphi_j, h_i). \tag{3.57}$$

The above operation defines the matrix $\mathbf{F}_\varphi$, with matrix elements $f_{j,i} \in (N, M)$. The maximum matrix element $f_{\max} = \max[f_{j,i}]_{N\times M}$ is determined and the matrix is normalized to unit,

$$\mathbf{R}_\varphi = \frac{1}{f_{\max}}\mathbf{F}_\varphi. \tag{3.58}$$

The elements of the matrix $r_{j,i} = R(\varphi_j, h_i)$ which are used for the investigation of increased backscattering areas, can vary within the range $0 \le r_{j,i} \le 1$; the areas where $r_{j,i} \to 1$ are the areas of highest heterogeneity.

To clarify the methodology of determining the heights of increased backscattering, let us consider experimental data obtained by the FSL scanning lidar in the vicinity of a wildfire near Missoula in Montana, in 2009. Scanning was made within the angular sector from $\varphi_{\min} = 10°$ to $\varphi_{\max} = 60°$ over a total number of slope directions $N_\varphi = 28$. In Fig. 3.51, the corresponding 28 normalized functions $R(\varphi_j, h_i)$ are shown as the gray curves with the thick black curve representing the resulting heterogeneity function $R_{\theta,\max}(h)$ versus height, defined as

$$R_{\theta,\max}(h) = \max\lfloor R(\varphi_1, h), R(\varphi_2, h), \dots, R(\varphi_j, h), \dots, R(\varphi_N, h)\rfloor. \tag{3.59}$$

The procedure for determining the maximum heights of areas of increased backscattering with scanning lidar is similar to that used for zenith-directed lidar. To determine the maximal height of the smoke-polluted atmosphere, the parameter χ, should be selected and matched with $R_{\theta,\max}(h)$. As in zenith one-directional profiling, the main issue is selection of the optimal level of χ. In the case of a poorly defined boundary, as is generally the case in smoke polluted atmospheres, the height h_{up}, where the selected parameter χ matches $R_{\theta,\max}(h)$, can dramatically depend on χ (Section 2.9). This observation is illustrated in Fig. 3.52, where the profile of the above function $R_{\theta,\max}(h)$ is analyzed. One can see that the selection of different χ yields significantly different heights of interest h_{up}. For example, the change of χ from 0.1 to 0.15 decreases the height h_{up} from 4581 to 3078 m; the change from 0.15 to 0.2 decreases it from 3078 to 2988 m; etc.

As in one-directional zenith searching (Section 2.9), it is sensible to define the upper smoke-plume boundary height $h_{\text{sm,max}}$ as the maximum height where smoke

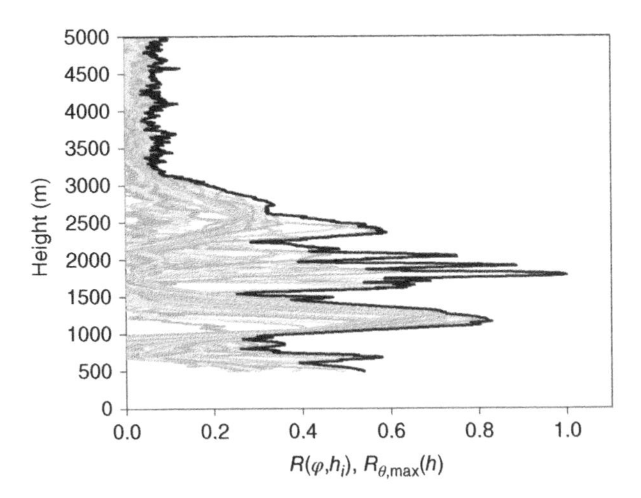

Fig. 3.51 An example of experimental data obtained with the FSL scanning lidar in smoke-polluted atmosphere. The gray curves represent the set of 28 functions $R(\varphi_j, h_i)$. The resulting heterogeneity function $R_{\theta,\max}(h)$ is shown as the thick black curve (Kovalev et al., 2011a).

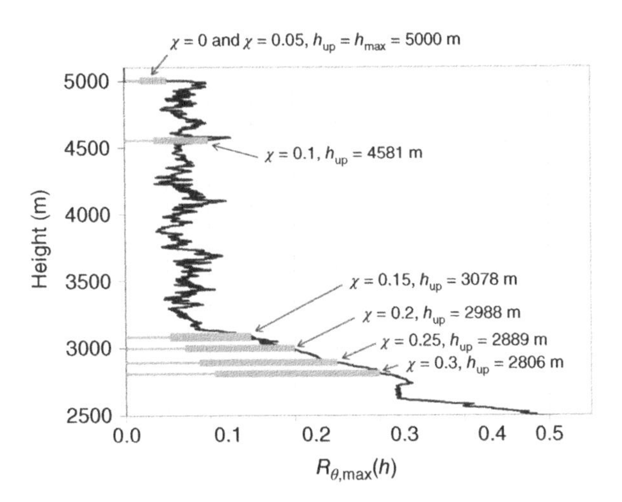

Fig. 3.52 The black curve is the same heterogeneity function $R_{\theta,\max}(h)$ as in Fig. 3.6.4, and the gray blocks illustrate the dependence of the height h_{up} on selected χ (Kovalev et al., 2011a).

aerosols are detectable in the presence of a noise component. This approach requires reliably distinguishing fluctuations caused by the smoke-plume heterogeneity from those caused by noise fluctuations in the examined function $R_{\theta,\max}(h)$. When determining the height of the smoke-plume upper boundary $h_{sm,\max}$, the digitized image defined in the previous subsection as AHHI is extremely helpful. However, to apply it in practice, the uncertain term, "increased backscatter gradient," should be quantified. This can be done by defining AHHI as a histogram, which shows the number of heterogeneity events $n(h)$ using χ as the quantifying criterion. Such a more concrete definition is clarified in Fig. 3.53. The thick black curve in the left side of the figure is the function $R_{\theta,\max}(h)$, the same as in the previous figures; the black-gray squares shows AHHI derived with $\chi = 0.15$. The maximal height h_{up}, where the minimal number of the heterogeneity events exceeds zero, in our case, $n(h) = 1$, is equal to 3078 m. The maximum number of heterogeneity events is fixed over two altitude ranges, $1000 - 1150\,\text{m}$ ($n = 28$) and $2350 - 2550\,\text{m}$ ($n = 22$), so that two separate layers at different heights are discriminated. Using AHHI histograms, one can determine the heights h_{up} that correspond to different χ. In the calculations, consecutive values of χ with the fixed step $\Delta\chi$ are utilized, so that $\chi_0 = 0$, $\chi_1 = \Delta\chi$, $\chi_2 = 2\Delta\chi$, ..., etc. For each discrete χ, the AHHI plot is built and the maximum height $h_{up}(\chi)$, where the number of heterogeneity events, $n(h) > 0$, is determined.

Fig. 3.54 illustrates the basic principle of determining the optimal level χ_{opt}, which allows quantifying the maximum smoke-plume height $h_{sm,\max}(\chi_{opt})$. The methodology of selecting such an optimal level χ_{opt} and determining the corresponding $h_{sm,\max}(\chi_{opt})$ is, in principle, the same as that used for zenith-directed lidar (Section 2.9.2). Initially, the level $\chi_0 = 0$ is selected. Because of the presence of nonzero instrumental noise the height, $h_{up}(\chi_0)$, determined as the maximal height where $n(h) > 0$, will be equal to the maximal profiling height h_{max}; in the case under consideration, $h_{up}(\chi_0) = h_{max} = 5000\,\text{m}$. Then the next level $\chi_1 = 0.05$ is analyzed. As one can see in the figure, this level does not change the initial height $h_{up} = 5000\,\text{m}$; in other words, the selected level $\chi_1 = 0.05$ is still below

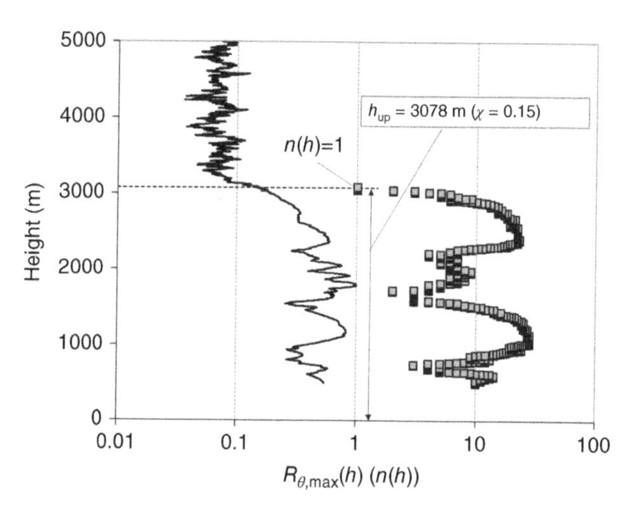

Fig. 3.53 The thick black curve is the function $R_{\theta,\max}(h)$, same as in Fig. 3.51. The black-gray squares show AHHI derived with $\chi = 0.15$. The corresponding maximal height, where the number of the heterogeneity events exceeds zero, [$n(h) = 1$], is $h_{up} = 3078\,\text{m}$ (Kovalev et al., 2011a).

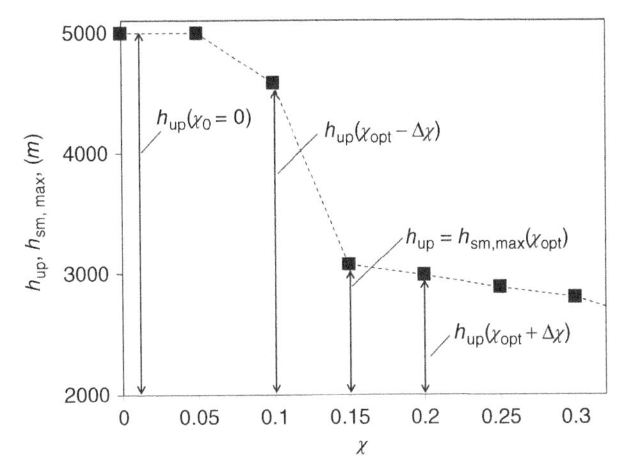

Fig. 3.54 Dependence of the height h_{up} on the selected χ for the heterogeneity function $R_{\theta,max}(h)$, shown in Fig. 3.51 (Kovalev et al., 2011a).

the interfering noise. Selecting then the larger levels equal to $\chi_2 = 0.1$, $\chi_3 = 0.15$, etc., one can plot the set of corresponding AHHI and determine the upper heights, $h_{up}(\chi_2)$, $h_{up}(\chi_3)$, etc. The objective of these operations is to establish the point where the heterogeneity event can be reliably discerned from noise and the maximal smoke-plume height $h_{sm,max}$ found.

As with one-directional profiling, determining the optimal level χ_{opt} and the corresponding height of the smoke plume $h_{sm,max}(\chi_{opt})$ may be based on the calculation of the differences between the adjacent heights $h_{up}(\chi_k)$ and $h_{up}(\chi_{k+1})$. As in Section 2.9, the established value of χ_{opt} should meet two conditions. First, the difference between the adjacent points, $h_{up} = h_{sm,max}(\chi_{opt})$ and the previous, $h_{up}(\chi_{opt} - \Delta\chi)$, should be maximal. Second, the next consecutive increase in χ, that is, the selection of χ equal to $\chi_{opt} + \Delta\chi$, then $\chi_{opt} + 2\Delta\chi$, etc., should result in only a minor decrease in the corresponding heights, $h_{up}(\chi)$. In our case, the decrease from $\chi = 0.1$ to 0.15 results in the largest decrease in $h_{up}(\chi)$ from 4581 to 3078 m (Fig. 3.54). The next consecutive increase of χ from 0.15 to 0.2, then from 0.2 to 0.25, etc. does not significantly reduce the extracted heights; for χ larger than 0.15, the difference between any pair of consecutive maximum heights is less than 100 m, that is, ~0.3% relative to the established smoke-plume height. Accordingly, the application of this principle to the analyzed data yields the optimal level $\chi_{opt} = 0.15$ and the corresponding maximal smoke-plume height $h_{sm,max}(\chi_{opt}) = 3078$ m. One should mention that to avoid significantly underestimating the maximum height, the value of χ_{opt} should be small, presumably, within the range, 0.1–0.2. This requirement puts some reasonable restrictions on the level of interfering noise in recorded signals.

Finally, let us consider an example of experimental data obtained by the FSL scanning lidar in the vicinity of the Kootenai Creek Fire in Montana in July–August 2009. The smoke-plume dynamics was monitored for the purpose of acquiring experimental data necessary for the evaluation of existing smoke-plume rise and dispersion models

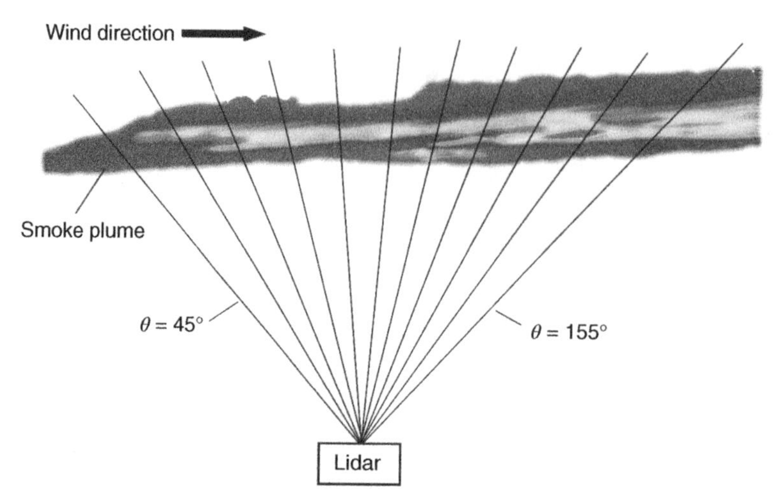

Fig. 3.55 Schematic of data collection with the FSL lidar during the Kootenai Creek Fire in Montana. The lidar performed vertical scanning in discrete azimuthal directions within the azimuthal sector 45°–155° (Kovalev et al., 2011a).

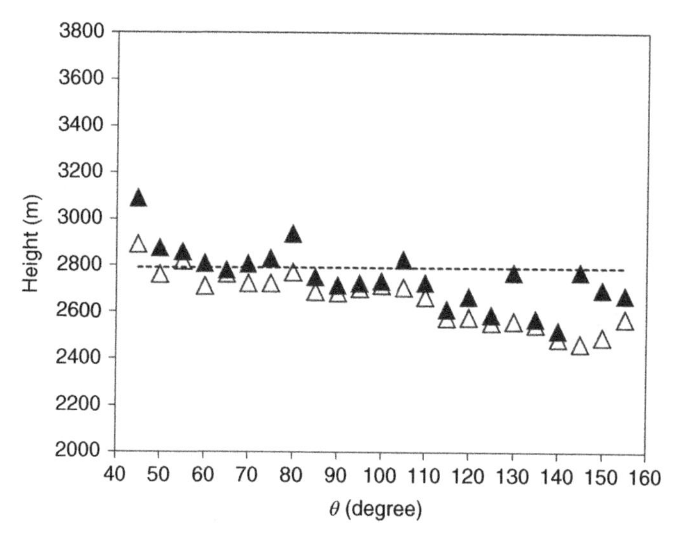

Fig. 3.56 Maximum heights of the horizontally extended smoke-plume at different azimuthal directions in the vicinity of the Kootenai Creek Fire. The horizontal dashed line indicates the smoke plume height determined from airborne measurements (Kovalev et al., 2011a).

for biomass fires (Urbanski et al., 2010). The schematic of the scanning lidar setup in the vicinity of the Kootenai Creek Fire is shown in Fig. 3.55. The fire occurred in a wild mountainous area from which the smoke plume spread in an easterly direction across the valley where the lidar site was chosen. The height of the lidar site was approximately 900 m below the height of the wildfire area. To obtain information on the spread of the smoke plume, lidar vertical scans were performed under 23 azimuthal directions, from $\theta = 45°$ to $\theta = 155°$, with an angular increment $\Delta\theta = 5^0$. Each scan yielded a vertical cross section of attenuated backscatter in the corresponding azimuthal direction. From these, the maximum smoke-plume heights were determined using dependencies discussed above.

An example of the maximum heights, obtained for the examined 23 azimuthal directions on August 27, 2009, from 12 : 09 to 12 : 27 local time, is shown in Fig. 3.56. Two consequent heights, $h_{sm,max}(\chi_{opt})$ and $h_{up}(\chi_{opt} + \Delta\chi)$, are shown as the filled and empty triangles, respectively. These adjacent levels characterize how well the upper boundary was defined. It follows from the figure that the maximal height of the smoke plume slowly decreases as the smoke moves away from the source. The increased difference between two consequent heights, which takes place for example, along azimuthal directions 130° and 145°, indicates that the upper smoke boundaries become less well defined as the smoke moves away from its source. The horizontal dashed line indicates the smoke-plume height determined from airborne measurements of aerosol concentration downwind of the Kootenai Creek fire (Urbanski et al., 2010).

BIBLIOGRAPHY

Adam M. Vertical versus scanning lidar measurements in horizontally homogeneous atmosphere. Appl. Opt. 2012;51:4491–4500.

Adam M. Notes on Rayleigh scattering in lidar signals. Appl. Opt. 2012a;51:2135–2149.

Adam M, Kovalev V, Wold C, Newton J, Pahlow M, Hao WM, Parlange MB. Application of the Kano-Hamilton multiangle inversion method in clear atmospheres. J. Atmos. Oceanic Technol. 2007;24:2114–2128.

Agishev RR, Comeron A. Spatial filtering efficiency of monostatic biaxial lidar: analysis and applications. Appl. Opt. 2002;41:7516–7521.

Ahmad SR, Bulliet EM. Performance evaluation of a laboratory-based Raman lidar in atmospheric pollution measurement. Opt. Laser Technol. 1994;26:323–331.

Althausen D, Oelsner P, Rohmer A, Baars H. Comparison of High Spectral Resolution Lidar with Raman lidar. In: Papayannis A, Balis D, Amiridis V, editors. *Proceedings of Twenty-Sixth International Laser Radar Conference*. Vol. 1, (Porto Heli, Greece, June 2012). 2012. p 43–46.

Anderson TA, Ogren JA. Determining aerosol radiative properties using the TSI 3563 integrating nephelometer. Aerosp. Sci. Technol. 1998;29:57–69.

Angström A. Techniques of determining the turbidity of the atmosphere. Tellus 1961;13(2):214–223.

Ansmann A. Ground-truth aerosol lidar observations: can the Klett solutions obtained from ground and space be equal for the same aerosol case? Appl. Opt. 2006;45:3367–3371.

Ansmann A, Riebesell M, Weitkamp C. Measurement of atmospheric aerosol extinction profiles with a Raman lidar. Opt. Lett. 1990;13:746–748.

Solutions in LIDAR Profiling of the Atmosphere, First Edition. Vladimir A. Kovalev.

Ansmann A, Wandinger U, Riebesell M, Weitkamp C, Michaelis W. Independent measurement of extinction and backscatter profiles in cirrus clouds by using a combined Raman elastic-backscatter lidar. Appl. Opt. 1992;31:7113–7131.

Baars H, Ansmann A, Engelmann R, Althausen D. Continuous monitoring of the boundary-layer top with lidar. Atmos. Chem. Phys. 2008;8:7281–7296.

Bakhvalov NS. *Numerical methods: analysis, algebra, ordinary differential equations, MIR*. Nauka; 1977, (In Russian).

Balin YS, Kavkyanov SI, Krekov GM, Rasenkov IA. Noise-proof inversion of lidar equation. Opt. Lett. 1987;12:13–15.

Balis DS, Papayannis A, Galani E, Marenco F, Santacesaria V, Hamonou E, Chazette P, Ziomas I, Zerefos C. Tropospheric lidar aerosol measurements and sun photometric observations at Thessaloniki, Greece. Atmos. Environ. 2000;34:925–932.

Balis DS, Amiridis V, Zerefos C, Gerasopoulos E, Andreae M, Zanis P, Kazantzidis A, Kazadzis S, Papayannis A. Raman lidar and sunphotometric measurements of aerosol optical properties over Thessaloniki, Greece during a biomass burning episode. Atmos. Environ. 2003;37:4529–4538.

Barlow R. *Statistics: A Guide to the Use of Statistical Methods in the Physical Sciences*. New York, NY: John Wiley & Sons, Inc; 1999.

Barret EW, Ben-Dov O. Application of lidar to air pollution measurements. J. Appl. Meteorol. 1967;6:500–515.

Berkoff TA, Welton EJ, Campbell JR, Scott VS, Spinhirine JD. Investigation of overlap correction techniques for the Micro-Pulse Lidar NETwork (MPLNET). IEEE Int. Geosci. Remote Sens. 2003:4395–4397.

Berkoff TA, Ji Q, Reid E, Valencia S, Welton EJ, Spinhirne JD. Analytically derived thermal correction to reduce overlap bias errors in Micro-Pulse Lidar data. In: Singh UN, editor. *Proceedings of SPIE **5984**, Lidar Technologies, Techniques, and Measurements for Atmospheric Remote Sensing*. 2005. p 59840R. The Society of Photo-Optical Instrumentation Engineers.

Bevington PR, Robinson DK. *Data Reduction and Error Analysis for the Physical Sciences*. 3rd ed. McGraw-Hill Higher Companies, Inc; 2003.

Beyerle G, McDermid S. Altitude range resolution of differential absorption lidar ozone profiles. Appl. Opt. 1999;38:924–927.

Biavati G, Donfrancesco G, Cairo F, Feist DG. Correction scheme for close-range lidar returns. Appl. Opt. 2011;50:5872–5882.

Bissonnette LR. Sensitivity analysis of lidar inversion algorithms. Appl. Opt. 1986;25:2122–2125.

Bissonnette LR. Multiple scattering of narrow light beams in aerosols. Appl. Phys. B 1995;60:315–323.

Bissonnette LR. Multiple-scattering lidar equation. Appl. Opt. 1996;35:6449–6465.

Bissonnette LR, Roy G, Poutier L, Cober SG, Isaac GA. Multiple-scattering lidar retrieval method: tests on Monte Carlo simulations and comparisons with *in situ* measurements. Appl. Opt. 2002;41:6307–6324.

Boers R, Melfi SH. Cold-air outbreak during MASEX: lidar observations and boundary layer model test. Boundary-Layer Meteorol. 1987;39:41–51.

Boyd JP. Chebyshev and Fourier Spectral Methods. Dover Publications; 2001.

Bristow M. Suppression of afterpulsing in photomultipliers by gating the photocathode. Appl. Opt. 2002;41:4975–4987.

Brooks IM. Finding boundary layer top: application of a wavelet covariance transform to lidar backscatter profiles. J. Atmos. Oceanic Technol. 2003;20:1092–1105.

Browell EV, Ismail S, Shipley ST. Ultraviolet DIAL measurements of O_3 profiles in regions of spatially inhomogeneous particulates. Appl. Opt. 1985;24:2827–2836.

Cadet B, Giraud V, Haeffelin M, Keckhut P, Rechou A, Baldy S. Improved retrievals of the optical properties of cirrus clouds by a combination of lidar methods. Appl. Opt. 2005;44:1726–1734.

Campbell JR, Hlavka DL, Welton EJ, Flynn CJ, Turner DD, Spinhirne JD, Scott VS, Hwang IH. Full-time, eye-safe cloud and aerosol lidar observation at atmospheric radiation measurement program sites: instruments and data processing. J. Atmos. Oceanic Technol. 2002;19:431–442.

Cattrall C, Reagan J, Thome K, Dubovik O. Variability of aerosol and spectral lidar and backscatter and extinction ratios of key aerosol types derived from selected Aerosol Robotic Network locations. J. Geophys. Res. 2005;110:D10S11.

Cheng AYS, Walton A, Chan CS, Chan RLM, Chan MH. Data inversion of eye-safe Micro-Pulse Mie lidar for internal boundary layer studies. In: Singh UN, editor. *Laser Radar Techniques for Atmospheric Sensing, Proceedings of SPIE* ***5575***. Maspalomas, Gran Canaria, Spain. The Society of Photo-Optical Instrumentation Engineers 2004; 67–74.

Collis, RTH Lidar. In: *Atmospheric Exploration by Remote Probes. Proceedings of the Scientific Meetings of the Panel on Remote Atmospheric Probing*, (Panel on Remote Atmospheric Probing, April 18-20 and May 16–17,1968). 1969;2:147-171.

Comerón A, Rocadenbosch F, López MA, Rodríguez A, Muñoz C, García-Vizcaíno D, Sicard M. Effects of noise on lidar data inversion with the backward algorithm. Appl. Opt. 2004;43:2572–2577.

Cook, D.A. Computers and M&S - Modeling and Simulation 101. Cross-Talk Jan. 2001 www.stsc.hill.af.mil/crosstalk/2001/01/cook.html.

Cook DA, Skinner JM. How to perform credible verification, validation, and accreditation for modeling and simulation. J. Defense Software Eng. 2005;2005:20–24.

Cooper DI, Eichinger WE. Structure of the atmosphere in an urban planetary boundary layer from lidar and radiosonde observations. J. Geophys. Res. 1994;99:22937–22948.

Cuesta J, Flamant PH, Flamant C. Synergetic technique combining elastic backscatter lidar data and sunphotometer AERONET inversion for retrieval by layer of particulate optical and microphysical properties. Appl. Opt. 2008;47:4598–4611.

Davis PA. The analysis of lidar signatures of cirrus clouds. Appl. Opt. 1969;8:2044–2102.

Davis KJ, Gamage N, Hagelberg CR, Kiemle C, Lenschow DH, Sullivan PP. An objective method for deriving atmospheric structure from airborne lidar observations. J. Atmos. Oceanic Technol. 2000;17:1455–1468.

De Tomasi F, Blanco A, Perrone MR. Raman lidar monitoring of extinction and backscattering of african dust layers and dust characterization. Appl. Opt. 2003;42:1699–1709.

Dho SW, Park YJ, Kong HJ. Experimental determination of the geometrical form factor in the lidar equation for inhomogeneous atmosphere. Appl. Opt. 1997;36:6009–6010.

DoD Modeling and Simulation (M&S) Glossary (1998) Defense Modeling and Simulation Office, Department of Defense, DoD 5000.59-M, www.dtic.mil/whs/directives/corres/pdf/500059m.pdf.

Donovan DP, Whiteway JA, Carswell AI. Correction for nonlinear photon-counting effects in lidar systems. Appl. Opt. 1993;32:6742–6753.

Durieux E, Fiorani L. Measurement of the lidar signal fluctuations with a shot-per-shot instrument. Appl. Opt. 1998;37:7128–7131.

Eberhard WL. Correct equations and common approximations for calculating Rayleigh scatter in pure gases and mixtures and evaluation of differences. Appl. Opt. 2010;49:1116–1130.

Ehret G, Kiemle C, Renger W, Simmet G. Airborne remote sensing of tropospheric water vapor with a near-infrared differential absorption lidar system. Appl. Opt. 1993;32:4534–4551.

Eloranta EW. Practical model for the calculation of multiply scattered lidar returns. Appl. Opt. 1998;37:2464–2472.

Eloranta EW. In: Weitkamp K, editor. High Spectral Resolution Lidar. In *Lidar: Range-Resolved Optical Remote Sensing of the Atmosphere*. Springer Series in Optical Sciences. New York: Springer-Verlag; 2005.

Eloranta EW, Razenkov IA, Garcia JP, Hedrick J. Observations with the University of Wisconsin arctic high spectral resolution lidar. In: Pappalardo G, Amodeo A, editors (Matera, Italy, 2004). *Proceedings of 22nd International Laser Radar Conference*. 2004. European Space Agency.

Elouragini S, Flamant PH. Iterative method to determine an averaged backscatter-to-extinction ratio in cirrus clouds. Appl. Opt. 1996;35:1512–1518.

Esselborn M, Wirth M, Fix A, Tesche M, Ehret G. Airborne high spectral resolution lidar for measuring particulate extinction and backscatter coefficients. Appl. Opt. 2008;47:346–358.

Ferguson JA, Stephens DH. Algorithm for inverting lidar returns. Appl. Opt. 1983;22:3673–3675.

Fernald, F. G. (1982) Comments on the analysis of atmospheric lidar observations. *Proceedings of 11th International Laser Radar Conference*, (Madison, University of Wisconsin, June 1982), 213-215.

Fernald FG. Analysis of atmospheric lidar observations: some comments. Appl. Opt. 1984;23:652–653.

Fernald FG, Herman B, Reagan JA. Determination of aerosol height distributions by lidar. J. Appl. Meteorol. 1972;11:482–489.

Ferrare RA, Melfi SH, Whiteman DN, Evans KD, Poellot M, Leifer R. Raman lidar measurements of aerosol extinction and backscattering 1. Methods and comparisons. J. Geophys. Res. 1998;103(D16):19,673–19,689.

Fiorani L, Calpini B, Jaquet L, Van den Bergh H, Durieux E. Correction scheme for experimental biases in differential absorption lidar tropospheric ozone measurements based on the analysis of shot per shot data samples. Appl. Opt. 1997;36:6857–6863.

Fiorani L, Armenante M, Capobianco R, Spinelli N, Wang X. Self-aligning lidar for the continuous monitoring of the atmosphere. Appl. Opt. 1998;37:4758–4764.

Flamant C, Pelon J, Flamant PH, Durant P. Lidar determination of the entrainment zone thickness and the top of the unstable marine atmospheric boundary layer. Boundary-Layer Meteorol. 1997;83:247–284.

Fujii T, Fukuchi T, editors. *Laser Remote Sensing*. Boca Raton, FL: Taylor & Francis Group; 2005.

Fujimoto T, Uchino O, Nagai T. Estimation of DIAL algorithms for stratospheric ozone. In: McCormick MP, editor. *Proceedings of the Seventeenth International Laser Radar Conference (ILRC17)*. (Chiba University, Sendai, Japan, 1994). 1994. p 392–395.

Godin S, Carswell AI, Donovan DP, Claude H, Steinbrecht W, McDermid IS, McGee TJ, Gross MR, Nakane H, Swart DPJ, Bergwerff HB, Uchino O, von der Gathen P, Neuber R. Ozone differential absorption lidar algorithm intercomparison. Appl. Opt. 1999;38:6225–6236.

Hair JW, Caldwell LM, Krueger DA, She C-Y. High-spectral-resolution lidar with iodine-vapor filters: measurement of atmospheric-state and aerosol profiles. Appl. Opt. 2001;40:5280–5294.

Hair JW, Hostetler CA, Cook AL, Harper DB, Ferrare RA, Mack TL, Welch W, Izquierdo LR, Hovis FE. Airborne high spectral resolution lidar for profiling aerosol optical properties. Appl. Opt. 2008;47:6734–6752.

Hamilton PM. Lidar measurement of backscatter and attenuation of atmospheric aerosol. Atmos. Environ. 1969;3:221–223.

Hawkins DM. The problem of overfitting. J. Chem. Inf. Comput. Sci. 2004;44:1–12.

Hennemuth B, Lammert A. Determination of the atmospheric boundary layer height from radiosonde and lidar backscatter. Boundary-Layer Meteorol. 2006;120:181–200.

Hooper WP, Eloranta EW. Lidar measurements of wind in the planetary boundary layer: the method, accuracy, and results from joint measurements with radiosonde and kytoon. J. Clim. Appl. Meteorol. 1986;25:990–1001.

Hughes HG, Ferguson JA, Stephens DH. Sensitivity of a lidar inversion algorithm to parameters relating atmospheric backscatter and extinction. Appl. Opt. 1985;24:1609–1613.

Hunt WH, Poultney SK. Testing the linearity of response of gated photomultipliers in wide dynamic range laser radar systems. IEEE Trans. Nucl. Sci. 1975;NS-22:116–120.

Ignatenko VM. Experimental determination of the lidar geometrical function. In: Gushchin GP, editor (Main Geophysical Observatory, Leningrad, 1991). Proceedings of the Main Geophysical Observatory. Vol. 499. 1985. p 91–95(in Russian).

Johnson W, Repasky KS, Carlsten JL. Micropulse differential absorption lidar for identification of carbon sequestration site leakage. Appl. Opt. 2013;52:2994–3003.

Kano M. On the determination of backscattering and extinction coefficient of the atmosphere by using a laser radar. Paper. Meteorol. Geophys. 1968;19:121–129.

Kaskaoutis DG, Kambezidis HD, Hatzianastassiou N, Kosmopoulos PG, Badarinath KVS. Aerosol climatology: dependence of the Angstrom exponent on wavelength over four AERONET Sites. Atmos. Chem. Phys. Discuss. 2007;7:7347–7397.

Kaul, B. V. (1977) *Laser Sensing the Aerosol Pollution in the Atmosphere*, Dissertation (Institute of Atmospheric Optics, Tomsk, U.S.S.R, 1976) (in Russian).

Kempfer U, Carnuth W, Lotz R, Trickl T. A wide-range ultraviolet lidar system for tropospheric ozone measurements: development and application. Rev. Sci. Instrum. 1994;65:3145–3164.

Kennedy MC, O'Hagan A. Bayesian calibration of computer models. J. R. Statist. Soc., Series B (Statistical Methodology) 2001;63:425–464.

Kinjo H, Kuze H, Sakurada Y, Takeuchi N. Calibration of the lidar measurement of tropospheric aerosol extinction coefficients. Jpn. J. Appl. Phys. 1999;38(1A):293–297.

Kiureghiana AD, Ditlevsen O. Aleatory or epistemic? Does it matter? Struct. Saf. 2009;31:105–112.

Klett JD. Stable analytical inversion solution for processing lidar returns. Appl. Opt. 1981;20:211–220.

Klett JD. Lidar inversion with variable backscatter/extinction ratios. Appl. Opt. 1985;24:1638–1643.

Kovalev VA. Near-end solution for lidar signals that includes a multiple scattering component. Appl. Opt. 2003;42:7215–7224.

Kovalev VA. Distortions of the extinction coefficient profile caused by systematic errors in lidar data. Appl. Opt. 2004;43:3191–3198.

Kovalev VA. Determination of slope in lidar data using a duplicate of the inverted function. Appl. Opt. 2006;45:8781–8789.

Kovalev VA, Eichinger WE. *Elastic Lidar. Theory, Practice, and Analysis Methods*. Hoboken, New Jersey: John Wiley & Sons, Inc; 2004. DOI: 10.1002/0471643173.

Kovalev VA, Kolgotin A. Comparison of alternative techniques for extracting the extinction coefficient and the lidar ratio from optical depth and backscatter coefficient profiles. In: *Proceedings of Twenty-Forth International Laser Radar Conference (ILRC24)* (Boulder, Colorado, June 2008). Organizing Committee; 2008. p SO1P-33.

Kovalev VA, McElroy JL. Differential absorption lidar measurement of vertical ozone profiles in the troposphere that contains aerosol layers with strong backscatter gradients: a simplified version. Appl. Opt. 1994;33:8393–8401.

Kovalev VA, Moosmüller H. Distortion of particulate extinction profiles measured with lidar in a two-component atmosphere. Appl. Opt. 1994;33:6499–6507.

Kovalev VA, Hao WM, Wold C, Adam M. Experimental method for the examination of systematic distortions in lidar data. Appl. Opt. 2007a;46:6710–6718.

Kovalev VA, Hao WM, Wold C. Determination of the particulate extinction-coefficient profile and the column-integrated lidar ratios using the backscatter-coefficient and optical-depth profiles. Appl. Opt. 2007b;46:8627–8634.

Kovalev V, Wold C, Hao WM, Nordgen B. Improved methodology for the retrieval of the particulate extinction coefficient and lidar ratio from the lidar multiangle measurement. In: Singh UN, Pappalardo G, editors. *Proceedings of SPIE **6750**, Lidar Technologies, Techniques, and Measurements for Atmospheric Remote Sensing*. 2007c. p 67501B-1–67501B-9.

Kovalev V, Hao WM, Wold C, Newton J, Latham DJ, Petkov A. Investigation of optical characteristics and smoke-plume dynamics in the wildfire vicinity with lidar. In: Hao WM, editor(SPIE, Bellingham, WA, 2008). *Remote Sensing of Fire: Science and Application, Proceedings of SPIE **7089***. 2008. p 708906-1–108906-8. The Society of Photo-Optical Instrumentation Engineers.

Kovalev VA, Wold C, Petkov A, Hao WM. Alternative method for determining the constant offset in lidar signal. Appl. Opt. 2009a;48:2559–2565.

Kovalev VA, Petkov A, Wold C, Urbanski S, Hao WM. Determination of smoke plume and layer heights using scanning lidar data. Appl. Opt. 2009b;48:5287–5294.

Kovalev V, Petkov A, Wold C, Hao WM. Determination of the smoke-plume heights with scanning lidar using alternative functions for establishing the atmospheric heterogeneity locations. In: *Proceedings of the 25th International Laser Radar Conference (5–9 July 2010, St.-Petersburg, Russia)*. Tomsk: Publishing House of IAO SB RAS; 2010. p 71–74.

Kovalev VA, Petkov A, Wold C, Hao WM. Lidar monitoring of regions of intense backscatter with poorly defined boundaries. Appl. Opt. 2011a;50:103–109.

Kovalev VA, Petkov A, Wold C, Hao WM. Modified technique for processing multiangle lidar data measured in clear and moderately polluted atmospheres. Appl. Opt. 2011b;50:4957–4966.

Kovalev VA, Petkov A, Wold C, Urbanski S, Hao WM. Essentials of multiangle data processing methodology for smoke polluted atmospheres. Rom. J. Phys. 2011c;56:520–529.

Kovalev VA, Wold C, Petkov A, Hao WM. Direct multiangle solution for poorly stratified atmospheres. Appl. Opt. 2012;51:6139–6146.

Kunkel KE, Eloranta EW, Shipley ST. Lidar observations of the convective boundary layer. J. Appl. Meteorol. 1977;16:1306–1311.

Kunz GJ. Transmission as an input boundary value for an analytical solution of a single-scatter lidar equation. Appl. Opt. 1996;35:3255–3260.

Kunz GJ, Leeuw G. Inversion of lidar signals with the slope method. Appl. Opt. 1993;32:3249–3256.

Lammert A, Bösenberg J. Determination of the convective boundary-layer height with laser remote sensing. Boundary Layer Meteorol. 2006;119:159–170.

Larchevêque G, Balin I, Nessler R, Quaglia P, Simeonov V, Bergh H, Calpini B. Development of a multiwavelength aerosol and water-vapor lidar at the Jungfraujoch Alpine Station (3580 m Above Sea Level) in Switzerland. Appl. Opt. 2002;41:2781–2790.

Lavrent'ev MM, Romanov VG, Shishatskii SP. *Ill-posed problems of mathematical physics and analysis*. In: *Translations of Mathematical Monographs*. Vol. 64. American Mathematical Society; 1980.

Lee HS, Schwemmer GK, Korb CL, Dombrowski M, Prasad C. Gated photomultiplier response characterization for DIAL measurements. Appl. Opt. 1990;29:3303–3315.

Liu B, Esselborn M, Wirth M, Fix A, Bi D, Ehret G. Influence of molecular scattering models on aerosol optical properties measured by high spectral resolution lidar. Appl. Opt. 2009;48:5143–5154.

Mao F, Gong W, Zhu Z. Simple multiscale algorithm for layer detection with lidar. Appl. Opt. 2011;50:6591–6598.

Marenco F, Santacesaria V, Bais AF, Balis D, Sarra A, Papayannis A, Zerefos C. Optical properties of tropospheric aerosols determined by lidar and spectrophotometric measurements (Photochemical Activity and Solar Ultraviolet Radiation campaign). Appl. Opt. 1997;36:6875–6886.

Mason JC, Christopher D. Chebyshev Polynomials. Florida: CRS Press LLC; 2003.

Matthies, H. G. (2007) Quantifying Uncertainty: Modern Computational Representation of Probability and Applications, *Extreme Man-Made and Natural Hazards in Dynamics of Structures*, 105–135.

Mattis I, Ansmann A, Müller D, Wandinger U, Althausen D. Multiyear aerosol observations with dual-wavelength Raman lidar in the framework of EARLINET. J. Geophys. Res. 2004;109:D13203.

McDermid IS, Godin MS, Lindqvist LO. Ground-based laser DIAL system for long-term measurements of stratospheric ozone. Appl. Opt. 1990;29:3603–3612.

McDermid IS, Walsh TD, Deslis A, White ML. Optical systems design for a stratospheric lidar system. Appl. Opt. 1995;34:6201–6210.

Measures RM. *Laser Remote Sensing*. New York: John Wiley & Sons; 1984.

Megie G, Menzies RT. Complementarity of UV and IR differential absorption lidar for global measurements of atmospheric species. Appl. Opt. 1980;19:1173–1183.

Melfi SH, Spinhirne JD, Chou SH, Palm SP. Lidar observations of vertically organized convection in the planetary boundary layer over the ocean. J. Clim. Appl. Meteorol. 1985;24:806–821.

Menut L, Flamant C, Pelon J, Flamant PH. Urban boundary-layer height determination from lidar measurements over the Paris area. Appl. Opt. 1999;38:945–954.

Mitev V, Matthey R, Makarov V. Compact micro-pulse backscatter lidar and examples of measurements in the planetary boundary layer. Rom. J. Phys. 2011;56:437–447.

Murayama T, Masonis SJ, Redemann J, Anderson TL, Schmid B, Livingston JM, Russell PB, Huebert B, Howell SG, McNaughton CS, Clarke A, Abo M, Shimizu A, Sugimoto N, Yabuki M, Kuze H, Fukagawa S, Maxwell-Meier K, Weber RJ, Orsini DA, Blomquist B, Bandy A, Thornton D. An intercomparison of lidar-derived aerosol optical properties with airborne measurements near Tokyo during ACE-Asia. J. Geophys. Res. 2003;108:8651.

Pahlow M, Kovalev VA, Ansmann A, Helmert K. Iterative determination of the aerosol extinction coefficient profile and the mean extinction-to-backscatter ratio from multiangle lidar data. In: Pappalardo G, Amodeo A, editors (Matera, Italy, 2004). *Proceedings of 22nd International Laser Radar Conference*, 2004. p 491–494. European Space Agency.

Pappalardo G, Amodeo A, Pandolfi M, Wandinger U, Ansmann A, Bösenberg J, Matthias V, Amiridis V, De Tomasi F, Frioud M, Iarlori M, Komguem L, Papayannis A, Rocadenbosch F, Wang X. Aerosol lidar intercomparisons in the framework of EARLINET. 3. Raman lidar algorithm for aerosol extinction, backscatter, and lidar ratio. Appl. Opt. 2004;43:5370–5385.

Pedhazur EJ, Shmelkin LP. *Measurement, Design, and Analysis: An Integrated Approach*. New York: Psychology Press, Tailor & Francis Group; 1991. p 819 p.

Pelon J, Megie G. Ozone monitoring in the troposphere and lower stratosphere: evaluation and operation of a ground-based lidar station. J. Geoph. Res. 1982;87:4947–4955.

Piironen P, Eloranta EW. Demonstration of a high-spectral-resolution lidar based on a iodine absorption filter. Opt. Lett. 1994;19:234–236.

Platt CMR. Lidar and radiometric observations of cirrus clouds. J. Atmos. Sci. 1973;30:1191–1204.

Platt CMR. Remote sounding of high clouds: I. Calculation of visible and infrared optical properties from lidar and radiometer measurements. J. Appl. Meteorol. 1979;18:1130–1143.

Pornsawad P, Böckmann C, Ritter C, Rafler M. Ill-posed retrieval of aerosol extinction coefficient profiles from Raman lidar data by regularization. Appl. Opt. 2008;47:1649–1661.

Powell DM, Reagan JA, Rubio MA, Erxleben WH, Spinhirne JD. ACE-2 multiple angle micro-pulse lidar observations from Las Galletas, Tenerife, Canary Islands. Tellus 2000;B52:652–661.

Proffitt MH, Langford AO. Ground-based differential absorption lidar system for day or night measurements of ozone throughout the free troposphere. Appl. Opt. 1997;36:2568–2585.

Razenkov IA, Eloranta EW, Hedrick JP, Holz RE, Kuehn RE, Garcia JP. A high spectral resolution lidar designed for unattended operation in the arctic. In: Bissonnette LR, Roy G, Vallee G, editors. *Proceedings of 21st International Laser Radar Conference*. Vol. 1, (Quebec, Canada, July 2002). 2002. p 57–60.

Reichardt J, Wandinger U, Klein V, Mattis I, Hilber B, Begbie R. RAMSES: German Meteorological Service autonomous Raman lidar for water vapor, temperature, aerosol, and cloud measurements. Appl. Opt. 2012;51:8111–8131.

Rocadenbosch F, Comerón A. Error analysis for the lidar backward inversion algorithm. Appl. Opt. 1999;38:4461–4474.

Rocadenbosch F, Comeron A, Pineda D. Assessment of lidar inversion errors for homogeneous atmospheres. Appl. Opt. 1998;37:2199–2206.

Rocadenbosch F, Comerón A, Albiol L. Statistics of the slope-method estimator. Appl. Opt. 2000;39:6049–6057.

Rocadenbosch F, Sicard M, Comeron A. Automated variable-resolution algorithm for Raman extinction retrieval. In: Pappalardo G, Amodeo A, editors. *Proceedings of the Twenty-Second International Laser Radar Conference (ILRC22)*, (Matera, Italy, July 2004). 2004. European Space Agency.

Rogers RR, Hair JW, Hostetler CA, Ferrare RR, Obland MD, Cook AL, Harper DB, Burton SP, Shinozuka Y, McNaughton CS, Clarke AD, Redemann J, Russell PB, Livingston JM, Kleinmam LI. NASA LaRC airborne high spectral resolution lidar aerosol measurements during MILAGRO: observations and validation. Atmos. Chem. Phys. Discuss. 2009;9:8817–8856.

Rogers RR, Hostetler CA, Hair JW, Ferrare RA, Liu Z, Obland MD, Harper DB, Cook AL, Powell KA, Vaughan MA, Winker DM. Assessment of the CALIPSO Lidar 532 nm attenuated backscatter calibration using the NASA LaRC airborne high spectral resolution lidar. Atmos. Chem. Phys. 2011;11:1295–1311.

Rothermal J, Jones W. Ground-based measurements of atmospheric backscatter and absorption using coherent CO_2 lidar. Appl. Opt. 1985;24:3487–3496.

Russell PB, Swissler TJ, McCormick MP. Methodology for error analysis and simulation of lidar aerosol measurements. Appl. Opt. 1979;18:3783–3797.

Sasano Y. Tropospheric aerosol extinction coefficient properties derived from scanning lidar measurements over Tsukiba, Japan, from 1990 to 1993. Appl. Opt. 1996;35:4941–4952.

Sasano Y, Nakane H. Significance of the extinction/backscatter ratio and the boundary value term in the solution for the two-component lidar equation. Appl. Opt. 1984;23:11–13.

Sasano Y, Shimizu H, Takeuchi N, Okuda M. Geometrical form factor in the laser radar equation: an experimental determination. Appl. Opt. 1979;18:3908–3910.

Sasano Y, Browell EV, Ismail S. Error caused by using a constant extinction/backscattering ratio in the lidar solution. Appl. Opt. 1985;24:3929–3932.

Sassen K, Dodd GC. Lidar crossover function and misalignment effects. Appl. Opt. 1982;21:3162–3165.

Shcherbakov V. Regularized algorithm for Raman lidar data processing. Appl. Opt. 2007;46:4879–4889.

Shimizu H, Sasano Y, Nakane H, Sugimoto N, Matsui I, Takeuchi N. Large scale laser radar for measuring aerosol distribution over a wide area. Appl. Opt. 1985;24:617–626.

Sicard M, Chazette P, Pelon J, Won JG, Yoon SC. Variational method for the retrieval of the optical thickness and the backscatter coefficient from multiangle lidar profiles. Appl. Opt. 2002;41:493–502.

Simeonov V, Larcheveque G, Quaglia P, Van den Bergh H, Calpini B. Influence of the photomultiplier tube spatial uniformity on lidar signals. Appl. Opt. 1999;38:5186–5190.

Spinhirne J. Monitoring of tropospheric optical properties by lidar. In: *Atmospheric Aerosols: The Optical Properties and Effects, NASA CP-2004*. NASA: Washington, DC; 1976.

Spinhirne J. Micro Pulse Lidar. IEEE Trans. Geosci. Remote Sens. 1993;31:48–55.

Spinhirne J. Eye-safe lidar provides full-time atmospheric monitoring. Photonic Spectra 1996;30(3):98–99.

Spinhirne JD, Reagan JA, Herman BM. Vertical distribution of aerosol extinction cross section and interference of aerosol imaginary index in the troposphere by lidar technique. J. Appl. Meteorol. 1980;19:426–438.

Spinhirne JD, Russel PB, Livingston JM. Slant-lidar aerosol extinction measurements and their relation to measured and calculated albedo changes. J. Clim. Appl. Meteorol. 1984;23:1204–1221.

Spinhirne J, Rall JAR, Scott VS. Compact eye safe lidar systems. Rev. Laser Eng. 1995;23:112–118.

Stelmaszczyk K, Dell'Aglio M, Chudzynski S, Stacewicz T, Wöste L. Analytical function for lidar geometrical compression form-factor calculations. Appl. Opt. 2005;44:1323–1331.

Stevens SS. Measurement, statistics, and the schemapiric view. Science 1968;161:849–856.

Sunesson JA, Apituley A, Swart DPJ. Differential absorption lidar system for routine monitoring of tropospheric ozone. Appl. Opt. 1994;33:7045–7058.

Takamura T, Sasano Y, Hayasaka T. Tropospheric aerosol optical properties derived from lidar, sun photometer, and optical particle counter measurements. Appl. Opt. 1994;33:7132–7140.

Taylor JR. *An Introduction to Error Analysis*. 2nd ed. Sausalito, CA: University Science Books; 1997.

Tikhonov AN, Arsenin VY. *Solutions of Ill-Posed Problems*. New York: Winston and Sons; 1977.

Tomine K, Hirayama C, Michimoto K, Takeuchi N. Experimental determination of the crossover function in the laser radar equation for days with a light mist. Appl. Opt. 1989;28:2194–2195.

Trefethen LN, Bau D. *Numerical Linear Algebra*. Philadelphia, PA: SIAM; 1997.

Uchino O, Tabata I. Mobile lidar for simultaneous measurements of ozone, aerosols, and temperature in the stratosphere. Appl. Opt. 1991;30:2005–2012.

Urbanski, S., V. Kovalev, W. M. Hao, C. Wold, and A. Petkov (2010) Lidar and airborne investigation of smoke plume characteristics: Kootenai Creek Fire case study. *Proceedings of the 25th International Laser Radar Conference* (IAO SB RAS, 2010), pp. 1051–1054.

Uthe EE, Livingston JM. Lidar extinction methods applied to observations of obscurant events. Appl. Opt. 1986;25:678–684.

Veretennikov VV, Abramochkin AI. Determination of the optical and microstructural characteristics of water droplet clouds in laser sensing taking into account multiple scattering. J. Atmos. Oceanic Technol. 2009;22:527–535. Pleiades Publishing, Ltd.

Viezee W, Uthe EE, Collis RTH. Lidar observations of airfield approach conditions: an exploratory study. J. Appl. Meteorol. 1969;8:274–283.

Volkov SN, Kaul BV, Shelefontuk DI. Optimal method of linear regression in laser remote sensing. Appl. Opt. 2002;41:5078–5083.

Vorontsov KV. Splitting and similarity phenomena in the sets of classifiers and their effect on the probability of overfitting. Pattern Recognition Image Anal. 2006;19:412–420.

Wandinger U. Raman lidar. In: Weitkamp K, editor. *Lidar, Range-Resolved Optical Remote Sensing of the Atmosphere*. Springer Series in Optical Sciences. New York: Springer-Verlag; 2005.

Wandinger U, Ansmann A. Experimental determination of the lidar overlap profile with Raman lidar. Appl. Opt. 2002;41:511–514.

Weinman JA. Effects of multiple scattering on light pulses reflected by turbid atmospheres. J. Atmos. Sci. 1976;33:1763–1771.

Weinman JA. Derivation of atmospheric extinction profiles and wind speed over the ocean from a satellite-borne lidar. Appl. Opt. 1988;27:3994–4001.

Whiteman DN. Application of statistical methods to the determination of slope in lidar data. Appl. Opt. 1999;15:3360–3369.

Winsberg E. Simulated experiments: methodology for a virtual world. Philos. Sci. 2003;70:105–125.

Wylie CR, Barret LB. *Advanced Engineering Mathematics*. 5th ed. McGraw-Hill, Inc; 1982.

Zege EP, Ivanov AP, Katsev IL. Determination of the extinction and scattering parameters of the water medium and the atmosphere by the analysis of erosion of the reflected pulse signal. Bull. Acad. Sci. U.S.S.R., Geophys. Ser. 1971;7:750–754, (in Russian).

Zege EP, Katsev IL, Polonsky IN. Analytical solution to LIDAR return signals from clouds with regard to multiple scattering. Appl. Phys. B 1995;60:345–354.

Zhao Z. Signal-induced fluorescence in photomultipliers in differential absorption lidar systems. Appl. Opt. 1999;38:4639–4648.

Zuev VE, Krekov GM. In: Zuev VE, editor. *Optical Models of the Atmosphere*. Leningrad: Gidrometeoizdat, Chapter 5; 1986. (in Russian).

Zuev VE, Makushkin YS, Marichev VN, Mitsel AA, Zuev VV. Lidar differential absorbing and scattering technique: Theory. Appl. Opt. 1983;22:3733–3741.

Zuev VE, Zadde GO, Kavkjanov SI, Kaul BV. Interpretation of lidar return signals from the regions of large optical depths. In: Zuev VE, editor. *Remote Sensing in the Atmosphere*. Novosibirsk: Nauka, U.S.S.R.; 1978. p 60–68(in Russian).

INDEX

Solutions in LIDAR Profiling of the Atmosphere, First Edition. Vladimir A. Kovalev.

Printed and bound by CPI Group (UK) Ltd, Croydon, CR0 4YY

16/04/2025

14658352-0002